S. Enenkel W. Stille

Antibiotics in the Tropics

Antibacterial Therapy with Limited Resources

With 16 Figures and 132 Tables

Springer-Verlag
Berlin Heidelberg New York
London Paris Tokyo

Dr. med. Sabine Enenkel
Department of Pediatrics

Professor Dr. Wolfgang Stille
Department of Internal Medicine

University Hospital,
Johann Wolfgang Goethe-Universität,
Theodor-Stern-Kai 7, 6000 Frankfurt/Main 70, FRG

ISBN-13: 978-3-540-18683-0 e-ISBN-13: 978-3-642-73276-8
DOI: 10.1007/978-3-642-73276-8

Library of Congress Cataloging-in-Publication Data. Enenkel, S. (Sabine), 1957–. Antibiotics in the tropics. Includes bibliographies and index. 1. Bacterial diseases – Developing countries – Chemotherapy. 2. Antibiotics – Developing countries. 3. Tropical medicine. I. Stille, W. (Wolfgang) II. Title. [DNLM: 1. Antibiotics – therapeutic use. 2. Bacterial Infections – drug therapy. 3. Tropical Medicine. QV 350 E56a]
RC115.E54 1988 616.9'2061 87-37652

2119/3140-543210

ALLE ZEIT WACH 1842

Preface

International textbooks on infectious diseases and antibacterial chemotherapy are usually written for readers in North America and Europe. In many ways, they are not appropriate for the problems encountered in developing countries. This book, in contrast, intends to define the rules of antibacterial chemotherapy practised under conditions of limited resources. It is meant for everyone concerned with the use of antibiotics in developing countries, including doctors, medical assistants, pharmacists, officials in health ministeries, and medical students.

Throughout the book, treatment recommendations are made for antibiotics from the WHO list of essential drugs.[1] For example, emphasis has therefore been put upon chloramphenicol as a stable, unexpensive and widely available oral agent suitable for the treatment of severe bacterial infections like septicemia and meningitis. So-called "international chemotherapy" with modern cephalosporins and acylaminopenicillins has been outlined for comparison.

Since it is the aim of the book to base treatment recommendations on data from developing countries, many data on the etiology of common bacterial infections in developing countries have also been included. Most of the data are from African, English-speaking developing countries, but references have been made to the literature on South East Asia, India or Papua New Guinea, where appropriate. On the other hand, pertinent data were not available in every instance, so that several statements and recommendations had to be made as "best guess". The authors are aware of these imperfections and will welcome comments from the readers.

The authors are very grateful to the Springer-Verlag for the help and support in the production of this book.

Frankfurt/Main, March 1988 S. Enenkel, W. Stille

[1] WHO Expert committee (1983) The use of essential drugs. Tech Rep Ser No. 685, WHO, Geneva.

Contents

Introduction 1

Part I. Principles of Antibiotic Therapy

1 The Range of Antibiotics Available 5

General Principles of Antibiotic Therapy –
The WHO List of Essential Drugs 5
The Main Antibiotics of the WHO List 10
Classical Penicillins 10
Benzylpenicillin (Penicillin G) and Its Oral Form
Phenoxymethylpenicillin 10
Repository Penicillins 11
Ampicillin 11
Penicillinase-Stable Penicillins 12
Tetracyclines 12
Chloramphenicol 13
Co-Trimoxazole and Sulfonamides 14
Co-Trimoxazole 14
Sulfonamides 15
Aminoglycosides 15
Gentamicin 16
Erythromycin 17
Metronidazole 17
Nitrofurantoin 18
Rifampicin 19
Further Antibiotics and Therapeutic Groups
(Not on WHO List) 19
Quinolones 19
Old Quinolones 20
Modern Quinolones 20
Acylaminopenicillins 21
Cephalosporins 21
Other Antibiotics 23
Vancomycin 23
Clindamycin 23
Imipenem 24
Amoxicillin/Clavulanic Acid (Augmentin) 24

2 *Choice of Antibiotics* 26

3 *Bacterial Resistance in Developing Countries* 34

4 *Administration of Antibiotics* 39
Topical Administration 40

5 *Dosage of Antibiotics* 41
Recommended Doses 41
Duration of Treatment 41
Tolerance 43
Dosage in Neonates, Infants, and Children 43
Pediatric Dosage 43
Premature and Full-Term Neonates and Infants During First Month of Life 44

6 *Side Effects and Contraindications* 45
Toxic Side Effects 45
Allergic Side Effects 47
Biological Side Effects 48

7 *Combinations of Antibiotics* 49

8 *Costs of Antibiotics* 51

Part II. Treatment of Infections and Infectious Diseases

9 *Septicemia* 55

10 *Infective Endocarditis* 71
Streptococcal Endocarditis 75
Endocarditis Due to Pneumococci or Group A Streptococci 76
Enterococcal Endocarditis 76
Staphylococcal Endocarditis 77
Staphylococcus Aureus 77
Staphylococcus Epidermidis 78
Endocarditis Due to Gram-Negative Bacteria 78
Culture-Negative Endocarditis 78
Prevention of Endocarditis 79

11 *Purulent Pericarditis* 81

12 Infections of the Central Nervous System 83

Bacterial Meningitis 83
Brain Abscess 101
Subdural Empyema 102

13 Infections of the Respiratory Tract 105

Acute Respiratory Infections in Childhood 105
Common Cold 108
Tonsillitis, Pharyngitis 108
Rheumatic Fever 109
Peritonsillar Abscess, Retropharyngeal Abscess, and Ludwig's Angina 110
Croup 110
Acute Bronchitis 111
Acute Exacerbation of Chronic Bronchitis 111
Bronchiolitis 112
Pneumonia 112
Treatment of Community-Acquired Pneumonia 119
Treatment of Special Types of Pneumonia 127
Staphylococcal Pneumonia 127
Measles Pneumonia 127
Klebsiella Pneumonia 129
Aspiration Pneumonia 129
Acute Abscess-Forming Pneumonia 129
Interstitial Pneumonias 130
Legionnaires' Disease 130
Secondary, Hospital-Acquired Pneumonias 130
Postoperative Pneumonia 131
Postinfluenza Pneumonia 131
Efficacy of Antimicrobial Agents in Different Types of Pneumonia 131
Empyema 131
Chronic Lung Abscess 134
Bronchiectasis 134
Whooping Cough 135
Diphtheria 136

14 Infections of the Gastrointestinal Tract 142

Diarrheal Diseases 142
Appendicitis 161
Peritonitis 163
Pyogenic Liver Abscess 165
Biliary Tract Infections 166

15 *Infections of the Musculoskeletal System* 170

Osteomyelitis 170
Septic Arthritis 178
Pyomyositis 183

16 *Skin and Wound Infections* 187

Common Bacterial Infections 187
Prophylaxis of Wound Infections 198
Indications of Prophylaxis 203
Modern Rules of Prophylaxis 203
Tetanus 205
Gas Gangrene 206

17 *Sexually Transmitted Diseases* 210

Gonorrhea 216
Other Sexually Transmitted Diseases 219

18 *Obstetric and Gynecological Infections* 222

Bartholinitis 222
Vulvovaginitis in Children 222
Vaginitis, Vulvitis in Adults 223
Types of Vaginal Disease 223
Pelvic Inflammatory Disease 225
Other Obstetric and Gynecological Infections 226
Septic Abortion 226
Puerperal Sepsis 226
Mastitis 227
Pyrexia During Labor 227

19 *Genitourinary Infections* 231

Urinary Tract Infections 231
Urethritis 240
Prostatitis, Epididymitis 240
Acute Epididymitis 241

20 *Infections of the Ear, Nose, and Throat* 243

External Otitis 243
Malignant External Otitis 243
Otitis Media 243
Acute Otitis Media 244
Chronic Suppurative Otitis Media 244
Mastoiditis 245

Sinusitis . . . 245
Stomatitis . . . 247
Viral Stomatitis . . . 247
Candida Stomatitis . . . 247
Necrotizing Ulcerative Gingivitis . . . 247
Noma (Cancrum Oris) . . . 248
Osteomyelitis of the Jaw . . . 248
Cervical Lymphadenitis . . . 249

21 *Eye Infections* . . . 251

Ulcerating Blepharitis . . . 251
Hordeolum . . . 251
Abscess and Cellulitis of Lid . . . 251
Orbital Cellulitis . . . 251
Acute Dacryocystitis . . . 252
Acute Bacterial Conjunctivitis . . . 252
Ophthalmia Neonatorum . . . 253
Trachoma . . . 254
Keratitis . . . 256
Bacterial Keratitis . . . 256
Herpetic Keratitis . . . 256
Fungal Keratitis . . . 257
Phlyctenular Keratitis . . . 257
Measles Keratitis . . . 257
Bacterial Endophthalmitis . . . 257

22 *Salmonella Infections* . . . 259

Typhoid Fever . . . 259
Nontyphoidal Salmonella Infections . . . 262

23 *Rickettsial Diseases* . . . 265

24 *Leptospirosis* . . . 269

25 *Brucellosis* . . . 271

26 *Relapsing Fever* . . . 274

27 *Anthrax* . . . 276

28 *Plague* . . . 278

29 *Listeria Infections* . . . 280

30 *Melioidosis* . . . 282

31 *Nonvenereal Treponematoses* . . . 284

32 Mycobacterial Infections 285

Tuberculosis 285
Leprosy 295
Buruli Ulcer 299

33 Actinomycosis 302

34 Nocardiosis 303

35 Rare Bacterial Diseases: Bartonellosis, Erysipeloid, and Ratbite Fever 305

36 Fungal Infections 306

Dermatophyte Infections 307
Tinea Capitis 307
Fungal Infections of the Nails 308
Tinea Versicolor 309
Candida Infections 309
Genital Candidiasis 309
Oral Candidiasis 310
Candida Esophagitis 310
Candida Enteritis 310
Candida Pneumonia 310
Candida Infections of the Urinary Tract 311
Candida Septicemia 311
Candida Infections of the Skin 311
Chronic Mucocutaneous Candidiasis 311
Cryptococcal Infection 312
Infections by Molds 312
Aspergillus Infections 312
Mycetoma 313
Chromomycosis 313
Rhinosporidiosis 314
Sporotrichosis 314
Subcutaneous Phycomycosis 314
Rhinoentomophthoromycosis 314
Histoplasmosis 315
Blastomycosis 315
Paracoccidioidomycosis 316
Coccidioimycosis 316
Systemic Treatment of Fungal Infections 316

37 Treatment of AIDS 319

Subject Index 322

Introduction

Modern antibacterial chemotherapy can be learned from a range of more or less sophisticated international textbooks [1-6]. General and specialized scientific journals frequently report on new developments and trends of antibiotic therapy. Scientists from the United States, Japan, and Western Europe meet frequently at international congresses like the Interscience Conference of Antibacterial Chemotherapy (ICAAC) or the International Congress of Chemotherapy (ICC). The abundant literature on antibacterial chemotherapy deals mainly with "international standards" of chemotherapy, according to which modern antibacterial chemotherapy of severe infections is carried out mainly with the newer cephalosporins, modern penicillins, and aminoglycosides. The principles of treatment with these modern agents are doubtlessly well adapted to the countries of origin of the scientists describing them, and accordingly the belief that these principles are internationally accepted is commonly held in these countries.

However, the antimicrobial agents needed to apply these principles in practice are not available to the great majority of people in the world. Apart from the modern chemotherapy practised in the United States, Western Europe, and Japan, there are at least two other types of chemotherapy. Antibacterial chemotherapy in the socialist block is carried out with standard antibiotics; supply is sufficient, but modern cephalosporins and penicillins are available only to relatively few selected patients. The chemotherapy practised under conditions of limited resources in developing countries also differs markedly. Limited supply and very strict cost-efficiency calculations necessitate different regimens of treatment.

Antibacterial chemotherapy in developing countries is not however, uniform. Three different types of treatment can be observed:

Type I: Chemotherapy for rich and powerful minorities, diplomats, and other foreigners. The standards of treatment are usually identical with those of chemotherapy in the United States, Western Europe, and Japan. There is a demand for the most modern and most active substances.
Type II: Basic chemotherapy for the poor majority. The antibiotics are usually supplied by the state hospitals. Only a limited number of standard substances are available.
Type III: An intermediate type. A varying number of patients can afford to buy antibiotics which are more effective than the standard substances supplied free by the state hospitals. The drugs are frequently purchased by relatives from a nearby dispensary and brought into the hospital.

Antibacterial chemotherapy for the poor majority (type II) and for patients who can spend some money (type III) are by far the most important of these types. Many states pay a remarkable percentage of their medical budget for antibiotics.

The prudent use of antibacterial agents is of very great value. In contrast to many other drugs, antibiotics are curative; with correct treatment, the patient is healed without relapse.

Despite the high prevalence of bacterial infections in developing countries, we were unable to find textbooks or even concise surveys on antibacterial chemotherapy in tropical areas. In contrast to the vast amount of data regarding antibiotics in the United States, Japan, and Europe, only very few data on antibiotics in clinical trials are available from developing countries. Our team combines sufficient knowledge of modern antibiotics, common infections, and also tropical medicine. Nevertheless, we underestimated the difficulties involved in writing this book. In the absence of pertinent data from developing countries, several statements and recommendations had to be made on the basis of "best guess." It is the aim of this book to define rules of antibacterial chemotherapy practised under limited conditions and with limited resources. In this context, it is interesting to note that some treatment policies, such as single-dose therapy, had been practised in tropical countries a long time before they were "invented" in Europe. There is a large demand for correctly carried out sophisticated or simple clinical trials concerning the use of antibiotics in the tropics.

References

1. Garrod LP, Lambert HP, O'Grady F (1981) Antibiotics and chemotherapy, 5th edn. Churchill Livingstone, Edinburgh
2. Greenwood D (1983) Antimicrobial chemotherapy. Baillière Tindall, London
3. Kucers A, McK Bennett N (1987) The use of antibiotics, 4th edn. Heinemann, London
4. Mandell GL, Douglas RG, Bennett JE (1985) Principles and practice of infectious diseases 2nd edn. Wiley, New York
5. Simon C, Stille W (1985) Antibiotika-Therapie in Klinik und Praxis, 6th edn. Schattauer, Stuttgart
6. Simon C, Stille W, Wilkinson PJ (1984) Antibiotic therapy in clinical practice. Schattauer, Stuttgart

Part I
Principles of Antibiotic Therapy

1 The Range of Antibiotics Available

General Principles of Antibiotic Therapy - The WHO List of Essential Drugs

Physicians in the USA and Europe are faced with an ever-increasing number of antibiotics and other antibacterial agents. (The same development can be observed in other fields of drug therapy.) For reasons of economy and practicability, hospital pharmacies no longer stock all the derivatives available on the market. A choice is usually made from groups of antibiotics similar in activity and antibacterial spectrum. Thus, the pharmacies set up lists of antibiotics tailored to the needs of the hospitals they are serving. Physicians working in the hospital are limited to the agents on these lists when prescribing antibiotics for treatment.

There is no doubt that antibacterial chemotherapy is more difficult than many other forms of modern therapy. Optimal treatment of bacterial infections requires knowledge of 30-35 agents (excluding tuberculosis). The great majority of bacterial infections are, however, treated with only 10 agents.

In Europe, antibacterial chemotherapy is usually practised as follows.

The majority of minor infections are treated with only four oral substances (alternatives in parentheses):

- Phenoxymethylpenicillin (propicillin)
- Amoxicillin (bacampicillin)
- Doxycycline
- Co-trimoxazole

An additional number of antibiotics are used for selected indications in ambulatory patients:

- Erythromycin
- Flucloxacillin (cloxacillin)
- Cefaclor (cefadroxil, cefalexin)
- Amoxicillin/clavulanic acid
- Clindamycin
- Metronidazole (ornidazole)
- Quinolones (norfloxacin, ofloxacin, pefloxacin, ciprofloxacin)
- Intramuscular penicillins
- Topical antibiotics

Treatment of hospitalized patients often requires parenteral antibiotics. Two penicillins, three cephalosporins, and one aminoglycoside are the standard derivatives:

- Benzylpenicillin
- Acylaminopenicillin (piperacillin, azlocillin, mezlocillin)

- Cefazolin
- Cefoxitin
- Cefotaxime (cefmenoxime, ceftizoxime)
- Gentamicin (tobramycin, sisomicin)

The following antibiotics are reserved for special indications (selected cases only):

- Cefotiam (cefuroxime, cefamandole)
- Ceftriaxon, cefoperazone, cefsulodin, ceftazidime, latamoxef, i.v. ampicillin, azlocillin, mezlocillin, imipenem, amikacin (netilmicin), parenteral ciprofloxacin

Fusidic acid, vancomycin, fosfomycin, and chloramphenicol are needed only for rare indications.

The lists of antibiotics used in European or North American chemotherapy do not take much account of costs, availability or practicability.

Selection of antibiotics for treatment of common bacterial infections in developing countries will necessarily result in different lists. The "List of Essential Drugs" (Table 1.1) published by WHO (hereinafter referred to as "WHO list") is designed especially for the needs of developing countries. This list, in its latest version from 1983 contains a broad range of antibiotics and antibacterial agents. Antibiotics currently leading on the European and North American market (cephalosporins and acylaminopenicillins) are not included.

It is clear that new antibiotics must be used for a certain time on a broad scale before they can be included in the WHO list. Very widely used antibiotics, e.g., tetracycline, cannot immediately be replaced by the better alternative, in this instance doxycycline. A number of important but expensive antibiotics are not included in the WHO list; obviously, cost must play a role in exclusion. Some important principles of chemotherapy, can be recognized by considering agents excluded from the list. For instance, only short-acting sulfonamides are included; the longer-acting sulfonamides, the poorly absorbed sulfonamides, and the triple sulfas (combinations of several sulfonamides) are not regarded as standard drugs. Combinations of antimicrobials are not included, with the exceptions of co-trimoxazole and thioacetazone + isoniazid. The large number of combinations of antibiotics with other drugs (expectorants, phenazopyridine) do not fulfil the requirements for standard treatment. The WHO list is remarkably free from "pharma junk." Historical or second-class antibiotics are also excluded from the list. Examples are old aminoglycosides (kanamycin), suboptimal macrolides (oleandomycin), fosfomycin, doubtful beta-lactams (mecillinam, ciclacillin), and old quinolones (nalidixic acid, cinoxacin). Substances which are not licensed in the country where they were developed have to be regarded with caution.

The total number of antibiotics in the WHO list should not be underestimated. Many antibiotics included are selected as "example of a therapeutic group" (Table 1.2).

There are also more or less equivalent alternatives to several antibiotics in the WHO list which are not specified as "examples of a therapeutic group" (Table 1.3). The more localized distribution of these agents could be the reason for this.

Table 1.1. Antibacterial and antifungal agents on the revised model list of essential drugs [8]

Main list	Complementary list	Route of administration, dosage forms, and strengths[a]
	Antibacterial drugs	
	Penicillins	
• ampicillin (4)		capsule or tablet, 250 mg, 500 mg (anhydrous) powder for oral suspension, 125 mg (anhydrous)/5 ml powder for injection, 500 mg (as sodium salt) in vial
benzylpenicillin		powder for injection, 0.6 g (=1 million IU), 3.0 g (=5 million IU) (as sodium or potassium salt) in vial
phenoxymethylpenicillin		tablet, 250 mg (as potassium salt) powder for oral suspension 250 mg (as potassium salt)/5 ml
procaine benzylpenicillin (7)		powder for injection, 1 g (=1 million IU), 3 g (=3 million IU)
benzathine benzylpenicillin (5)		injection 1.44 g benzylpenicillin (=2.4 million IU)/5 ml in vial
	Other antibacterial drugs	
• chloramphenicol (7)		capsule, 250 mg powder for injection, 1 g (as sodium succinate) in vial
• cloxacillin		capsule, 500 mg (as sodium salt) powder for injection, 500 mg (as sodium salt) in vial
erythromycin		capsule or tablet, 250 mg (as stearate or ethylsuccinate) oral suspension, 125 mg (as stearate or ethylsuccinate)/5 ml powder for injection, 500 mg (as lactobionate) in vial
• gentamicin (4)		injection, 10 mg, 40 mg (as sulfate)/ml in 2-ml vial
• metronidazole		tablet, 200–500 mg injection, 500 mg in 100 ml suppository, 500 mg, 1 g
spectinomycin (8)		powder for injection, 2 g (as hydrochloride) in vial
• sulfadimidine (4)		tablet, 500 mg oral suspension, 500 mg/5 ml injection, 1 g (sodium salt) in 3-ml ampoule
• sulfamethoxazole + trimethoprim (4)		tablet, 100 mg + 20 mg, 400 mg + 80 mg
• tetracycline (4)		capsule or tablet, 250 mg (hydrochloride)
	• amikacin (B, C) (4)	injection, 250 mg (sulfate)/ml in 2-ml ampoule
	doxycycline (B) (5, 6)	capsule or tablet, 100 mg (as hydrochloride) injection, 100 mg (as hydrochloride)/5 ml in ampoule
	nitrofurantoin (A, B) (4, 7)	tablet, 100 mg

[a] When the strength is specified in terms of a selected salt or ester, this is mentioned in brackets: when it refers to the active moiety, the name of the salt or ester in brackets is preceded by the word "as".

Table 1.1 (continued)

Main list	Complementary list	Route of administration, dosage forms, and strengths[a]
	Antileprosy drugs	
clofazimine		capsule, 100 mg
dapsone		tablet, 50 mg, 100 mg
rifampicin		capsule or tablet, 150 mg, 300 mg
	ethionamide (B)	tablet, 125 mg, 250 mg
	protionamide (B)	tablet, 125 mg
	Antituberculosis drugs	
ethambutol		tablet, 100-500 mg (hydrochloride)
isoniazid		tablet, 100-300 mg
pyrazinamide		tablet, 500 mg
rifampicin		capsule or tablet, 150 mg, 300 mg
streptomycin (4)		powder for injection, 1 g (as sulfate) in vial
thioacetazone + isoniazid		tablet, 50 mg + 100 mg, 150 mg + 300 mg
	Antifungal drugs	
amphotericin B		powder for injection, 50 mg in vial
griseofulvin (8)		tablet or capsule, 125 mg, 250 mg
nystatin		tablet, 500000 IU pessary, 100000 IU
	flucytosine (B) (4, 8)	capsule, 250 mg infusion, 2.5 g in 250 ml
	Antifungal drugs (dermatological)	
benzoic acid + salicylic acid		ointment or cream, 6% + 3%
• miconazole		ointment or cream, 2% (nitrate)
nystatin		ointment or cream, 100000 IU/g
	Antiinfective drugs (dermatological)	
• neomycin + • bacitracin		ointment, 5 mg neomycin sulfate + 500 IU bacitracin zinc/g
	Disinfectants	
• chlorhexidine		solution, 5% (gluconate) for dilution
• iodine		solution, 2.5%
	Antiinfective agents (ophthalmological)	
silver nitrate		solution (eye drops), 1%
sulfacetamide		eye ointment, 10% (sodium salt) solution (eye drops), 10% (sodium salt)
• tetracycline		eye ointment, 1% (hydrochloride)

[a] See p. 7.

Explanatory Notes. In many instances various drugs could serve as alternatives to those on the list. In these cases, the substance selected provides an *example of a therapeutic group* and is distinguished by being preceded by a symbol (•). It is imperative that this should be understood when drugs are selected at national level, since the choice is then influenced by the comparative cost and availability of equivalent products.

Examples of Acceptable Substitution. • Sulfadimidine: any other short-acting systemically active sulfonamide unlikely to cause crystalluria.

Table 1.1 (continued)

Numbers in parentheses following the drug names indicate:
(1) Drugs subject to international control under the Single Convention on Narcotic Drugs (1961) and the Convention on Psychotropic Substances (1971);
(2) Specific expertise, diagnostic precision or special equipment required for proper use;
(3) Greater potency;
(4) In renal insufficiency, contraindicated or dosage adjustments necessary;
(5) To improve compliance;
(6) Special pharmacokinetic properties for purpose;
(7) Adverse effects diminish benefit risk ratio;
(8) Limited indications or narrow spectrum of activity;
Letters in parentheses following the drug names indicate the reasons for the inclusion of *complementary drugs:*
(A) When drugs in the main list cannot be made available;
(B) When drugs in the main list are known to be ineffective or inappropriate for a given individual;
(C) For use in rare disorders or in exceptional circumstances.

Table 1.2. Alternatives to antibacterial and antifungal agents marked in the WHO list as "examples of a therapeutic group"

Agents in the WHO list	Alternatives
Ampicillin	Amoxicillin, bacampicillin, pivampicillin, talampicillin
Chloramphenicol	Thiamphenicol
Cloxacillin	Oxacillin, dicloxacillin, flucloxacillin, nafcillin
Gentamicin	Tobramycin, sisomicin, dibekacin
Metronidazole	Ornidazole, tinidazole
Sulfadimidine	Sulfadiazine, sulfafurazole, sulfamethizole
Tetracycline	Oxytetracycline, minocycline, doxycycline
Amikacin	Netilmicin
Miconazole (topical)	Clotrimazole, and many other azole derivatives
Sulfamethoxazole + trimethoprim (≙ co-trimoxazole)	Sulfamoxole + trimethoprim, sulfametrole + trimethoprim, sulfadiazin + trimethoprim, sulfadiazin + tetroxoprim

Table 1.3. Alternatives to antibiotics in the WHO list that are not distinguished as "examples of a therapeutic group"

Antibiotics in the WHO list	Alternatives
Phenoxymethylpenicillin	Propicillin, azidocillin, phenethicillin
Procaine penicillin	Clemizol penicillin
Nitrofurantoin	Hydroxymethylnitrofurantoin
Erythromycin	Josamycin, spiramycin, oleandomycin

The antibiotics in the WHO list allow effective treatment of the majority of common bacterial infections; a wide range of pathogens is covered.

Additional agents are needed for treatment of infections due to *Pseudomonas aeruginosa* (excluding urinary tract infections), resistant enterobacteria, and methicillin-resistant staphylococci, which are often nosocomial infections. It seems likely that modern cephalosporins (cefotaxime), acylaminopenicillins, and also ketoconazole would be included in the list as supplementary drugs if the problems of costs could be overcome (e.g., special manufacturers prices for WHO). The introduction of new quinolones (norfloxacin, ofloxacin, ciprofloxacin), with their broad spectrum of activity - gram-positive cocci and enterobacteria, including salmonellae and shigellae - will be of great interest for developing countries. The main disadvantage of this interesting group of antibiotics is the contraindication in children.

The WHO list is valuable as a model list which, modified to local conditions, can be used for provision of standard antibiotic therapy. In the chapters on treatment of infections, the use of additional antibiotics will be discussed, when appropriate. The differences between "international standard therapy" and "therapy under conditions of limited resources" lie often not so much in efficacy as in the avoidance of side effects. A classical example of this is the use of chloramphenicol in typhoid fever.

The Main Antibiotics of the WHO List

It is necessary to summarize the position of the main antibiotics. Extensive details can be found in the various textbooks of antibiotics [1-7].

Classical Penicillins

Nomenclature of Penicillins. Classical benzylpenicillin is meant by the term 'penicillin'. The term 'penicillins' means the group of all penicillin derivatives.

Benzylpenicillin (Penicillin G) and Its Oral Form Phenoxymethylpenicillin

Alternatives: Propicillin, azidocillin.

Activity. Very active against streptococci, pneumococci, peptostreptococci, meningococci, corynebacteria, spirochaetes, *Borrelia, Actinomyces,* clostridia, and many gram-negative anaerobes (except *Bacteroides fragilis*). Less active against enterococci, *Listeria,* and *Bacillus* spp. Many strains of gonococci and *Staphylococcus aureus* are now resistant.

Pharmacokinetics. Short half-life, mainly renal elimination.

Dosage. Due to the good tolerance, a broad range of daily doses is possible (1-4 g phenoxymethylpenicillin, orally, 4 - 10 - 30 MU (megaunits) benzylpenicillin i.v.). Penicillins can be given for very long times.

Main Indications. Infections by sensitive strains. Tonsillitis, erysipelas, dental infection, rheumatic fever, lobar pneumonia, ENT infections, treatment of skin and wound infection outside the hospital. Pelvic inflammatory disease, meningitis in adults, endocarditis, *Borrelia* infection, lues, leptospirosis.

Main Side Effects. Allergy, neurotoxicity in high doses of benzylpenicillin and other penicillins.

Repository Penicillins

Repository forms of penicillin still play an important role in the tropics. They have the advantage of long half-lives. The blood and tissue levels are quite low; only highly sensitive bacteria (pneumococci, streptococci, *Treponema*) can be treated with these products.

Procaine penicillin gives sufficient levels for 1 day. With benzathine penicillin, there are low levels for 2-3 weeks. Repository forms of penicillin have the risk of prolonged severe allergies and of shock by accidental intravenous injection. The increasing risk of AIDS and hepatitis is an argument for preferring oral penicillin, if possible.

Ampicillin

Alternatives with a better oral absorption are amoxicillin, bacampicillin, and pivampicillin.

Activity. Spectrum of activity identical with benzylpenicillin. Additional higher activity against *Hemophilus influenzae, Bordetella pertussis, Streptococcus faecalis,* other enterococci, *Listeria.* The majority of strains of *E. coli, Salmonella, Shigella, Proteus mirabilis* are sensitive, but clinical use may lead to resistant strains. Ampicillin is not stable to beta-lactamases, so penicillin-resistant gonococci and staphylococci, but also penicillin-resistant enterobacteria and *Bacteroides,* are not inhibited by ampicillin.

Pharmacokinetics. Similar to classical penicillins. Short half-life and mainly renal elimination.

Dosage. Due to the good tolerance, a broad range of daily doses is possible; from 1-4 g orally or i.v.

Main Indications. Infections by sensitive strains, where phenoxymethylpenicillin cannot be used. Bronchitis, ENT infection, uncomplicated urinary tract infections, meningitis in children, pelvic inflammatory disease. Intravenously together with gentamicin in enterococcal endocarditis.

Main Side Effects. Allergy, increased frequency of benign skin reactions. Selection of resistant gram-negative bacteria when used extensively in hospital. Generally, the old and incompletely absorbed ampicillin should be replaced by the nearly completely absorbed amoxicillin or bacampicillin.

Penicillinase-Stable Penicillins (Anti-Staphylococcal Penicillins)

Cloxacillin

Alternatives: Oxacillin, dicloxacillin, flucloxacillin, nafcillin

Activity. Effective against penicillinase-producing staphylococci, less active than benzylpenicillin against penicillin-sensitive staphylococci, streptococci and other gram-positive pathogens.

Pharmacokinetics. Short half-life, oral and parenteral administration, metabolized, incomplete renal excretion.

Dosage. Oral absorption is better on an empty stomach (1 h before or 2-4 h after meal), 2-4 g daily, divided in 4(−6) doses, parenterally 3-4(−10) g in 4-6 hourly injections or short infusions.

Main Indications. Infections with penicillinase-producing staphylococci.

Main Side Effects. Similar to benzylpenicillin.

Tetracyclines

The classical derivative is tetracycline. Nearly identical are oxytetracycline, demeclocycline, and methacycline. Doxycycline and minocycline have much better absorption, longer half-life, and higher activity.

Activity. The spectrum of activity within the tetracycline group is practically identical, with moderate to good sensitivity against streptococci, pneumococci, gonococci, meningococci, *Listeria, Actinomyces, Hemophilus, Yersinia, Brucella, Vibrio, Campylobacter, Leptospira, Mycoplasma, Chlamydia,* and *Rickettsia.* There is a worldwide increasing tendency toward resistance against tetracycline, even in formerly sensitive strains like pneumococci. A remarkable percentage of staphylococci, *E. coli, Klebsiella,* and *Salmonella* spp. are now resistant against tetracycline.

Pharmacokinetics. Incomplete absorption of tetracycline and oxytetracycline. Half-life of tetracycline about 8 h. Doxycycline is nearly completely absorbed and has a half-life of 15 h; a single dose is sufficient to achieve effective concentrations for 24 h. Good tissue penetration, but poor CSF penetration. Excretion mainly via the kidneys (except for doxycycline).

Dosage. The range of dosage is small. The standard dosage consists of 1-2 g/day tetracycline or oxytetracycline, 100-200 mg/day doxycycline. Intravenous forms have certain additional risks and should be used in exceptional cases only.

Main Indications. Infections by sensitive strains. Important for the treatment of chronic bronchitis, interstitial pneumonia, nonspecific urethritis, brucellosis, tularemia, plague, leptospirosis, lymphogranuloma inguinale, trachoma, cholera, rickettsiosis, melioidosis, *Listeria* infections. May be used in wound infection, urinary tract infection, enteritis, syphilis, and gonorrhea. Certain forms of severe acne respond well to long-term treatment with tetracyclines. The high amount of resistance is relevant in the treatment of wound infections.

Main Side Effects. Intestinal intolerance, liver damage after overdosage, photosensitivity. The teeth of small children may be stained yellow irreversibly. Therefore, tetracyclines should not be used in children before the end of the 6th year. Local irritation is common with the i.v. forms.

Chloramphenicol

The original substance is very bitter, stable, and poorly water-soluble. Chloramphenicol palmitate is tasteless and can be used in syrup. Parenteral forms contain the water-soluble but inactive chloramphenicol succinate; it must be activated by enzymes. Thiamphenicol is the less active analogue with a similar antibacterial spectrum.

Activity. Broad-spectrum activity against most gram-positive and gram-negative bacteria, but also rickettsiae, chlamydiae, *Mycoplasma,* spirochetes, and *Bacteroides.* Resistance is not uncommon in *Staphylococcus aureus* and gram-negative bacteria. Resistant strains of *Hemophilus influenzae* are rare.

Pharmacokinetics. Nearly complete oral absorption. The serum levels of oral chloramphenicol are higher than the same dose of i.v. form. Half-life 3 h, excretion, mainly renal.

Dosage. The dosage range is 1.5-3 g/day in three to four divided doses. Children: 50-80 mg/kg/day. The total dose should generally be limited to 25-30 g for adults (700 mg/kg for children). Topical chloramphenicol is frequently used in ophthalmic preparations (1%).

Indications. Chloramphenicol is now used very restrictively in Europe and the USA. Typhoid fever and gram-negative meningitis *(Hemophilus, E. coli)* are the main indications. The position of the cheap and stable chloramphenicol is far more important in developing countries, where it plays a considerable role in the treatment of other severe bacterial infections. It can be used in the treatment of peritonitis, pneumonia, septicemia, osteomyelitis, pyomyositis, brain abscess, intraocular infections, meningitis.

Side Effects. The restricted use of chloramphenicol in Europe is due solely to side effects. In a very small percentage of the patients, dangerous irreversible pancytopenia may arise several weeks after the treatment. The frequency increases with higher total dose, but blood disorders may arise even after short courses of treatment. The dangerous pancytopenia should not be confused with the quite common depression of erythropoiesis during the treatment. Other side effects are rare. Mild gastrointestinal symptoms like loose stools are quite common, but not dangerous. Allergies are very rare. In newborn infants, overdose may lead to "grey syndrome" due to toxic accumulation. The effects of chloramphenicol on hematopoiesis result in a general restriction of its use; chloramphenicol should not be given in trivial infections which can be treated safely with less dangerous agents. Great caution should be used in patients with blood diseases like aplastic anemia or in combinations with other hematotoxic drugs.

Thiamphenicol has a more marked acute effect on hematopoiesis, but irreversible bone marrow aplasia has not been described. Short-course treatment with thiamphenicol is commonly used for therapy of gonorrhea.

Co-Trimoxazole and Sulfonamides

Co-Trimoxazole

Co-trimoxazole is a fixed combination of trimethoprim plus the sulfonamide sulfamethoxazole. Combinations with other sulfonamides are equivalent but not superior. Tetroxoprim is an analogue to trimethoprim.

Activity. The two agents act as a synergistic combination with a higher activity than the single components. Co-trimoxazole has a broad spectrum of activity (gram-positives except clostridia, gram-negatives except *Pseudomonas*). Resistance can be found in staphylococci, pneumococci, *Hemophilus*, gonococci, *E. coli*, and other gram-negative bacteria. Co-trimoxazole is also active against malaria and *Pneumocystis* infection.

Pharmacokinetics. The pharmacokinetics of the two components are nearly identical. Half-life 10 h, mainly excreted in urine.

Dosage. Standard dose 1.92 g/day (1.6 g sulfamethoxazole + 0.32 g trimethoprim). Long-term treatment possible. A parenteral form is available.

Main Indications. Infections of urinary and respiratory tract, ENT infections, prostatitis, urethritis. Drug of choice for the treatment of typhoid fever and infective enteritis (shigellosis, cholera, salmonellosis). Active in brucellosis, infections by *Nocardia, Mycobacterium marinum,* chancroid. Common use as prophylaxis in leukemia. Less active in staphylococcal infections (wound infections), streptococcal infections (tonsillitis, erysipelas), anaerobic infections (peritonitis, empyema). Drug of choice for the treatment of *Pneumocystis* pneumonia in very high dosage.

Side Effects. Allergy not infrequent due to the sulfonamide component. Severe forms (Lyell's syndrome) possible. Depression of hematopoiesis possible, but rarely severe. Nephrotoxicity may arise in dehydration or underlying renal diseases. Gastric complaints frequent. The parenteral form may cause local phlebitis.

Sulfonamides

Sulfonamides can be classified by their half-life. The leading derivatives from a great bulk of mainly historical sulfonamides are:
- *Short-acting:* Sulfafurazole, sulfadimidine
- *Medium-acting:* Sulfadiazine, sulfamoxole, sulfamethoxazole
- *Long-acting:* Sulfmethoxydiazine
- *Ultra-long-acting:* Sulfadoxine, sulfalene
- *Nonabsorbable:* Sulfaguanole, sulfaguanidine.

Activity. Sulfonamides are only weakly active with a delayed action. They have a quite broad range of activity. Good activity only against streptococci, pneumococci, meningococci. Lower activity against gonococci, *Chlamydia, E. coli, Shigella.* Marked resistance during therapy or after extensive use. A high percentage of common bacteria are now resistant to sulfonamides.

Pharmacokinetics. Good oral absorption (except nonabsorbable sulfonamides). Half-life differing from 8 h in short-acting sulfonamides to 64 h in ultra-long-acting derivatives. The pharmacokinetics of different sulfonamides are quite different.

Dosage. Sulfonamides are used at quite different daily doses:
- Short-acting: 4-6 g
- Medium-acting: 1 g
- Long-acting: 0.5 g
- Ultra-long-acting: 0.2 g (1-2 g/week)
- Nonabsorbable: 4 g

Main Indications. Sulfonamides are inferior to modern antibiotics. The use of these cheap and stable drugs is justified only if better alternatives are not available. The best activity can be achieved in cystitis in females. Sulfonamides are also active in bronchitis and in streptococcal and pneumococcal infections and have been used for chlamydial infections (trachoma). Nonabsorbable sulfonamides played a great role in former times for the treatment of infective enteritis; the majority of *Shigella* strains are now resistant to sulfonamides. Medium- to long-acting sulfonamides are the drug of choice in the treatment of toxoplasmosis (together with pyrimethamine).

Side Effects. Side effects are frequent: marked gastrointestinal intolerance with short-acting sulfonamides. Allergy frequent, sometimes severe. Fatal Lyell's syndrome may occur with long-acting sulfonamides. Renal damage due to crystalluria with some older derivatives. Suboptimal old derivatives may have a very poor tolerance.

Aminoglycosides

Gentamicin is the main derivative of the group of aminoglycosides. Practically identical are sisomicin, tobramycin, and dibekacin. Amikacin and netilmicin are active against some gentamicin-resistant strains. Classical aminoglycosides like kanamycin, neomycin, and paromomycin are inferior to gentamicin. Streptomycin still has a certain position in the treatment of tuberculosis.

Aminoglycosides are very stable; there is no loss of activity in prolonged hot storage.

Gentamicin

Activity. Gentamicin is active against most strains of *E. coli, Klebsiella, Proteus, Pseudomonas,* and *Staphylococcus.* Moderate activity against gonococci, *Hemophilus, Salmonellae.* Streptococci, *Bacteroides,* and clostridia are relatively resistant. Marked synergistic activity with beta-lactams against *Pseudomonas,* enterococci, *Klebsiella.* Increasing resistance in hospitals after prolonged frequent use of gentamicin for years.

Pharmacokinetics. The pharmacokinetics of all aminoglycosides are very similar. No oral absorption. Half-life 120 min, mainly renal elimination. Marked accumulation in renal insufficiency.

Dosage. The daily dose depends on the underlying bacterium. Dose range of 2–3 mg/kg/day to 5 mg/kg/day in two divided doses. Usually patients receive 2 times daily 80–120 mg gentamicin. General USA recommendations are higher than European recommendations. Renal insufficiency necessitates a reduction of dosage.

Main Indications. Gentamicin is now used in Europe mainly to achieve synergistic effects together with beta-lactams. This is necessary in patients with septicemia, severe gram-negative infections, peritonitis, osteomyelitis, or endocarditis. Monotherapy with gentamicin alone in severe gram-negative infections leads to poor results despite in vitro sensitivity; it is justified only in pyelonephritis. The combination of an acylaminopenicillin with gentamicin or tobramycin is the optimal treatment for severe *Pseudomonas* infection. Combinations with metronidazole, clindamycin, or cephalosporins may be used in mixed infections (peritonitis, pelvic inflammatory disease). Topical gentamicin is valuable in eye infections, burns, and infected wounds. Sophisticated local therapies like polymethylmethacrylate (PMMA) beads or gentamicin in bone cement are of increasing importance in Europe.

Side Effects. The modern aminoglycosides all have comparable ototoxicity and nephrotoxicity. Vestibular damage is common in excessive dosage or accumulation and is usually irreversible. Nephrotoxicity may lead to oliguria and tubular necrosis. Combination with old cephalosporins (cephaloridine), and certain diuretics (ethacrynic acid) should be avoided. Allergic reactions (rash, urticaria) are uncommon. Neurotoxicity may occur in combination with anesthetics or muscle relaxants or after intraperitoneal use. The use of aminoglycosides in pregnancy carries the risk of ototoxicity in the newborn.

Amikacin should be used only if gentamicin is ineffective. It has a higher daily dosage (1–2 g/day) and higher toxicity.

Erythromycin

Erythromycin is the main agent of the group of macrolide antibiotics, the other members of which (spiramycin, oleandomycin, josamycin) are inferior to erythromycin. Several forms are available with different oral absorption and metabolism.

Activity. Active against the main gram-positive bacteria (staphylococci, pneumococci, streptococci), gonococci, *Bordetella, Chlamydia, Mycoplasma.* Less active against *Campylobacter, Legionella,* spirochaetes, meningococci, and *Corynebacterium* spp. Resistance is common in *Staphylococcus aureus* and *Hemophilus.*

Pharmacokinetics. Incomplete absorption after oral administration. The most reliant form is erythromycin ethylsuccinate. Poor absorption is possible.

Dosage. The daily dose is 1-2 g for adults in two to four divided doses (40 mg/kg/day in children).

Indications. Main use in acute respiratory infections, otitis media, skin infections, erythrasma. Drug of choice in legionellosis. Alternative to penicillin in allergic patients. Effective in trachoma, lymphogranuloma inguinale, urethritis, whooping cough.

Side Effects. General good tolerance, but quite frequently mild gastrointestinal disorders. Allergies are rare. Erythromycin estolate (rarely other forms) frequently leads to allergic intrahepatic cholestasis with pain and jaundice. Extended use of erythromycin in hospital frequently leads to dissemination of resistant staphylococci; it should be avoided.

Metronidazole

Metronidazole is a member of the group of nitroimidazoles. Other substances from this group are alternatives, the main agents being tinidazole and ornidazole.

Activity. Metronidazole is active only against anaerobic bacteria (except *Propionibacterium* and *Actinomyces*). Most important is the activity against *Bacteroides* species, anaerobic cocci, and clostridia. High activity against ameba, *Trichomonas, Giardia.*

Pharmacokinetics. Good oral absorption, very good tissue penetration. Excretion in the urine, unchanged and as red-brown metabolites.

Dosage. The recommended dosage is 1.0-1.5 g in two to four divided doses. Higher doses are necessary for amebic infection. Lower doses are sufficient for trichomoniasis and giardiasis.

Indications. Mixed anaerobic infections like aspiration pneumonia, pyogenic liver abscess or other intra-abdominal abscesses, peritonitis. The exclusive activity against anaerobes requires a combination with other antibiotics to cover a concomitant infection with enterobacteria and streptococci (aminoglycoside, ampicillin, tetracycline). Intravenous metronidazole is effective in ulcerative stomatitis, gingivitis, periodontitis, and noma; in these conditions it should be given together with benzylpenicillin. Often used for prophylaxis in major colonic surgery. Active in pseudomembranous enterocolitis (inferior to vancomycin) and Crohn's disease. Drug of choice for the treatment of amebic infections and other intestinal parasites. Metronidazole is often used for the treatment of vaginitis.

Side Effects. Metronidazole and the other nitroimidazoles are carcinogenic and mutagenic in the laboratory; they should not be given for trivial infections or in pregnancy. Marked intolerance to alcohol after metronidazole. Some patients complain of an unpleasant metallic taste and gastrointestinal upset. After high dosage and prolonged therapy, central nervous disorders (paresthesia, dizziness, ataxia, reversible neuritis) and also reversible granulocytopenia may arise. Ornidazole and tinidazole have a reduced dosage, lower metabolism, and somewhat better tolerance.

Nitrofurantoin

Nitrofurantoin is the main derivative of the quite toxic group of nitrofurans.

Activity. Most pathogens of urinary tract infections (*E. coli, Klebsiella,* enterococci, staphylococci); inactive against *Pseudomonas* and *Proteus.*

Pharmacokinetics. Rapid absorption and urinary excretion.

Dosage. Standard dosage for adults: 150–300 mg/day in three divided doses.

Indications. Nitrofurantoin was in former times widely used for the treatment of urinary tract infections. It is active in noncomplicated cystitis. It should not be used in invasive urinary tract infections like pyelonephritis.

Side Effects. Nitrofurantoin has poor acute tolerance; nausea, vomiting, and gastric complaints are common. Accumulation in renal insufficiency may lead to severe, irreversible paralysis. Allergies are relatively frequent and may be severe with high mortality from shock, hepatitis, allergic pneumonitis, asthma, and lung fibrosis. Nitrofurantoin is teratogenic and mutagenic in the laboratory and should not be given during pregnancy. In general, the relation between activity and side effects is very poor in nitrofurantoin. It should be given in the treatment of urinary tract infections only if other agents are unavailable. The only residual indication in Europe is constituted by some selected cases of chronic obstructive urinary tract infection. Long-term or repeated use carries a considerable risk of severe side effects and even death.

Rifampicin

Rifampicin (rifampin in the USA) is the only relevant semisynthetic ansamycin antibiotic. It is the most active antituberculous agent and also has a certain position in common bacterial infections.

Activity. Very active against all gram-positive bacteria and *Neisseria, Legionella,* chlamydiae, and mycobacteria. Most gram-negative rods are relatively insensitive, except *Hemophilus influenzae.* Primary resistance is rare, but secondary resistance develops rapidly. Rifampicin should therefore be used only in combination.

Pharmacokinetics. Good oral absorption; mainly biliary elimination. Good tissue penetration.

Dosage. Standard dosage of 600 mg/day in one dose or in two divided doses.

Indications. Rifampicin is the most active agent in the combination therapy of tuberculosis and leprosy. Due to resistance, broad use in other infections should be avoided. Nevertheless, it is highly active in chronic staphylococcal infections (osteomyelitis, endocarditis, abscesses). It must be combined in these indications. Rifampicin is active in the prophylaxis of meningococcal infections. Despite high activity it should not be given for the treatment of gonorrhea.

Side Effects. Frequent increase of transaminases, with fatal acute hepatitis possible. Monitoring of transaminases is desirable. Gastrointestinal intolerance, allergy, and granulocytopenia are rare. Repeated use may occasionally cause hypersensitivity with hemolysis, interstitial nephritis, and acute tubular necrosis. Rifampicin may interfere with many other drugs which are metabolized in the liver (coumarin, tolbutamide, methadone, ketoconazole, anticoagulants). It is mutagenic and must be avoided in pregnancy.

Further Antibiotics and Therapeutic Groups (Not on WHO List)

Quinolones

A new group of chemotherapeutic agents has been used increasingly since 1982 in Europe and Japan. These fluorinated derivatives of nalidixic acid are usually, albeit incorrectly, called "quinolones". Since some of the agents are not quinoline derivatives (nalidixic acid, enoxacin), the term "gyrase inhibitors" is preferred in some countries. Quinolones can be divided into old and modern derivatives:

Modern quinolones		*Old quinolones*	
Ciprofloxacin	Norfloxacin	Nalidixic acid	Piromidic acid
Ofloxacin	Enoxacin	Pipemidic acid	Oxolinic acid
Pefloxacin		Cinoxacin	

Old Quinolones

The old quinolones have poor activity and suboptimal pharmacokinetics with partially extreme metabolism. They were used only to treat urinary tract infections. Nalidixic acid is, at least in India, a common agent for treating shigellosis.

Modern Quinolones

Activity. The modern fluorinated quinolones have a greatly enlarged spectrum of activity in comparison to the old derivatives. They are also 100-1000 times more active. The differences in antibacterial activity within this group of fluoquinolones are not very marked; generally, ciprofloxacin has the highest activity. The modern quinolones cover a very broad spectrum of bacteria. They have very high activity against nearly all aerobic gram-negative bacteria (rods and cocci). Generally, their activity against gram-positive bacteria and *Bacteroides* species is somewhat less, but still sufficient. Resistance may arise, but up to now is a clinical problem only in chronic *Pseudomonas* infection.

Pharmacokinetics. The differences in pharmacokinetics result mainly from different absorption and metabolism. Ofloxacin displays nearly complete absorption and only minor metabolites, in contrast to the other agents. The tissue penetration of these agents is generally good.

Dosage. The dosage is 800 mg/day for norfloxacin, 500-1500 mg/day for ciprofloxacin, and 200-400-800 mg/day for ofloxacin.

Indications. The main indications for quinolones are infections by gram-negative bacteria. They are of very high value for the treatment of complicated urinary tract infection. Norfloxacin is licensed only for this indication. Quinolones may be used for gram-negative wound infection. Their activity in respiratory tract infection is mainly due to their high activity against *Hemophilus influenzae*. Their activity against pneumococci may be insufficient. Quinolones have a very high activity against *Neisseria* species in the single-dose treatment of gonorrhea. Ciprofloxacin, ofloxacin, and pefloxacin are effective in the treatment of typhoid fever. They can also be used for the treatment of *Salmonella* and *Shigella* enteritis, and are active against chancroid. Ciprofloxacin and ofloxacin also have a certain activity against tuberculosis.

Side Effects. The tolerance of quinolones is good. They are non-allergenic and have only minor side effects (gastrointestinal intolerance, slight CNS toxicity like dizziness, unrest, headache). The major disadvantage is derived from animal experiments. All quinolones, including the old derivatives, lead to considerable cartilage damage in experimental animals. There is doubt whether this is relevant to man, but as a general precaution, quinolones are not allowed in children and pregnant and lactating women.

Outlook. Many further derivatives of this group are in development. It is hoped that agents may be found with an even higher activity against gram-positive bacteria; maybe a perfect agent for the treatment of tuberculosis will emerge. The heat resistance of quinolones is an advantage in the tropics, and in contrast to modern penicillins and cephalosporins, they are quite cheap. Quinolones will probably have a considerable influence on the practical treatment of bacterial infections in tropical countries.

Acylaminopenicillins

Acylaminopenicillins are a group of ampicillin derivatives with enlarged spectrum. The main agents are: azlocillin, mezlocillin, and piperacillin.

Activity. Like ampicillin, but better activity against *Pseudomonas, Klebsiella, Proteus.* Inactive against penicillinase-producing staphylococci.

Pharmacokinetics. Similar to parenteral ampicillin, but with a higher biliary excretion.

Dosage. Wide range from 3-15-20 g/day i.v. in two to four divided doses, depending on site and severity of infection.

Indications. Severe infections by *Pseudomonas, Klebsiella, Proteus, Serratia,* usually together with gentamicin. In Europe, often used in complicated urinary tract infection, pelvic inflammatory disease, biliary and abdominal infections, and septicemia. The combination of acylaminopenicillins with cefotaxime is important for the empirical treatment of infections in leukemia.

Side Effects. Comparable to other penicillins: rash, allergy, occasionally reversible granulocytopenia.

Cephalosporins

Cephalosporins are a group of quite different derivatives which are commonly separated into first, second, and third generations.

Group	*Main agents*
First generation	Cefazolin, cefazedone (inferior and historical derivatives are cephalothin, cephaloridine)
Second generation	Cefamandole, cefuroxime, cefotiam, cefoperazone
Cefoxitingroup	cefoxitin, cefmetazole, cefotetan, latamoxef
Third generation	Cefotaxime, cefmenoxime, ceftizoxime, ceftazidime, ceftriaxone
Oral cephalosporins	Cefalexin, cefradin, cefadroxil, cefaclor

Activity. The different groups of cephalosporins have quite different activities. The *first-generation* cephalosporins are active against streptococci (except enterococci), pneumococci, staphylococci, *E. coli, Proteus mirabilis, Klebsiella.* The activity of oral cephalosporins is comparable but somewhat less than first-generation cephalosporins. *Second-generation* cephalosporins have a more or less enlarged spectrum of activity against resistant gram-negative rods (*Klebsiella, Proteus vulgaris, Hemophilus,* gonococci). Cefoxitin and related derivatives are active, too, against beta-lactamase-producing *Bacteroides fragilis. Third-generation* cephalosporins have a very high activity against all enterobacteria. They are active to a lesser extent against staphylococci and against *Pseudomonas.* They are very active against *Hemophilus* and *Neisseria* spp. Ceftazidime has a remarkable activity against *Pseudomonas* spp.

Pharmacokinetics. Oral cephalosporins are completely absorbed with a rapid and complete renal excretion. Most cephalosporins have comparable pharmacokinetics, with a short half-life and mainly renal excretion. Ceftriaxone has a prolonged half-life with marked biliary excretion.

Dosage. Quite similar in all parenteral cephalosporins: 3–4–6 g/day in two or three divided doses according to site and severity of infection. Ceftriaxone 2–4 g, oral cephalosporins 3–6 g.

Indications. Cephalosporins are very important antibiotics for clinical use in Europe, US and Japan. Oral cephalosporins are used mainly for pediatric infections and mild respiratory and wound infections. *First-generation* cephalosporins are useful for the treatment of wound infections and community-acquired pneumonias and for perioperative prophylaxis. *Second-generation* cephalosporins are useful for less severe secondary infections in the hospital. Cefuroxime has sufficient activity against *Hemophilus* and advantages in severe pediatric infections. Cephalosporins of the cefoxitin group are useful for the treatment of pelvic inflammatory disease. *Third-generation* cephalosporins are important for the treatment of severe infections by resistant enterobacteria (pneumonia, septicemia, cellulitis). They are frequently used in severe infections together with aminoglycosides (gentamicin) or acylaminopenicillins (piperacillin, azlocillin). Cefotaxime and ceftriaxone are active in the treatment of typhoid, meningitis, severe *Hemophilus* infection, and osteomyelitis. *Second- and third-generation* cephalosporins can be used for the single-dose treatment of gonorrhea with high efficiency. Ceftriaxone may have certain advantages under conditions of limited resources due to its extended half-life.

Side Effects. Comparable to penicillins: rash, allergy, occasionally reversible granulocytopenia, pain after i.m. injection. The methyltetrazole side chain may lead to hypoprothrombinemia (cefmetazole, cefotetan, cefamandole, cefoperazone, cefmenoxime). The expensive and not very stable cephalosporins have disadvantages for broad use under conditions of limited resources. They may be used in selected cases involving highly resistant bacteria or in infants, where other antibiotics cannot be given.

Other Antibiotics

Vancomycin

Vancomycin is the leading derivative of the glycopeptide antibiotics. Several improved derivatives (teichoplanin) are under development.

Activity. Only gram-positive bacteria (staphylococci, streptococci, clostridia, *Corynebacterium*); resistance practically absent. No activity against any gram-negative bacteria.

Pharmacokinetics. Oral absorption absent. Half-life 6 h; elimination exclusively renal with marked accumulation in renal insufficiency.

Dosage. Adult dose: 1-2 g as two very slow infusions. Oral administration (0.125 g every 6-8 h in enterocolitis).

Main Indications. Severe staphylococcal infections (septicemia, endocarditis, osteomyelitis), mainly by methicillin-resistant strains. Very active in pseudomembranous enterocolitis due to *Clostridium difficile.*

Side Effects. Frequent vascular reactions (flush and hypotension due to rapid infusion). Real allergic reactions (rash, urticaria) may occur. Ototoxicity mainly in renal insufficiency.

Clindamycin

Clindamycin is an improved derivative of lincomycin with higher activity and better absorption.

Activity. Mainly staphylococci, pneumococci, streptococci (except enterococci). Somewhat lower activity against *Bacteroides,* fusobacteria, *Actinomyces,* peptostreptococci, clostridia, and *Campylobacter.* No activity against enterococci, *Neisseria,* or any aerobic gram-negative bacilli.

Pharmacokinetics. Good oral absorption. Half-life 3 h; principally renal elimination, mainly as metabolites.

Dosage. The recommended dosage is 0.6-1.2 g orally in three to four divided doses; a parenteral form is available. Certain degree of accumulation in severe hepatic and renal disease.

Indications. Severe staphylococcal or anaerobic infections like empyema, lung abscess, severe oral infection. In peritonitis, intra-abdominal abscesses, and severe pelvic infections, clindamycin should be combined with other antibiotics (gentamicin, ampicillin) to cover enterobacteria and enterococci. Clindamycin is very effective in methicillin-resistant staphylococci. It is the drug of choice in the treatment of staphylococcal osteomyelitis in Europe. It may be used for streptococcal infections in allergy to penicillin.

Side Effects. Loose stools in 5%-20% of cases. Sometimes severe pseudomembranous enterocolitis by *Clostridium difficile* may arise, which necessitates treatment with vancomycin or metronidazole. Rarely allergy and rise of transaminases. Phlebitis common in the i.v. form. Rapid i.v. injection must be avoided.

Imipenem

Imipenem is the first derivative of the new group of penems - beta-lactam antibiotics which are neither penicillins nor cephalosporins. The formulation of imipenem is a combination with cilastatin to inhibit renal cleavage of the antibiotic. Imipenem and cilastatin sodium are present in the ampoules in a 1:1 ratio.

Activity. Imipenem has a very broad spectrum of activity, including almost all gram-positive organisms and gram-negative bacteria. It is stable to all beta-lactamases; only some strains of *Pseudomonas cepacia, Streptococcus faecium,* and methicillin-resistant staphylococci are not affected by imipenem.

Pharmacokinetics. Similar to penicillin with a short half-life and mainly renal elimination.

Dosage. The usual daily dose is 3 times 0.5 g imipenem (ampoules) together with the same amount of cilastatin.

Main Indications. Imipenem is being used increasingly in Europe and the USA for severe bacterial infections, mainly in intensive care patients, leukemia, and severe wound infections. It has a high clinical activity in mixed infections which include *Bacteroides fragilis.* Often it is used if other forms of treatment have been ineffective. It is active in infections by *Mycobacterium avium-intracellulare,* but not in tuberculosis.

Side Effects. Similar to other beta-lactams: allergies (rashes), eosinophilia (local reactions). Rapid injection should be avoided. The use of the unstable and expensive imipenem will be restricted to very few selected cases in developing countries.

Amoxicillin/Clavulanic Acid (Augmentin)

Activity. In comparison to amoxicillin extended spectrum of activity against certain beta-lactamase-producing bacteria (e.g. staphylococci, *Hemophilus,* enterobacteria).

Pharmacokinetics. Oral and parenteral administration, half-life about 1 h, renal excretion of amoxicillin higher than of clavulanic acid.

Dosage. Orally 1-2 tablets of 375 mg (250 mg amoxicillin and 125 mg clavulanic acid) or 1 tablet of 625 mg (500 mg amoxicillin and 125 mg clavulanic acid) 3 times a day. Parenterally 600 mg (500 mg amoxicillin and 100 mg clavulanic acid) or 1.2 g (1 g amoxicillin and 200 mg clavulanic acid) both 3 times a day.

Main Indications. Infections with amoxicillin-resistant bacteria, the beta-lactamases of which are inhibited by clavulanic acid.

Main Side Effects. Known side effects of amoxicillin, in addition about 10% of patients suffer from nausea, vomiting, and diarrhea predominately due to the clavulanic acid.

References

1. Garrod LP, Lambert HP, O'Grady F (1981) Antibiotics and chemotherapy, 5th edn. Churchill Livingstone, Edinburgh
2. Gilman AY, Goodman LS, Rawl L, Murad S (eds) (1985) Goodman and Gilman's the pharmacological basis of therapeutics, 7th edn. Macmillan, New York
3. Greenwood D (1983) Antimicrobial chemotherapy. Baillière Tindall, London
4. Kucers A, McK Bennett N (1987) The use of antibiotics, 4th edn. Heinemann, London
5. Mandell GL, Douglas RG, Bennett JE (1985) Principles and practice of infectious diseases, 2nd edn. Wiley, New York
6. Simon C, Stille W (1985) Antibiotika - Therapie in Klinik und Praxis, 6th edn. Schattauer, Stuttgart
7. Simon C, Stille W, Wilkinson PJ (1984) Antibiotic therapy in clinical practice. Schattauer, Stuttgart
8. WHO Expert Committee (1983) The use of essential drugs. Tech Rep Ser No. 685, WHO, Geneva

2 Choice of Antibiotics

The decision to start treatment with an antibiotic depends largely on reliable clinical diagnosis. Pooling the information from anamnesis, complaints, prior therapy, symptoms, and direct clinical examinations, the majority of acute infections can be localized. Routine bacteriological facilities are available only in a minority of places in the world, and in developing countries therapy of infection usually has to be carried out without sophisticated bacteriology. Gram-stains of pus, sputum, and CSF fluid may give important information in selected cases. In general, the antibiotic treatment has to be carried out without further bacteriological results.

The selection of antibiotics depends largely on the potential spectrum of pathogens in the actual infection. Typical clinical situations (pyelonephritis, osteomyelitis) have quite different typical spectra of pathogens. The mode of action of the antibiotic, its pharmacokinetic properties, the site and severity of the infections, and the patient's age and hepatic and renal function influence the choice. Other important factors are the ease of administration and the therapeutic range of the antibiotic. Nonmedical factors are the availability and the price of the antibiotic. The selection of the appropriate antibiotic thus depends on sensitivity, pharmacokinetic parameters, clinical efficacy, and side effects. The clinical efficacy of the more important antibiotics against common and rare bacterial pathogens is summarized in Tables 2.1-2.3. Only a few types of infection are seen frequently in clinical practice, and for all of these it is possible to define a treatment of choice. The empiric treatment of common bacterial infections with antibiotics taken from the WHO list only is presented in Table 2.4. In contrast, Table 2.5 presents the directed treatment of common bacterial infections according to the pathogen. Treatment of fungal infections is displayed in Table 2.6.

In the absence of culture results, the response to initial antibiotic treatment plays a more important role in the evaluation of common bacterial infections. The distinction into responsive and nonresponsive infections is appropriate when the etiological agent cannot be identified. Practically, this means that if pneumonia, for instance, does not respond to treatment with benzylpenicillin or phenoxymethylpenicillin, pneumococci can almost certainly be excluded as the pathogens. Treatment of a nonresponsive pneumonia differs markedly from the initial treatment, since other pathogens are now much more likely to be the causative agent.

Bacteriological examination has a much more important position in the evaluation of nonresponsive infections. In certain clinical conditions, there is a high probability of highly resistant strains (chronic urinary tract infections, burns, chronic osteomyelitis, infected ulcers). Bacteriological cultures are more desirable in these circumstances.

Table 2.1. Clinical use of antibiotics on the WHO list

Bacterial species	Benzylpenicillin	Ampicillin	Cloxacillin	Tetracycline	Chloramphenicol	Gentamicin, tobramycin	Amikacin	Erythromycin	Co-trimoxazole	Sulfonamide	Metronidazole	Rifampicin	Nitrofurantoin[a]
Corynebacterium diphtheriae	●	+	+	⊕	+	+	+	⊕	+	+	○	+	○
Streptococci	●	+	+	+	+	+	+	⊕	+	+	○	+	○
Pneumococci	●	+	+	+	+	○	○	⊕	⊕	+	○	+	○
Enterococci	+	●	○	⊕	+	○	○	⊕	+	○	○	○	⊕
Staphylococci	⊕	+	●	+	+	+	+	⊕	+	○	○	⊕	+
Gonococci	●	+	+	⊕	+	+	+	⊕	+	+	○	⊕	+
Meningococci	●	+	+	+	⊕	+	+	⊕	+	+	○	⊕	+
Hemophilus influenzae	○	●	○	●	●	+	+	⊕	⊕	+	○	⊕	○
Legionella pneumophila	○	○	○	+	○	○	○	●	○	○	○	●	○
Brucella	○	+	○	●	⊕	+	+	○	⊕	○	○	⊕	○
Bacteroides fragilis	○	○	○	⊕	⊕	○	○	+	+	○	●	+	○
Bacteroides melaninogenicus	●	⊕	○	⊕	+	○	○	+	+	○	●	+	○
Mycobacterium tuberculosis	○	○	○	+	○	+	+	○	○	○	○	●	○
Treponema pallidum	●	+	+	⊕	+	○	○	+	○	○	○	○	○
Listeria	⊕	●	○	+	+	+	+	+	+	○	○	+	○
Clostridia	●	+	+	⊕	+	○	○	⊕	+	○	○	+	○
Pseudomonas aeruginosa	○	○	○	+	+	●	●	○	○	○	○	○	○
Escherichia coli	○	⊕	○	⊕	⊕	⊕	⊕	○	●	+	○	○	+
Klebsiella pneumoniae	○	○	○	⊕	+	●	●	○	●	+	○	○	+
Enterobacter aerogenes	○	○	○	⊕	+	●	●	○	●	+	○	○	+
Proteus vulgaris	+	+	○	+	+	●	●	○	●	+	○	○	○
Proteus mirabilis	+	●	○	+	+	●	●	○	●	+	○	○	○
Salmonella typhi	+	⊕	○	+	●	+	+	○	●	+	○	○	○
Salmonella typhimurium	+	●	○	⊕	⊕	+	+	○	●	+	○	○	○
Shigella	+	●	○	⊕	⊕	+	+	○	●	+	○	○	○
Rickettsiae	○	○	○	●	●	○	○	+	○	○	○	⊕	○
Mycoplasma pneumoniae	○	○	○	●	⊕	○	○	⊕	○	○	○	⊕	○
Chlamydia psittaci	○	○	○	●	⊕	○	○	+	○	○	○	⊕	○
Chlamydi trachomatis	○	○	○	●	⊕	○	○	●	⊕	+	○	⊕	○

● Very active, first choice for treatment; ⊕ good activity, second line antibiotic; + active but only recommended in special cases or not at all; ○ inactive.

[a] Use only in UTI.

Table 2.2. Clinical use of the more important antibiotics not on the WHO list

Bacterial species	Azlocillin	Mezlocillin	Piperacillin	Cefazolin	Cefuroxime	Cefoxitin	Cefotaxime	Ceftazidime	Latamoxef	Augmentin	Clindamycin	Fusidic acid	Norfloxacin	Ofloxacin	Ciprofloxacin	Vancomycin	Imipenem
Corynebacterium diphtheriae	+	+	+	+	+	+	+	+	+	+	+	+	○	+	+	+	+
Streptococci	+	+	+	+	⊕	+	+	+	+	+	⊕	+	○	+	+	+	⊕
Pneumococci	+	+	+	+	⊕	+	+	+	+	+	⊕	+	○	+	+	+	⊕
Enterococci	⊕	●	⊕	○	○	○	○	○	○	+	○	○	+	+	+	⊕	⊕
Staphylococci	+	+	+	⊕	⊕	⊕	+	○	○	●	●	⊕	+	⊕	⊕	●	●
Gonococci	+	+	+	+	●	⊕	●	⊕	⊕	⊕	○	+	●	●	●	○	●
Meningococci	+	+	+	+	⊕	+	⊕	⊕	+	+	○	+	+	⊕	⊕	○	⊕
Hemophilus influenzae	●	+	●	+	⊕	+	●	●	⊕	●	○	○	+	●	●	○	●
Legionella pneumophila	○	○	○	○	○	○	+	+	+	○	○	○	○	+	+	○	+
Brucella	+	+	+	+	+	+	+	+	+	○	○	○	○	⊕	⊕	○	+
Bacteroides fragilis	⊕	+	⊕	○	○	●	○	○	●	+	●	○	○	+	+	○	●
Bacteroides melaninogenicus	+	+	+	○	+	●	+	+	⊕	+	●	○	○	+	+	○	⊕
Mycobacterium tuberculosis	○	○	○	○	○	○	○	○	○	○	○	○	○	+	+	○	○
Treponema pallidum	+	+	+	⊕	⊕	+	+	+	+	+	○	○	○	○	○	○	+
Listeria	+	⊕	⊕	○	○	○	○	○	○	⊕	+	○	○	+	+	○	+
Clostridia	+	+	+	+	+	+	+	+	+	+	+	+	○	+	+	⊕	+
Pseudomonas aeruginosa	●	⊕	●	○	○	○	+	●	+	○	○	○	⊕	●	●	○	●
Escherichia coli	⊕	●	●	⊕	●	●	●	⊕	⊕	⊕	○	○	●	●	●	○	●
Klebsiella pneumoniae	○	+	⊕	+	⊕	●	●	⊕	⊕	⊕	○	○	●	●	●	○	●
Enterobacter aerogenes	○	○	+	+	⊕	⊕	⊕	⊕	●	○	○	○	●	●	●	○	●
Proteus vulgaris	⊕	●	●	+	⊕	●	●	●	●	+	○	○	●	●	●	○	●
Proteus mirabilis	●	●	●	⊕	⊕	●	●	●	●	⊕	○	○	●	●	●	○	●
Salmonella typhi	+	+	+	+	+	+	⊕	+	+	+	○	○	+	●	●	○	+
Salmonella typhimurium	+	+	+	+	+	+	⊕	+	+	+	○	○	●	●	●	○	⊕
Shigella	+	+	+	+	+	+	+	+	+	+	○	○	●	●	●	○	+
Rickettsiae	○	○	○	○	○	○	○	○	○	○	○	○	+	+	+	○	○
Mycoplasma pneumoniae	○	○	○	○	○	○	○	○	○	○	○	○	+	+	+	○	○
Chlamydia psittaci	○	○	○	○	○	○	○	○	○	○	○	○	+	+	+	○	○
Chlamydia trachomatis	○	○	○	○	○	○	○	○	○	○	○	○	+	⊕	⊕	○	○

● Very active, first choice for treatment; ⊕ good activity, second line antibiotic; + active but only recommended in special cases or not at all; ○ inactive.

Table 2.3. Clinical use of the more important antibiotics against less common pathogens

Bacterial species	Benzylpenicillin	Ampicillin	Cefazolin	Cefoxitin	Cefotaxime	Streptomycin	Tetracycline	Chloramphenicol	Erythromycin	Kanamycin	Gentamicin	Polymyxin	Sulfonamide	Co-trimoxazole	Quinolones
Acinetobacter	∅	±	∅	∅	±	±	⊕⊕	∅	±	+	+	+	∅	+	+
Actinomyces israelii	⊕	+ +	+ +	+ +	+	±	±	±	±	∅	∅	∅	+	+	
Aeromonas hydrophila	∅	±	∅	+	+	±	+ +	+ +	∅	±	±	+	+	+	+ +
Bacillus anthracis	⊕⊕	+	+ +	+	+	+	+	+	+	+	+	∅	±	+	+
Bordetella pertussis	∅	⊕⊕	∅	∅	+	±	+ +	+ +	⊕⊕	±	±	±	∅	+	+
Borrelia recurrentis	+	+	+	+	+	+	⊕⊕	+	+	+	+	∅	∅	∅	+
Campylobacter jejuni	+	+	+	+	+	+	⊕⊕	⊕⊕	⊕⊕	+	⊕⊕	∅	∅	+	+
Citrobacter	∅	+	+	∅	+	+	+	+ +	∅	+ +	⊕	+	+	⊕	+ +
Enterobacter cloacae	∅	+	+	±	±	+	+ +	+ +	∅	+ +	⊕⊕	+	+	+	+ +
Erwinia	∅	±	±	+	⊕⊕	+	+	+ +	∅	+ +	+	+	∅	+	+ +
Erysipelothrix rhusiopathiae	⊕⊕	+	+	+	+	∅	+	+	+	∅	∅	∅	∅		+
Francisella tularensis	∅	∅	∅	∅	+	⊕⊕	+ +	+ +	∅	+	+	+	∅	±	+
Fusobacterium spp.	⊕⊕	+	+	+	+	+	+	+	+	∅	∅	∅	+	+	+
Hemophilus ducreyi	∅	+	+	+	+	+	+	+	+	+	+	∅	⊕	⊕⊕	+ +
Hafnia	∅	+	+	+	⊕⊕	+	+ +	+ +	∅	+ +	⊕⊕	+	+	+	+
Leptospira	⊕	+	+	+	+	±	⊕⊕	±	+	∅	∅	∅	∅	∅	
Moraxella lacunata	∅	∅	+	+	+	+	⊕⊕	+ +	∅	+	+	∅	∅	±	+
Nocardia asteroides	∅	∅	∅	∅	∅	∅	⊕⊕	+	∅	∅	+	∅	⊕⊕	⊕⊕	
Pasteurella multocida	⊕	+	+	+	+	±	⊕⊕	+	+ +	∅	+	+ +	+	+	+ +
Providencia	∅	∅	∅	⊕	⊕⊕	∅	∅	±	∅	+	+ +	∅	∅	⊕	+ +
Ps. cepacia	∅	∅	∅	∅	+	∅	∅	+	∅	∅	∅	∅	+	+	+
Ps.mallei	∅	∅	∅	∅	+	+	⊕⊕	+	∅	+	+	∅	+	⊕	+
Ps. maltophilia	∅	∅	∅	∅	+	∅		+	∅	∅	∅	∅	∅	⊕	+
Ps. pseudomallei	∅	∅	∅	∅	+	∅	⊕	⊕	∅	∅	∅	∅	+	+	+
Serratia marcescens	∅	∅	∅	⊕	⊕⊕	∅	∅	∅	∅	⊕	⊕	∅	∅	⊕	+ +
Streptobacillus moniliformis	⊕⊕	+	+	+	+	+	⊕⊕	+	+	+	+	∅	⊕	+	
Vibrio cholerae	∅	+	+	+	+	±	⊕⊕	+	+	+	+	∅	⊕	⊕	+
Y. enterocolitica	∅	∅	∅	+	+	+	⊕⊕	⊕⊕	∅	+	+	+	+	+	+
Y. pestis	∅	∅	∅	∅	+	⊕⊕	+ +	+ +	∅	+	+	∅	+	+	+
Y. pseudotuberculosis	∅	∅	∅	+	+	+	⊕⊕	⊕⊕	∅	+ +	+	+	+	+	+

⊕⊕ antibiotic of choice; + + very active; + moderately active; ± doubtful activity; ∅ inactive; no sign: unknown.
Ps., Pseudomonas, Y., Yersinia.

Table 2.4 see p. 32

Table 2.5. Common pathogens, common infections, and antibiotic therapy

Name and synonyms	Usual site of occurrence	Typical infections	WHO list		Non-WHO list, good activity	Poorly effective	Inactive
			Drug of choice	Second choice			
Staphylococcus aureus Coagulase-positive staphylococci	Skin, upper respiratory tract	Boils, infected wounds, mastitis, purulent parotitis, suppurative pneumonia, food poisoning, infections associated with foreign bodies, osteomyelitis, pyomyositis	Cloxacillin, when sensitive benzylpenicillin	Erythromycin, chloramphenicol (only in severe infections!)	Clindamycin, vancomycin, cefazolin, fusidic acid, imipenem	Ampicillin, tetracycline, aminoglycosides (only in combination)	Metronidazole
Staphylococcus epidermidis, coagulase-negative staphylococci, *Staphylococcus albus*	Always on the skin, nasal mucosa	Occasional cause of endocarditis, infections associated with foreign bodies (e.g. valve implants)	Same as *Staphylococcus aureus*, but often more resistant				
Streptococcus pyogenes, beta-hemolytic streptococci, group A streptococci	Throat	Erysipelas, scarlet fever, angina, rheumatic fever, puerperal sepsis, cellulitis, septicemia	Benzylpenicillin, phenoxymethylpenicillin	Erythromycin, ampicillin, tetracycline	Cephalosporins, clindamycin	Co-trimoxazole	Metronidazole, aminoglycosides
Streptococcus pneumoniae, pneumococci	Upper respiratory tract	Lobar pneumonia, bronchitis, sinusitis, corneal ulcer, meningitis, empyema, septicemia, otitis media	Benzylpenicillin, phenoxymethylpenicillin	Ampicillin, erythromycin, chloramphenicol (in meningitis), co-trimoxazole	Cephalosporins, clindamycin, vancomycin	Tetracycline, quinolones	Metronidazole, aminoglycosides
Group B streptococci, *Streptococcus agalactiae*	Genital tract, intestines, cause of animal infections	Neonatal septicemia and meningitis, gynecological infections, pyelonephritis	Benzylpenicillin, phenoxymethylpenicillin	Ampicillin	Cephalosporins	Tetracycline	Metronidazole, aminoglycosides
Group D streptococci, enterococci, *Strep. faecalis, Strep. faecium*, faecal streptococci, etc.	Intestines, urethra, vagina	Urinary infections, mixed infections of intestinal origin, septicemia, endocarditis	Ampicillin	Erythromycin, benzylpenicillin in high dosage	Vancomycin, mezlocillin, imipenem	Aminoglycosides (only in combination), tetracycline	Clindamycin, metronidazole, cephalosporins
Other aerobic streptococci, viridans and nonhemolytic streptococci	Upper respiratory tract, intestines	Subacute bacterial endocarditis	Benzylpenicillin	Ampicillin	Cefazolin, vancomycin, clindamycin	Tetracycline, aminoglycosides	Metronidazole
Anaerobic streptococci, peptostreptococci	Mouth, intestines, vagina	Mixed infections of intestinal or genital origin, dental infections, brain and lung abscesses	Benzylpenicillin, phenoxymethylpenicillin	Erythromycin, clindamycin, tetracycline, metronidazole	Cefoxitin	Aminoglycosides	

Escherichia coli	Intestines, possibly also mouth and vagina; common in animals too	Urinary infections, pyelonephritis, neonatal meningitis, cholangitis	Ampicillin, co-trimoxazole	Chloramphenicol, tetracyclines	Cephalosporins, mezlocillin, quinolones	Sulfonamides	Narrow spectrum antibiotics, metronidazole
Organisms of the *Klebsiella Enterobacter* group	Respiratory tract, intestines	As with *E.coli*, also *Klebsiella pneumoniae*	Co-trimoxazole	Chloramphenicol, gentamicin, amikacin	Cefotaxime, mezlocillin, quinolones, imipenem	Tetracycline	Ampicillin, metronidazole
Salmonella species (nontyphoidal salmonellae)	Animals; intestinal carriers in animals and men	Enteritis, food poisoning, septicemia, abscesses, osteomyelitis, meningitis in infants	Co-trimoxazole, amoxicillin	Chloramphenicol	Quinolones, mecillinam	Tetracycline, aminoglycosides	Cefazolin, metronidazole
Proteus mirabilis	Intestines, sewage, dirt	Urinary infections	Ampicillin, co-trimoxazole	Gentamicin, amikacin, chloramphenicol	Quinolones, mezlocillin, cephalosporins	Tetracycline	Metronidazole
Proteus vulgaris, Pr. morganii, Pr. rettgeri	Intestines, sewage, dirt	Urinary infections and occasionally in ulcers, burns, wound infections, chronic otitis	Co-trimoxazole	Gentamicin, amikacin	Quinolones, mezlocillin, cefotaxime, cefoxitin, imipenem	Ampicillin, tetracycline, chloramphenicol	Metronidazole
Pseudomonas aeruginosa	Rarely on skin or mucous membranes, but common in sewage and dirt, sometimes in intestines	Wound infections, particularly burns, chronic otitis, urinary infections, septicemia, chronic bronchitis, ecthyma gangrenosum, umbilical infections	Tobramycin, gentamicin	Amikacin	Azlocillin, piperacillin, quinolones, ceftazidime, imipenem	Co-trimoxazole	Ampicillin, cefazolin, metronidazole, tetracycline, chloramphenicol
Hemophilus influenzae	Respiratory tract	Chronic bronchitis, bronchopneumonia, otitis media, sinusitis, conjunctivitis, meningitis, septicemia, epiglottitis	Ampicillin	Tetracycline, co-trimoxazole, chloramphenicol (in meningitis)	Cefotaxime, cefuroxime, mezlocillin, quinolones	Benzylpenicillin, aminoglycosides	Metronidazole
Bacteroides melaninogenicus	Upper respiratory tract, rare in intestines	Dental sepsis, lung abscess, empyema, brain abscess	Benzylpenicillin, phenoxymethylpenicillin	Tetracycline, metronidazole	Cephalosporins, clindamycin, mezlocillin	Ampicillin, erythromycin	Co-trimoxazole, aminoglycosides, quinolones
Bacteroides fragilis	Mouth, intestines	Mixed infections of intestinal origin, appendicitis, pylephlebitis, septic thrombophlebitis, genital infections, abscesses	Metronidazole	Chloramphenicol	Clindamycin, cefoxitin, latamoxef, imipenem	Rifampicin	Co-trimoxazole, aminoglycosides, quinolones

Table 2.4. Choice of antibiotics from the WHO list

Common bacterial infection	Antibiotics											
	Penicillin	Ampicillin	Cloxacillin	Gentamicin	Erythromycin	Chloramphenicol	Tetracycline	Metronidazole	Sulfonamide	Co-trimoxazole	Amikacin	Others
UTI, noncomplicated	5	2	6	3	6	4	3	6	5	①	4	Nitrofurantoin 4
UTI, complicated	6	3	6	2	6	2	3	6	5	①	3	—
UTI, recurrent	5	2	6	4	6	4	3	6	3	①	4	Nitrofurantoin 3
Urethritis	3	2	6	4	5	4	2	6	5	①	4	Spectinomycin 3
PID	2	①	6	4	5	4	2	3	5	2	4	—
Tonsillitis	①	3	4	4	2	4	4	6	5	5	5	—
Sinusitis/otitis	①	2	4	5	3	4	2	6	5	2	5	—
Acute bronchitis (adults)	2	①	4	4	3	4	①	6	5	①	4	—
Pneumonia, lobar (adults)	①	2	3	4	3	4	2	6	5	3	4	—
Pneumonia, interstitial (adults)	5	5	6	5	2	4	①	6	5	3	5	Rifampicin 3
Chronic bronchitis	2	①	5	5	3	4	①	6	3	①	5	—
LRTI (children)	①	2	4	5	3	4	4	6	5	2	5	—
Pneumonia (child, hospital)	①	4	2	4	3	2	4	6	6	3	4	—
Sinusitis/otitis (children)	2	①	4	5	3	4	4	6	5	3	5	—
Wound infection (outpatient)	①	4	2	3	3	4	3	5	5	3	4	—
Wound infection (hospital)	4	4	①	2	3	4	5	4	5	3	4	—
Erysipelas	①	2	3	6	2	4	3	6	5	4	6	Procaine penicillin

UTI, urinary tract infection; PID, pelvic inflammatory disease. LRTI, Lower respiratory tract infection.
1, active: drug of choice; 2, active: second choice; 3, active: third choice; 4, active: use only in exceptional cases; 5, weakly active: activity only in special forms; 6, inactive.
[a] Penicillin = Benzylpenicillin, procaine penicillin, phenoxymethylpenicillin.

Table 2.6. Treatment of fungal infections

Fungal infections	Usual site of occurrence	Typical infections	Topical therapy			Systemic therapy	
			Drug of choice	Second choice	Active	Oral	Parenteral
Candida albicans	Skin, mouth, intestines	Candida stomatitis, (thrush), vaginitis, balanitis, occasional pneumonia, esophagitis, septicemia. Common in skin infections, e.g. as intertrigo and nail infections	Clotrimazole, miconazole, other azoles	Amphotericin B, nystatin	Ciclopiroxolamine, povidone iodine, Castellani paint, naftifin	Ketoconazole, flucytosine, fluconazole (investigational)	Amphotericin B, miconazole
Aspergillus fumigatus	Environment, dust, debris	Aspergilloma, pneumonia, sinusitis, otitis	Amphotericin B	Miconazole, other azoles	–	Itraconazole (investigational), flucytosine	Amphotericin B
Dermatophytes	Skin, animals	Ringworm, tinea, infection of nails	Whitfields ointment (6% benzoic acid + 3% salicylic acid)	Clotrimazole, miconazole, other azoles, tolnaftate	Haloprogin, natamycin	Griseofulvin, ketoconazole	Not necessary
Pityrosporum infection	Skin	Pityriasis versicolor	2.5% selenium sulfide	Miconazole, clotrimazole	Sulphur preparations	Ketoconazole	Not necessary
Cryptococcus neoformans	Environment, bird excreta	Pneumonia, meningitis, sinusitis	—— Not advisable ——			Flucytosine, fluconazole (investigational)	Amphotericin B

3 Bacterial Resistance in Developing Countries

The problems of bacterial resistance currently encountered in developing countries are to a certain degree comparable with the conditions met in Europe and the USA in 1950–1970; however, there are also remarkable differences. In developing countries there is usually a nonavailability of bacteriological services. Resistance is most often registered as clinical failure. Where a bacteriological laboratory is available, more often than not only simple routine tests can be performed; even widespread problems of bacterial resistance may thereby either remain undetected or appear as local problems only. Generally, bacterial resistance is of more concern in developing countries than in developed countries. Infections with resistant *Salmonella, Shigella* or *Neisseria gonorrhoeae,* for instance, are far more frequent. Resistance to common, widely available antibiotics such as the oral agents on the WHO list can not easily be overcome by the use of expensive parenteral antibiotics [6, 10].

Distribution of resistant bacteria is favored by many of the conditions prevailing in a tropical hospital. Overcrowding, crossinfection, nonsterile dressings, suboptimal disinfection, and other failures of hygiene are the main factors in a remarkably high level of resistance [3, 4]. Poor kitchens and inadequate food hygiene may contribute to the problem [23]. The high proportion of ambulatory treatment, short stay in the hospital, and short duration of treatment should antagonize the spread of resistant strains. However, a recent report from Brazil pointed out that hospitals may in fact act as reservoirs for infections with multiple resistant *Salmonella typhimurium* in the community [24]. According to common opinion, abuse of antibiotics, self-treatment, short duration of treatment, and minimal dosage all favor the development of resistance [10]. In general, these influences are certainly overrated in respect to developed countries, where the use of antibiotics in animal breeding contributes to a large extent to resistance. The widespread self-treatment with often poorly effective agents in inadequate dosage which is prevalent in many tropical cities is considered a major factor in the development of bacterial resistance in developing countries. Lack of sanitation and hygiene then favor the spread of resistant strains. The potential for development of resistance differs greatly among different groups of antibiotics. Certain agents, such as erythromycin, tetracyclines, and sulfonamides, carry a considerable risk of secondary resistance during treatment. With penicillins and cephalosporins, however, secondary resistance is nearly absent. Clinical nonresponse and emergence of resistant strains are usually due to the selection of other resistant species. For instance, *E. coli* in a wound infection is rapidly eliminated but is replaced by *Pseudomonas*. Selection of resistant strains under treatment with antibiotics plays a far more important role than the development of resistance in a previously sensitive strain.

Certain, sometimes highly resistant strains are often waterborne. The common sources of infection are usually certain fluids, aerosols, moistures or topical preparations. Very dangerous are contaminated infusions, which will lead to septic shock in a high percentage of patients. The main waterborne bacteria are: *Serratia marcescens, Klebsiella pneumoniae, Enterobacter* spp., *Pseudomonas aeruginosa* and other *Pseudomonas* species, and *Acinetobacter* spp.

In an outbreak or even a local cluster of these infections, investigations should be performed to find the common mechanism.

Carriers among the hospital personnel may play an important role in the distribution of certain bacterial infections. This is mainly so with *Staphylococcus aureus* and group A streptococci. Finger abscesses, boils, and nasal discharges are the main vectors from staff member to patient.

Some types of resistance (oxacillin resistance and oxacillin tolerance of *Staphylococcus aureus,* ampicillin resistance of *Hemophilus influenzae,* penicillinase production of *Neisseria gonorrhoeae*) are not easy to find and will be overlooked in a bacteriological laboratory if only routine tests are performed. Sensitivity testing of anaerobic bacteria requires sophisticated equipment; the results are nevertheless doubtful and are not consistent with the clinical course.

Only a limited number of bacterial species vary very much in their sensitivity to antibacterial agents. The main problems of resistance relate to staphylococci, enterobacteria and *Pseudomonas* spp.:

Staphylococcus aureus
Staphylococcus epidermidis
Enterobacteria
- *Escherichia coli*
- *Klebsiella pneumoniae*
- *Enterobacter* spp.
- *Proteus* spp.
- *Serratia* spp.

Pseudomonas aeruginosa

Certain bacterial species are only rarely resistant to common antibacterial agents, but show an increasing tendency toward resistance in developing countries:

- *Neisseria gonorrhoeae* (penicillinase-producing)
- *Streptococcus pneumoniae* (multiple resistance)
- *Shigella* spp. (multiple resistance)
- *Salmonella* spp. (multiple resistance)
- *Vibrio* cholerae (multiple resistance)
- *Hemophilus influenzae* (beta-lactamase-producing)

Penicillinase-producing strains of *Neisseria gonorrheae* (PPNG) have led to a major outbreak of resistant gonorrhea worldwide within the last decade. Originating from Southeast Asia, these strains are now present all over the world. As a practical consequence, the former standard therapy of gonorrhea with benzylpenicillin or ampicillin is now unreliable.

For many decades, anibiotic resistance was nearly absent in pneumococci. Only rarely could low-grade resistance to penicillin be found. Resistance to other antibiotics was uncommon, with the exception of sulfonamides and tetracycline. In 1977, however, multiple antimicrobial resistance in pneumococci was detected in South Africa [14]. There was a remarkable spread of the resistant strains in hospital patients, but surveys in the community rarely identified resistant strains. The management of highly resistant pneumococcal infections may lead to significant difficulties. The strains are resistant to all beta-lactam antibiotics except imipenem and also to other common antibiotics. Treatment has to be performed with chloramphenicol or with unusual combinations like erythromycin + rifampicin, or even with vancomycin or imipenem.

Penicillin resistance of the intermediate type of *Streptococcus pneumoniae* has been reported from several other tropical regions [8, 11, 28]. The problem of pneumococcal resistance endangers the widespread use of oral penicillins in the treatment of respiratory infections. The emergence of such strains must be considered in any local strategies for the treatment of respiratory infections.

Resistance of *Hemophilus influenzae* to ampicillin has become a problem in the United States. The resistance, due to beta-lactamases, may be overlooked by routine sensitivity tests. The frequency of ampicillin-resistant *Hemophilus* strains in Europe is low. Ampicillin-resistant *Hemophilus* strains have been reported from the tropics [5, 22, 25, 26]. These strains are mainly a problem in the treatment of infantile meningitis. The drug of choice in Europe is now cefotaxime or related cephalosporins; chloramphenicol is the alternative treatment.

There are only very few data concerning resistance profiles of enterobacteria and staphylococci from developing countries [21, 27]. Apparently, there are considerable problems of resistance in large hospitals. A general increase of resistance to common antibiotics was reported from Nigeria [1]. A high level of general resistance is also known from Dakar, Senegal [16], where sensitivity to ampicillin was found in only 35% of all isolates of *E. coli*. The sensitivity to tetracycline was even lower, namely 20%. A high level of resistance was also reported from Addis Abeba [13]. Amongst blood culture isolates, sensitive strains were the exception [12].

There are scarce but impressive reports of outbreaks by resistant bacterial infections in tropical hospitals. Epidemics of cholera have been reported, from Tanzania and Bangladesh, in which an initially sensitive strain became resistant within a few months due to R factors [17]. Multiresistant *Salmonella typhimurium* led to an important outbreak in children in Brazil [24]; hospital outbreaks of resistant salmonella infections have, however, occurred worldwide within the last decade, though particularly in developing countries.

There are also alarming reports on resistance to co-trimoxazole or trimethoprim from Chile, Bangkok, and India. Of the strains of *E. coli*, 40%–64% were resistant to trimethoprim or co-trimoxazole [18, 30].

There are several anecdotal reports of nosocomial infections by different common bacteria from tropical countries. In one clear-cut analysis of nosocomial infections, an increase of the nosocomial infection rate from 1.9% to 3.4% was reported [2]. A high number of nosocomial infections with common pathogens was reported from Nigeria [1]. There are reports on hospital-acquired *Serratia* infections [7, 9, 19]. The use of sophisticated modern equipment in the hospital may be associated with secondary infections. A large outbreak of *Klebsiella* septicemia was observed in Colombo due to contamination of intravenous fluids [15].

References

1. Alausa OK, Montefiore D (1978) Bacterial infections, sensitivity patterns and chemotherapy among hospital patients in the tropics. Scand J Infect Dis 10: 295-302
2. Anyiwo CE, Daniel SO, Ogunbi OO, Aromolaran GO (1980) Nosocomial infections - a continuing danger to patients at Lagos University Teaching Hospital. Public Health 94: 229-234
3. Anyiwo CE, Ogunbi O (1981) Controlling infection in an operating theatre. Niger Med Practit 2: 11-13
4. Bauer C (1982) Disinfectant usage and hygiene practices at Port Moresby General Hospital. Papua New Guinea Med J 25: 248-252
5. Bernard-Bonnin A-C, Ekoe T (1985) Les meningites purulentes de l'enfant a Yaounde: aspects epidemiologiques et prognostiques. Ann Soc Belge Méd Trop 65: 59-68
6. Burke JP, Levy SB (1985) Summary report on worldwide antibiotic resistance: international task forces on antibiotic use. Rev Infect Dis 7: 560-564
7. Chakravarti A, Mandal A, Sharma KB (1981) An outbreak due to multiple drug resistant *Serratia marcescens* in a children's hospital. Indian J Med Res 74: 196-201
8. Denis FA, Greenwood BD, REy JL, Prince-David M, Mboup S, Lloyd-Evans N, Williams K, Benbachir I, El Ndaghri N, Hansman D, Omanga V, Krubwa K, Duchassin M, Perrin J (1983) Etude multicentrique des sérotypes de pneumocoques en Afrique. Bull WHO 61: 661-669
9. Fall M, Kuakuvi N, Castets M, Diadhiou F, Martin L (1976) Infections néo-natales à *Serratia marcescens*. Bull Soc Méd Afr Noire Lang Franc 21: 57-64
10. Farrar WE (1985) Antibiotic resistance in developing countries. J Infect Dis 152: 1103-1106
11. Gratten M, Naraqi S, Hansman D (1980) High prevalence of penicillin-insensitive pneumococci in Port Moresby, Papua New Guinea. Lancet II: 192-195
12. Gedebou M, Tassew A, Azene G (1984) Blood culture isolates from an Addis Abeba hospital frequency and antibiotic sensitivities. East Afr Med J 61: 190-196
13. Gedebou M, Tassew A, Azene G (1983) Frequency and resistance patterns of bacterial isolates from surgical patients in a teaching hospital in Addis Abeba. Trop Geogr Med 35: 133-138
14. Jacobs MR, Koornhof HJ, Robins-Browne RM, Stevenson CM, Vermaak ZA, Freiman I, Miller GB, Witcomb MA, Isaaecson M, Ward JI, Austrian R (1978) Emergence of multiply resistant pneumococci. N Engl J Med 299: 735-740
15. Koopman JS, Fighetti de Olave A (1980) An extensive outbreak of *Klebsiella pneumoniae* bacteremias from in-use contamination of i.v. bottles. Bull Pan Am Health Organ 14: 185-192
16. Menard M (1982) Résultats des 5691 antibiogrammes pratiqués en 1979, 1980 et 1981 à l'Hòpital Principal de Dakar. Dakar Méd 27: 595-602
17. Mhalu FS, Mtango FDE, Msengi AE (1984) Hospital outbreaks of cholera transmitted through close person-to-person contact. Lancet II: 82-84
18. Murray BE, Alvarado T, Kim K-H, Vorachit M, Jayanetra P, Levine M, Prenzel I, Fling M, Elwell L, McCracken GH, Madrigal G, Odio C, Trabulsi LR (1985) Increasing resistance to trimethoprim-sulfamethoxazole among isolates of *Escherichia coli* in developing countries. J Infect Dis 152: 1107-1113
19. Muyembe TL, Makulu MU, Vandepitte J (1977) Considérations sur une épidémie d'infections hospitalières à *Serratia* observées aux cliniques universitaires de Kinshasa-Zaire. Méd Trop 37: 531-536
20. Nnochiri E (1970) The pattern of hospital-acquired infections in the teaching hospitals of East and West Africa. East Afr Med 47: 639-645
21. Nnochiri E (1975) Medical microbiology in the tropics. Oxford University Press, London
22. Odugbemi TO, Fadahunsi HO, Ogunbi O (1977) Some aspects of *Haemophilus influenzae* meningitis in Lagos. Niger Med J 7: 133-137
23. Price TV (1977) Studies of bacterial populations in the kitchens of the University of Papua New Guinea. Papua New Guinea Med J 20: 59-64
24. Riley LW, Ceballos BSO, Trabulsi LR, Fernandes de Toledo MR, Blake PA (1984) The significance of hospitals as reservoirs for endemic multiresistant *Salmonella typhimurium* causing infection in urban Brazilian children. J Infect Dis 150: 236-241
25. Scragg JN, Appelbaum PC, Govender DA (1978) The spectrum of infection and sensitivity of organisms isolated from African and Indian children in a Durban hospital. Trans R Soc Trop Med Hyg 72: 325-328

26. Simasathien S, Duangmani C, Echeverria P (1980) *Haemophilus influenzae* type B resistant to ampicillin and chloramphenicol in an orphanage in Thailand. Lancet II: 1214-1217
27. Vargas de la Rosa R, Peniche A, Martinez Ma del C (1982) Tendencias y otras características de las infecciones intrahospitaliarias en el Hospital de Pediatria del CMN, IMSS, 1977-1980. Rev Med IMSS 20: 543-555
28. Wamola IA, Mirza NB, Nsanzumuhire H (1981) Penicillin resistant pneumococcal meningitis in Kenyatta National Hospital, Nairobi, Kenya. East Afr Med J 58: 12-17
29. WHO Scientific Working Group (1983) Antimicrobial resistance. Bull WHO 61: 383-394
30. Young H-K, Jesudason MV, Koshi G, Amyes SGB (1986) Trimethoprim resistance amongst urinary pathogens in South India. J Antimicrob Chemother 17: 615-621

4 Administration of Antibiotics

The administration of an antibiotic depends largely on the clinical picture, the condition of the patient, and the availability. Most antibiotics can be given orally and parenterally. Modern beta-lactams (many penicillins and cephalosporins) and aminoglycosides have to be injected because they are not absorbed in the intestine. Only a few substances (phenoxymethylpenicillin, norfloxacin, oral esters of ampicillin) are not available in parenteral form. Antibiotics can be divided into several groups according to their absorption.

Intravenous administration of antibiotics generally leads to higher levels than oral intake; the only exception is chloramphenicol. Intravenous infusions of antibiotics may lead to severe overhydration in small children or in cardiac insufficiency. In serious infections, treatment should be started parenterally to achieve rapidly high blood and tissue concentrations. Once the patient starts to improve, treatment should be continued by mouth, if possible.

The need to give *parenteral antibiotics* to outpatients may be reduced in future. The availability of oral antibiotics with high absorption should be an argument against unnecessary parenteral antibiotics, and the worldwide increase of AIDS is another important reason for restricting them.

Oral absorption may be impaired by vomiting, diarrhea, cardiac insufficiency, or malabsorption. Oral administration may be impossible in unconscious or non-cooperative patients.

Table 4.1. Absorption of antibiotics and antibacterial agents

I. Optimal absorption: injection not necessary if patient is able to swallow	*III. Oral absorption sufficient but poor absorption possible*
Chloramphenicol	Erythromycin and other macrolides
Amoxicillin/bacampicillin	Tetracycline/oxytetracycline
Cefalexin/cefaclor/cefadroxil	Lincomycin
Ofloxacin	
Doxycycline/minocycline	*IV. Parenteral administration necessary: oral absorption insignificant or absent*
Co-trimoxazole	Aminoglycosides
Nitrofurantoin	Vancomycin
Metronidazole	Cephalosporins (except cefalexin group)
II. Sufficient but incomplete oral absorption	Azlocillin/mezlocillin/piperacillin
Ampicillin	Benzylpenicillin
Phenoxymethylpenicillin	Aztreonam
Cloxacillin	Imipenem
Clindamycin	
Rifampicin	*V. Antibiotics only for topical use*
Ciprofloxacin/pefloxacin/norfloxacin	Neomycin/kanamycin
	Bacitracin

Intramuscular injections may lead to problems. The absorption of the antibiotic can be poor in shock, or due to injection into subcutaneous fat. The blood levels are inferior to those achieved by i.v. injection. Many antibiotics have poor local tolerance (tetracyclines, erythromycin, cephalosporins); they may lead to remarkable inflammations, pain, and even abscesses. Intramuscular injection should be preferred with aminoglycosides.

If there is a necessity of combination therapy, antibiotics should not be mixed in the same ampoule or vial; inactivation may be possible. Parenteral antibiotics should never be mixed with other substances.

Topical Administration

The topical use of antibiotics has very limited value. The penetration into tissue is very poor. Instillation into local abscesses or empyema does not reach the bacteria in all parts of the infection. Quite often the absorption from local instillation is very rapid. In general, systemic administration of an antibiotic is far more reliant than local instillation. Local instillation may be used as an adjuvant to systemic treatment. Widespread topical treatment carries the risk of development of resistance, selection of resistant strains, and increased allergy in patients and the nursing staff. As a general principle, beta-lactam antibiotics should not be used for topical treatment due to the high risk of allergy.

Indications for superficial topical antibacterial therapy are: impetigo, conjunctivitis and other external eye infections, vaginitis, and external otitis.

Topical use of local antibacterials is useless and may be even dangerous in pharyngitis and tonsillitis, despite the great number of such preparations on the market. The topical treatment of infected wounds should be done with local disinfectants (chlorhexidine, povidone-iodine, etc.) rather than with antibiotics. Usually the cleaning and rinsing effects of antibiotic instillations are more important than the direct antimicrobial activity. Sophisticated forms of local treatment (gentamicin-PMMA-chains, gentamicin-containing bone cement etc.) are useful in modern orthopedic surgery. Instillations of antibiotics may be useful - in addition to systemic therapy - in deep eye infections, empyema, and certain rare forms of meningitis *(Pseudomonas!)*.

The local therapy of chronic ulcers may be useful, but carries risk of colonization with highly resistant bacteria. Intratracheal gentamicin is effective in the treatment and prophylaxis of gram-negative pneumonia in intensive care patients. Instillation of antibiotics in the urinary bladder using a catheter is ineffective, expensive, and potentially dangerous.

5 Dosage of Antibiotics

The principal objective in giving an antibiotic is to achieve a concentration at the site of infection which is sufficient to kill or inhibit the growth of the bacteria present. The dosage required to accomplish this object cannot always be defined in fixed terms; it often has to be adjusted to meet the special features of the individual case. Important factors which need to be taken into account are the sensitivity of the causative organism, the pharmacokinetics and tolerance of the antibiotic in relation to the patient's age and illness, and the site of infection. The dose recommendations made by the manufacturers are sometimes also influenced by commercial considerations, with preference for underdosage (less expensive) or twice daily dosage (more convenient). Overdosage is less frequently recommended.

Recommended Doses

Conventional daily doses of standard and additional antibacterial agents are summarized in Tables 5.1 and 5.2. There are considerable variations in the recommended doses between different countries. Thus, the standard daily dose of ampicillin is 0.75-1.5 g in Great Britain and Japan, but 3-4 g in the USA and West Germany. Clinical efficacy in *Hemophilus influenzae* or *E. coli* infections cannot be achieved with a low dose. At low dosage, ampicillin resembles phenoxymethylpenicillin; high dosage is required in order to achieve the desired activity against *H. influenzae* or *E. coli.* The daily dosages in Table 5.1 and 5.2 refer to middlesized Europeans with an average body weight of 70 kg. They must be modified for ethnic groups with smaller average size and weight.

Duration of Treatment

International textbooks usually recommend treatment for 10-14 days. Poor experience with weakly active sulfonamides, common opinions on resistance, and the goal of optimal treatment contribute to this recommendation. During the last decade, variations in the duration of treatment have been increasingly advocated. Certain infections (gonorrhea, urethritis, cystitis, lues, chancroid, shigellosis) can be treated by a single shot or single dose of an appropriate antibiotic.

Longer courses of treatment are needed for chronic infections (tuberculosis, fungal infections, etc.) and also in septicemic illnesses with a tendency to relapse or recur (e.g., staphylococcal septicemia, brucellosis, endocarditis). Prophylaxis against recurrent infection is particularly recommended in rheumatic fever. Extended courses of antibiotics are often necessary in patients with impaired immu-

Table 5.1. Daily dosage of essential antimicrobial agents (WHO list 1983)

Antibiotic	Adults	Children (except newborns)	Route
Benzylpenicillin	1-5 (-30) MU	0.04-0.1 (-1) MU/kg	i.m., i.v.
Procaine penicillin	0.6-1.2-2 MU	0.025-0.05 MU/kg	i.m.
Benzathine penicillin	1.2-2.4 MU (single dose)	0.6-1.2 MU (single dose)	i.m.
Phenoxymethylpenicillin	1.5 (-8) MU	0.05 (-0.1) MU/kg	p.o.
Ampicillin	3-4 g	100-150 (-200) mg/kg	p.o.
	1.5-2 (-20) g	100 (-400) mg/kg	i.m., i.v.
Amoxicillin	1-1.5 (-3) g	50 (-100) mg/kg	p.o.
Bacampicillin	2.4 g	60 mg/kg	p.o.
Tetracycline	1-1.5 (-2) g	20-30 mg/kg	p.o.
Doxycycline	0.1-0.2 g	2-4 mg/kg	p.o., i.v.
Minocycline	0.2 g	4 mg/kg	p.o., i.v.
Chloramphenicol	2-3 (-4) g	50-75 (-100) mg/kg	p.o., i.v., i.m.
Erythromycin	1-2 (-3) g	30-40 (-80) mg/kg	p.o., i.v.
Gentamicin, tobramycin	160-240 (-320) mg	2-3 (-5) mg/kg	i.m., slow i.v. injection
Amikacin	1-2 g	15 mg/kg	i.m., slow i.v. injection
Nitrofurantoin	(0.1-) 0.2-0.3 g	(2-) -5 mg/kg	p.o.
Short-acting sulfonamides (e.g., sulfadimidine, sulfisoxazole)	4-6 g	0-1 year: 1-2 g 1-6 years: 2-3 g 6-12 years: 3-4 g	p.o.
Co-trimoxazole	(0.9-) 1.9 (-2.8) g	20-30 mg/kg	p.o.
Metronidazole (anaerobes)	1.5-2.25 g	20-40 mg/kg	p.o., i.v.

MU, megaunit.

Table 5.2. Daily dosage of important supplementary antimicrobial agents

Antibiotic	Adults	Children (except the newborns)	Route
Azlocillin, mezlocillin, piperacillin	4-6-15 (-20) g	100-200 (-300) mg/kg	i.v., i.m.
Cefazolin, cefoxitin, cefuroxime, cefamandole, cefotiam, cefmetazole, cefotaxime, ceftazidime	3-6 g	60-150 mg/kg	i.v., i.m.
Latamoxef	2-4 g	50-100 mg/kg	i.v., i.m.
Ceftriaxone	1-2 (-4) g	50 (-100) mg/kg	i.v., i.m.
Imipenem	1.5-3 g	50 mg/kg	i.v.
Oral cephalosporins	2-4 g	50-100 mg/kg	p.o.
Clindamycin	0.6-1.2 (-2) g	10-20 mg/kg	p.o., i.m., i.v.
Fusicid acid	1.5 (-3) g	20-30 mg/kg	p.o.
Vancomycin	1-2 g	20-40 mg/kg	i.v.
Nalidixic acid	4 g	Contraindicated	p.o.
Pipemidic acid	0.8 g	Contraindicated	p.o.
Norfloxacin	0.8 g	Contraindicated	p.o.
Ofloxacin	0.2-0.4 g	Contraindicated	p.o.
Pefloxacin	0.8 g	Contraindicated	p.o., i.v.
Ciprofloxacin	0.5-1.5 g	Contraindicated	p.o., i.v.

nity (leukemia, immune deficiency diseases, etc.), who tend to relapse when treatment is discontinued.

Tolerance

If dose-related side effects are to be avoided, the maximum daily and total dosage should not as a rule be exceeded. The penicillins, and to a lesser extent, the cephalosporins, clindamycin, fusidic acid, and rifampicin, have such a broad therapeutic range that dose-related side effects are extremely uncommon. With most other antibiotics, however, side effects regularly occur when the maximum recommended dosage is exceeded.

A moderate dose is generally sufficient for less severe infections or when the causative organism is very sensitive, in which case the risk of dose-related side effects is less. If the infection does not respond, the dose can be increased up to the maximum recommended. With some antibiotics such as the tetracyclines, however, increasing the oral dose does not result in proportionately higher blood concentrations because gastrointestinal absorption is already taking place at the maximal rate.

The side effects of some antibiotics, such as cycloserine in tuberculosis and amphotericin B in systemic fungal infections, can be minimized by starting with a low dose and increasing it progressively to the therapeutic dose or the limit of tolerance.

Dosage in Neonates, Infants, and Children

Pediatric Dosage

Pediatric dosage may be based on the average body surface area of children. The following rules have been derived:

Age (years)	*Fraction of adult dosage*
¼	⅙
½	⅕
1	¼
3	⅓
7	½
12	⅔

In clinical practice, doses of antibiotics are usually doses according to body weight. The usual daily doses for children given in Tables 5.1 and 5.2 apply to infants and small children. Applying the rules of body weight to older children could result in excessive doses, higher even than those recommended for adults. For this reason, surface area should be the determining factor for schoolchildren. Children between 6 and 9 years thus receive one-half, and children between 10 and 12 years about two-third of the dosage recommended for adults.

Premature and Full-Term Neonates and Infants During First Month of Life

At this age, the excretion of antibiotics, whose main route of elimination is via the kidneys, is delayed because of renal immaturity; the peak blood levels decline more slowly, and the dose intervals may need to be 2-3 times longer than with older children. Individual variations in the degree of renal immaturity among premature and full-term babies must be taken into account. Special care is necessary in the first few days of life. Accumulation is less common after the first week. In particular, potentially toxic antibiotics should be given only in very low doses, or at extended dose intervals. It may be helpful to assay the serum concentration immediately before the next dose (the trough or residual level). Commercial kits now available for radioimmunoassay and enzyme immunoassay of aminoglycosides require only very small samples of blood. Those antibiotics, such as penicillin, which are generally nontoxic can be given without serious risk in the normal dosage for infants (Table 5.3).

Table 5.3. Dosage in newborns up to the end of the first month of life

Antibiotic	Daily dosage	Route of administration	Dose interval
Benzylpenicillin	0.06-0.5 MU/kg	i.v., i.m.	8-12 h
Procaine penicillin	0.05 MU/kg	i.m.	24 h
Phenoxymethylpenicillin	0.05 MU/kg	p.o.	8-12 h
Cloxacillin	50-100 (-200) mg/kg	p.o., i.m., i.v.	8-12 h
Ampicillin	100-200 mg/kg	i.m., i.v.	8-12 h
Azlocillin, mezlocillin, piperacillin	100-200 mg/kg	i.m., i.v.	8-12 h
Cefoxitin, cefmetazole, cefuroxime, cefamandole, cefotaxime, ceftazidime	60-100 mg/kg	i.m., i.v.	8-12 h
Ceftriaxone	50 mg/kg	i.m., i.v.	12-24 h
Oral cephalosporins	50-100 mg/kg	i.m., i.v.	8-12 h
Gentamicin, tobramycin	5 mg/kg (term infant, 1st week)	i.m., slow i.v. injection or infusion	12 h
	5 (-7.5) mg/kg (term infant, 2nd week onwards)		8-12 h
	3-3.5 mg/kg (<35 weeks gestational age)[a]		12-24 h
	2.5 mg/kg (<28 weeks gestational age)		24 h
Amikacin, kanamycin	15 (-20) mg/kg	i.m., slow i.v. injection or infusion	8-12 h
Erythromycin	30 mg/kg	p.o.	8-12 h
Clindamycin	15 (-20) mg/kg	p.o., i.v.	8-12 h
Chloramphenicol	25 mg/kg (1st and 2nd week) 50 mg/kg (3rd and 4th week)	p.o., i.v.	8-12 h
Metronidazole (anaerobic bacteria)	15 mg/kg (1st week) 20 mg/kg (2nd week onwards)	i.v.	8-12 h

MU, megaunit.

[a] A loading dose of 5 mg/kg/day in two divided doses is recommended for the first day of treatment.

6 Side Effects and Contraindications

Side effects of antibiotics can be divided into three main types: toxic, allergic, and biological. The main side effects of standard and additional antibacterial agents are summarized in Tables 6.1 and 6.2.

Toxic Side Effects

Modern antibiotics are usually well tolerated. Nevertheless, human toxicity can be divided into different types:

1. Acute toxicity (amphotericin B: nephrotoxicity)
2. Chronic toxicity (chloramphenicol: aplastic anemia)
3. Toxicity due to overdosage (penicillins: seizures)
4. Genetically determined toxicity (sulfonamides: 6-GPD-deficiency; isoniazid: slow vs fast acetylators)
5. Toxicity due to degradation products or inappropriate storage (tetracyclines)

Only few types of toxicity frequently limit the use of an antibiotic. The remarkable ototoxicity and nephrotoxicity of aminoglycosides requires restriction of daily dosage and limits the duration of treatment (preferably 7–10 days). Bone marrow aplasia caused by chloramphenicol is rare (1 in 10000) but very dangerous. It is usually irreversible. The hepatotoxicity of rifampicin should be monitored by regular liver function tests, but does not usually lead to permanent hepatic damage. Toxicity of certain agents in animals may lead to restrictions in clinical use, even if toxicity in humans is not yet known. Examples of this are the oncogenicity and teratogenicity of metronidazole and other nitroimidazoles. Their use in pregnancy and for treatment of minor infections cannot be recommended. The damage of articular cartilage of young animals led to the strict warning against quinolones in children.

Some types of side effects are frequently discussed but are apparently overestimated. This is true for the discoloration of teeth in infants by tetracyclines. The risk is apparently very small with doxycycline. Marked intolerance to alcohol is a side effect of metronidazole and other nitroimidazoles as well as of cephalosporins with an *N*-methyltetrazole side chain (cefamandole, latamoxef, cefoperazone, cefmenoxime). This side effect can easily be prevented by avoidance of alcohol during treatment.

Cumulation of an antibiotic due to decreased elimination may result in toxic side effects. Cumulation of aminoglycosides in renal failure frequently causes

Table 6.1. Important side effects of antimicrobial agents (WHO list)

Antimicrobial	Side effects							Contraindicated (except in life-threatening disease)		
	Allergic	Hematotoxicity	Nephrotoxicity	Hepatotoxicity	Neurotoxicity	Biological	Other	First trimester of pregnancy	Neonatal period	Severe renal failure
Benzylpenicillin	+ +				+	±				
Procaine penicillin	+ +				+		1			
Benzathine penicillin	+ +				+		1			
Phenoxymethylpenicillin	+ +					±				
Ampicillin	⊕⊕		±		±	+	3			
Cloxacillin	+ +		±		±		1,2			
Erythromycin	±			+[a]		±	1,2,3,5			
Tetracyclines	±	±	±	+		+	2,3	×	×	×
Chloramphenicol	±	⊕⊕			±	+	3	×	×	
Spectinomycin	±		+		±		1	×	×	×
Gentamicin, tobramycin	±		+		⊕⊕	±	4	×		×
Amikacin	+		+		⊕⊕	±	4	×	×	×
Sulfonamides	+ +	±	±	±			3		×	×
Co-trimoxazole	+ +	+	+	+		+	3	×	×	×
Metronidazole	±	±			+	+	3,9	×	×	
Nitrofurantoin	⊕⊕	±		+	⊕⊕		3,6	×	×	×

Key to side effects: + +, relatively frequent; +, occasional; ±, rare; ⊕⊕, principal complication and often the limiting factor for therapy; 1, local intolerance with i.m. or s.c. injection; 2, venous irritation with i.v. injection; 3, gastrointestinal intolerance; 4, histamine release; 5, development of secondary resistance; 6, pneumonia or pulmonary fibrosis; 7, bleeding tendency; 8, damage of growing cartilage; 9, alcohol intolerance; ×, contraindicated.

[a] As estolate.

cochlear damage. Other antibiotics may also cumulate but to a lesser extent (penicillins, cephalosporins, tetracyclines). Hepatic failure limits the use of antibiotics with predominantly biliary elimination (rifampicin, fusidic acid, cefoperazone, ceftriaxone). In general, biliary elimination of an antibiotic should not be regarded as an advantage. Nonspecific side effects such as gastrointestinal intolerance or flatulence are quite common.

Table 6.2. Important side effects of antimicrobial agents (not on WHO list)

Antimicrobial	Side effects							Contraindicated (except in life-threatening disease)		
	Allergic	Hematotoxicity	Nephrotoxicity	Hepatotoxicity	Neurotoxicity	Biological	Other	First trimester of pregnancy	Neonatal period	Severe renal failure
Carbenicillin, ticarcillin	++	±			+	+	2,7			×
Azlocillin, mezlocillin, piperacillin	++	+			±	+	2			
Cephaloridine	+	±	⊕⊕			+	1			×
Cefazolin	+	±	±			±	1,2			
Cefoxitin	+	±				±	1			
Cefmetazole	+	±				±	1,9			
Cefotaxime	+	±				±	1			
Latamoxef	+	±				±	1,7,9	?	?	
Cefoperazone	+	±		?		±	1,3,7,9	?	?	
Ceftriaxone	+	±		?		±	1,3,7	?	?	
Ceftazidime	+	±	?			±	1			
Imipenem	+		±		±	±		×		×
Clindamycin	±			±		⊕⊕	3			
Fusidic acid	±			+			1,3,5			
Vancomycin	++	±	±		+		1,2	×	×	×
Nalidixic acid	±			±	+	++	3,5,8	×	×	×
Norfloxacin	±		±		±	+	3,8	×	×	×
Ofloxacin	±				±	+	3,8	×	×	×
Ciprofloxacin			±			+	3,8	×	×	×
Pefloxacin	+		±		±		3,8	×	×	×

Key to side effects: ++, relatively frequent; +, occasional; ±, rare; ⊕⊕, principal complication and often the limiting factor in therapy; 1, local intolerance with i.m. or s.c. injection; 2, venous irritation with i.v. injection; 3, gastrointestinal intolerance; 4, histamine release; 5, development of secondary resistance; 6, pneumonia or pulmonary fibrosis; 7, bleeding tendency; 8, damage of growing cartilage; 9, alcohol intolerance; ×, contraindicated; ?, no data, but suspicion.

Allergic Side Effects

The frequency of allergic side effects is quite high in some antibiotics. Hypersensitivity to penicillins and cephalosporins leads frequently, in 2%-5%-8% of cases to fever, rashes, urticaria, and eosinophilia. Anaphylactic shock, which is fatal in about 10% of cases, is fortunately rare. Allergic reactions to penicillins or cephalosporins occur after the first dose if the drug has been taken by the patient previously, or late, after 9-11 days, if it is taken for the first time. Hypersensitivity occurs more often after parenteral or topical than after oral administration. Patients who are allergic to penicillin should be given cephalosporins with caution, since the reaction in case of cross-allergy (reported in approximately 10% of cases) is usually of a milder type. Certain penicillins (ampicillin!) apparently carry an increased risk of rashes. Hypersensitivity to sulfonamides may cause severe skin reactions, including Lyell's syndrome. Erythromycin (mainly the estolate) may cause allergic intrahepatic cholestasis. Allergic side effects of tetracyclines, quinolones, and chloramphenicol are rare.

Biological Side Effects

Biological side effects are commonly observed during antibacterial therapy but are usually not severe. Overgrowth of yeasts is frequently the cause of stomatitis, vaginitis, and occasionally esophagitis. Superinfection by *Klebsiella* spp., *Serratia* spp., *Staphylococcus aureus,* or *Pseudomonas aeruginosa* may be a consequence of treatment with certain broad-spectrum antibiotics (ampicillin, tetracycline). The most severe biological side effect is the induction of pseudomembranous enterocolitis. Many antibiotics (tetracycline, ampicillin, clindamycin) alter the intestinal flora such that overgrowth with toxigenic strains of *Clostridium difficile* may occur. The result is more or less severe acute ulcerative colitis with fever, diarrhea, and sometimes bloody stools. The treatment of choice is oral vancomycin (0.5 g/day). Less expensive, but also less effective, is metronidazole. Withdrawal of the causative agent is most important in pseudomembranous enterocolitis, as indeed in all other side effects.

7 Combinations of Antibiotics

As a general principle, an infection should be treated with a single antibiotic. Combinations of antibiotics are only occasionally justified. The main bacteriological reasons for the use of combinations are:

1. To extend the spectrum of activity
2. To delay the development of resistance
3. To achieve synergy

Synergy is present if one antibiotic is potentiated by another. This can be achieved in different ways, such as;

1. Blockade of sequential enzymes in the bacterial metabolic pathway (co-trimoxazole)
2. Blockade of enzymes produced by bacteria (e.g., beta-lactamases by clavulanic acid)
3. Enhancement of penetration of one antibiotic (e.g., gentamicin) by another (e.g., penicillin)
4. Effect on different binding proteins (double beta-lactam combinations).

Synergy in vitro is not necessarily relevant for the situation in the patient. The same is true for antagonism.

Combined antibiotic therapy is advised for the following indications:

Bacteriological Reasons for Combination Treatment

1. Benzylpenicillin or ampicillin alone have inadequate bactericidal activity for severe enterococcal infections. Reliable bactericidal activity in enterococcal endocarditis can only be obtained by combining the penicillin with an aminoglycoside (gentamicin).
2. The combination of gentamicin with azlocillin is superior to treatment with a single agent in severe *Pseudomonas* infection. The combination of a cephalosporin with gentamicin is more effective than single-agent treatment in severe *Klebsiella* infections.
3. Combined treatment delays the development of resistance in tuberculosis.
4. The combination of amphotericin B and flucytosine achieves better clinical results in severe fungal infection *(Candida, Aspergillus).* The dose of amphotericin B can be reduced in this combination.
5. Trimethoprim and sulfonamides inhibit sequential stages of the metabolic pathway in bacteria. This leads to a remarkable synergistic activity on a number of relevant bacterial species.

Clinical Reasons for Combination Treatment

1. Mixed infections (peritonitis, bronchiectasis, empyema) must be treated by antibiotic combinations. Usually, no single agent which covers the streptococci, *E. coli,* and *Bacteroides* species is available.
2. Endocarditis. Infection is difficult to eradicate from heart valves. Bactericidal combinations of antibiotics are always necessary.
3. Initial therapy in an acute life-threatening clinical situation. A combination of two or more antibiotics may be necessary to cover the relevant bacteria.
4. Infections associated with foreign bodies can only be eliminated by bactericidal antibiotic combinations, if at all. The majority of such infections fail to resolve until the foreign material is removed.
5. Severe impairment of host defences (leukemia, immunosuppression). Mixed infection and frequent change of pathogens are common. Antibiotic combinations with a very broad spectrum are favored.

Risks of Combination Treatment

Sophisticated combination therapy affords a lot of special knowledge. Many errors are possible:

1. Underdosage of one or both components because of reliance on presumed synergistic activity.
2. Combinations of two antibiotics which are not fully effective at the site of infection because of their different pharmacological properties.
3. Cumulative toxicity of two antibiotics with related side effects (amikacin + gentamicin).
4. Use of antagonistic combinations (penicillin + tetracycline).

Many fixed combinations of antibiotics are available worldwide, but they are generally undesirable. Usually they are not combinations of two optimal agents in the right dosage. The only such drug which is internationally accepted is the fixed combination of trimethoprim and sulfamethoxazole (= co-trimoxazole). The general recommendation is to abstain from crude combinations. The same is true for commercial preparations of antibiotics with other drugs (expectorants, analgesics, mucolytics).

8 Costs of Antibiotics

The prices of antibiotics may vary remarkably from country to country. Usually, the classical standard antibiotics are more expensive in Europe. The prices in Germany are relatively high. The prices of antibiotics relative to each other, however, are similar all over the world. Tables 8.1 and 8.2 show the prices for daily treatment with the main oral and parenteral antibiotics in Germany.

Table 8.1. Prices of oral antibiotics in Germany 1987 (DM 1 ≈ US $ 0.60)

Oral antibiotic	Daily dose for adults	German price (DM)
Phenoxymethylpenicillin	1.0–2.0 g	2.60–7.85
Ampicillin	2.0 g	4.10–6.60
Amoxicillin	1.5 g	6.00–10.70
Cloxacillin	2.0 g	—[a]
Flucloxaxillin	1–2 g	6.90–13.85
Amoxicillin/clavulunic acid	1.5 g	19.50
Cefradine	2.0 g	13.10
Cefalexin	1.5 g	9.95
Cefaclor	1.5–3 g	14.65–29.30
Tetracycline	1–2 g	1.40–4.10
Doxycycline	0.1 g	0.90–5.35
Minocycline	0.2 g	7.60
Erythromycin	1–2 g	4.15–11.80
Clindamycin	0.6–1.2 g	10.35–20.65
Fusidic acid	1.5 g	17.15
Nitrofurantoin	0.2–0.4 g	1.50–3.00
Nalidixic acid	4.0 g	13.70
Ofloxacin	0.4 g	8.70
Norfloxacin	0.8 g	7.00
Ciprofloxacin	0.5 g	7.00
Short-acting sulfonamide	4.0 g	0.50–2.40
Long-acting sulfonamide	0.5 g	1.00
Co-trimoxazole	1.92 g	1.15–2.55
Trimethoprim	0.4 g	3.50
Chloramphenicol	1.5–2 g	3.70–8.75
Metronidazole	1.2 g	4.75
Rifampicin	0.6 g	7.85
Isoniazid	0.3 g	0.19
Ethambutol	15 mg/kg	3–4
Pyrazinamide	20–30 mg/kg	3–4
Amphotericin B	0.8 g	19.40
Flucytosine	100–200 mg/kg	37–75
Griseofulvin	0.5–1 g	2.10–5.10
Ketoconazole	0.2 g	3.40

[a] Not available on the German market.

Table 8.2. Prices of parenteral antibiotics for hospital use in Germany 1987 (DM 1 ≈ US $ 0.60)

Parenteral antibiotic	Daily dose for adults	German price (DM)
Benzylpenicillin	4-10 megaunits	1.50-10.00
Procaine penicillin	0.6-4.8 g	7.00-55.00
Azlocillin	6-15 g	122.00-304.00
Mezlocillin	6-15 g	93.00-199.00
Piperacillin	6-16 g	106.00-270.00
Cephalothin	4-12 g	58.00-143.00
Cefazolin	1-6 g	19.00-107.00
Cefradine	2-8 g	54.00-197.00
Cefamandole	3-6 g	74.00-115.00
Cefuroxime	2.25-4.5 g	56.00-107.00
Cefoxitin	3-6 g	85.00-169.00
Cefotaxime	2-12 g	62.00-361.00
Cefsulodin	3-6	190.00-374.00
Ceftazidime	1-6 g	56.00-319.00
Ampicillin + cloxacillin	2-6 g	30.00-53.00
Gentamicin	0.24-0.48 g	61.00-113.00
Tobramycin	0.24-0.48 g	66.00-132.00
Netilmicin	0.3-0.45 g	68.00-102.00
Amikacin	0.5-1.0 g	81.00-158.00
Injectable tetracycline	0.5 g	12.40
Vancomycin (i.v.)	2 g	354.00
Ciprofloxacin	0.4 g	162.60

The prices for standard agents can be far lower in developing countries and there may be remarkable variations. No relevant comparative price lists for all regions of the world are available. The local price level may be added to Tables 8.1 and 8.2 by the reader.

The overall cost of antibiotics depend largely on the duration of treatment. More expensive modern agents like quinolones may lead to cure in 1-3 days, while older agents were necessary for 10-14 days. Generally, the total cost of treatment has to be considered. Ideally, of course, the selection of antibiotics should be based purely on the best interest of the particular patient, even if the cost is extremely high, e.g., in the case of development of a chronic debilitating process (osteomyelitis, empyema) from an acute infection. Under conditions of extremely limited resources, however, it may be relevant to calculate how many patients can be successfully treated with, say, 100 tablets of co-trimoxazole or, more generally, for a cost of $ 10.

Part II
Treatment of Infections and Infectious Diseases

9 Septicemia

Incidence and Etiology

Little information is available concerning the incidence and etiology of septicemia in developing countries. Incidence rates vary. Those reported from Nigeria, Jamaica, and Kuwait in three studies were 37.1, 16.1, and 10.9 cases, respectively, per 1000 hospital admissions [2, 12, 17]. A retrospective analysis of all deaths on the medical wards of the same Nigerian hospital from 1960 to 1973 attributed 1.3% of the total deaths to septicemia [1]. This is almost certainly an underestimate because conditions such as pneumonia, urinary tract infection, and pyomyositis, which are often associated with secondary septicemia, were classified according to the respective organ systems. The same argument applies to studies of the pattern of hospital admissions to adult medical wards, where only 0.26%-0.29% of cases were diagnosed as septicemia [7, 25].

Hospital-based studies on the incidence of neonatal septicemia report figures of 6.1 per 1000 live births in Benin [21], 9.0 per 1000 live births in Ibadan [10], and 7.7 per 1000 live births in Santiago de Chile [33]. However, only 15 cases of septicemia were diagnosed in a series of 19000 live births in a hospital in Antofagasta, Chile, giving an incidence of 0.7 per 1000 live births [33].

Apparently, many instances of septicemia remain undiagnosed in the tropics, where the facilities of a microbiological laboratory are only very rarely available. But the clinical diagnosis of septicemia is difficult to confirm even in well-equipped hospitals when antibiotics have been given prior to admission. There is, however, no doubt that septicemia is one of the leading causes of neonatal mortality in developing countries [8, 30]; reasons include the lack of prenatal care, poor hygiene at birth, and traditional practices of cutting, tying, and covering the umbilical cord. The high proportion of low birth weight infants also contributes to high mortality from septicemia [10].

Table 9.1 summarizes the findings of several studies of the etiology of septicemia in developing countries. More detailed information regarding the age distribution of different agents and the case-fatality ratios is given in Tables 9.2 and 9.3; these data are taken from a prospective study conducted in Nigeria [2].

The proportion of staphylococcal septicemia is strikingly high in three of four African studies compared to recent USA/European figures (see Tables 9.1 and 9.4). This reminds one of the pattern encountered in the USA and Europe from 1950 to 1960, when widespread use of penicillin and sulfonamide eliminated the streptococcal and pneumococcal septicemias predominant in the preantibiotic era (Table 9.4).

Table 9.1. Etiology of septicemia in developing countries

Institution/town [Reference]	Mpilo, Zimbabwe [9]	Ibadan UH, Nigeria [2]	Addis Abeba, Ethiopia [14]	Johannesburg UH, South Africa [5]	Kingston UH, Jamaica [17]	Kuwait University, Kuwait [12]
Period studied	1968–1977	Jan.–June 1976	1976–1981	1982	1982–1983	1982–1983
Total numbers of cases	2746	146	293	319 (337 isolates)	222 (233 isolates)	214 (225 isolates)
Age group	All ages	All ages	Adults	Children (0-120 months)	All ages	All ages excluding neonates
Departments included	Several institutions	Medicine and surgery	Medicine, surgery, obstetrics, gynecology	General pediatrics	Medicine, surgery, obstetrics, gynecology, pediatrics, and others	Medicine and surgery
Bacteria isolated (%)						
Streptococcus pneumoniae	6.9	1.4	3.1	22.6	12.4	2.2
Beta-hemolytic streptococci	0.9	0.6		3.3[a]	10.8[b]	1.8
Streptococcus viridans	2.0	1.4	3.1	0		0.9
Enterococci	4.3	2.7		0.6	5.6	4.9
Staphylococcus aureus	22.9	28.1	29.0	6.2	11.6	9.3
Escherichia coli	29.6	9.6	12.6	12.8	8.2	19.6
Klebsiella Enterobacter spp.		15.8	8.2	8.3	16.7	17.3
Proteus spp.	1.6	4.8	2.0	1.2	ND	4.4
Pseudomonas spp.	4.2	8.9	5.1	2.4	ND	6.2
Candida spp.	0.3	0	0	0.3	ND	0
Salmonella typhi	17.4	16.4	27.3	3.9		8.0
Other *Salmonella* spp.	4.3	7.5	4.1	16.6	8.2	8.5
Hemophilus influenzae	1.0	0	0	14.0	6.9	2.2
Neisseria meningitidis	2.0	1.4	0	4.7	ND	0
Other	2.6	1.4	5.5	3.3	19.8[c]	14.1
Mortality (%)	ND	33.6	ND	23.0	18.0	ND

UH, University Hospital; ND, no data.
[a] Group B streptococci. [b] Includes 5.2% group B streptococci. [c] Includes 13.3% enterobacteriaceae other than *E. coli* and *Klebsiella* spp.

Table 9.2. Etiology of septicemia in 146 patients grouped on the basis of age (Ibadan, Nigeria) [2]

Bacteria isolated	Age group *(n)*					All age groups *(n)*	Mortality	
	Under 1 month	1-11 months	1-14 years	15-49 years	Over 50 years		*(n)*	(%)
Salmonella typhi	1	1	10	12	0	24	1	4.2
Salmonella spp.	1	3	6	1	0	11	2	18.2
Klebsiella spp.	14	2	2	4	1	23	14	60.9
Proteus spp.	2	0	1	3	1	7	1	14.3
Escherichia coli	5	4	2	2	1	14	4	28.6
Pseudomonas aeruginosa	6	3	1	3	0	13	7	53.8
Hemophilus influenzae	0	1	1	0	0	2	1	50.0
Bacteroides fragilis	0	0	0	0	1	1	0	0
Staphylococcus aureus	15	4	11	9	2	41	17	41.4
Staphylococcus epidermidis	0	0	0	1	0	1	0	0
Streptococcus pneumoniae	0	0	1	0	1	2	1	50.0
Enterococci	1	2	1	0	0	4	0	0
Nonhemolytic streptococci	1	0	0	1	0	2	0	0
Group A streptococci	0	1	0	0	0	1	1	100
Total gram-negative bacteria	29	14	23	25	4	95	30	31.6
Total gram-positive bacteria	17	7	13	11	3	51	19	37.6
Total	46	21	36	36	7	146	49	33.6
Mortality *(n)*	19	7	8	12	3			
(%)	(41.3)	(33.0)	(22.2)	(33.3)	(42.8)			

Table 9.3. Mortality in septicemia in 146 patients grouped on the basis of initial diagnosis (Ibadan, Nigeria) [2]

Initial diagnosis	Patients *(n)*	Mortality	
		(n)	(%)
Pyrexia of undetermined origin	21	4	19.1
Infections of urinary tract	7	5	71.4
Infections of gastrointestinal tract (other than typhoid)	7	3	42.9
Typhoid	14	3	21.4
Infections of respiratory tract	5	3	60.0
Infections of central nervous system	11	3	27.3
Infections of cardiovascular system	3	1	33.3
Infections of musculoskeletal system	8	0	0
Neonatal sepsis and septicemia	34	14	41.1
Septicemia (children and adults)	23	7	30.0
Infections associated with malignancies	4	4	100,0
Other infections	9	2	22.2

Table 9.4. Comparison of etiology of septicemia in Boston and Frankfurt

Institution	Boston City Hospital			Frankfurt University Hospital	
[Reference]		[8]		[31]	[28]
Period studied	1935	1955	1972	1960-1969	1980-1982
Total number of cases	291	574	591	315	487
Age group	All ages			Adults	
Departments included	Medicine and surgery			Medicine	
Bacteria isolated (%)					
Streptococcus pneumoniae	32.5	12.5	15.3	4.6	1.0
Beta-hemolytic streptococci	17.2	0.7	6.4	1.0[a]	1.4[a]
Streptococcus viridans	13.6	8.9	8.9	14.6	6.0
Enterococci	0	3.9	5.4	7.9	6.4
Staphylococcus aureus	21.4	35.4	12.3	12.0	22.4
Escherichia coli	8.8	11.1	13.7	21.6	20.3
Klebsiella Enterobacter spp.	0	8.7	13.3	3.5	14.0[b]
Proteus spp.	1.9	10.7	4.5	3.8	3.3
Pseudomonas spp.	0.3	2.3	3.1	3.8	9.0
Candida spp.	0	2.5	4.2	0.6	2.2
Salmonella typhi	3.4	0	0	10.2	8.2
Other *Salmonella* spp.	0	0.2	0.5		
Hemophilus influenzae				1.3	0
Neisseria meningitidis				1.3	0
Other	0.8	4.7	12.4	13.4[c]	5.8
Mortality (%)	57.7	36.4	31.0	ND	ND

ND, no data.
[a] Group A streptococci.
[b] Including *Citrobacter* and *Serratia*.
[c] Including 6% mixed infections.

Staphylococcal septicemias are, worldwide, frequently associated with wound infections, skin infections, osteomyelitis, pyogenic arthritis, and staphylococcal pneumonia. Typical tropical portals of entry are minor traumas, tungiasis, and tropical ulcers, all of which are abundant in people who walk barefoot. Other known portals of entry in the tropics include infected mosquito bites, guinea worm infections, snakebite ulcers, the painless, penetrating ulcers of the feet in leprosy, and domestic burns, often following epileptic fits. Pyomyositis, a condition only rarely seen in temperate climates, is also frequently associated with staphylococcal septicemia.

A recent study of staphylococcal bacteremia from Zimbabwe stresses that 60 of a total of 107 cases occurred in newborns and children under 2 years of age [19]. This reflects the age distribution of the general population. Prematurity, protein-energy malnutrition and measles were common predisposing factors. Of note is that primary pneumonia was the most common focus of bacteremia, namely in 43 patients. The skin ranked second as focus of infection, accounting for 20 cases [19].

Table 9.5. Neonatal septicemia in the tropics and subtropics, etiology *(n)* and mortality (%, in parentheses)

Place [Reference] Bacteria isolated	Ibadan, Nigeria [10]	Benin, Nigeria [21]	Zaria, Nigeria [32]	Chandigarh, India [6]	Amman, Jordan [20]	Riyadh, Saudi Arabia [27]	Total
Gram-positive							
Staphylococcus aureus	22 (31.8)	18 (27.8)	38 (36.8)	14 (42.9)	18 (55.6)	14	124
Staphylococcus epidermidis	6	3	18	3	19	6	45
Beta-hemolytic streptococci	0	1	0	1	0	2[a]	4
Enterococci	5	0	8	5	0	0	18
Streptococcus viridans	1	0	0	6	0	0	7
Nonhemolytic streptococci	3	3	0	0	0	0	6
Streptococcus pneumoniae	0	5	0	0	6	0	11
Other	0	0	0	0	4	0	4
Subtotal gram-positive	37	30	64	29	47	22	219
Mixed infections	2	0	0	14	0	0	16
Gram-negative							
Escherichia coli	18 (61.0)	25 (48.0)	26 (61.5)	14 (85.7)	7 (42.9)	12	102
Klebsiella Enterobacter	24 (25.0)	9	12 (33.3)	7 (57.1)	11 (45.5)	9	82
Pseudomonas spp.	4 (75.0)	5	5 (20.0)	10 (90.0)	4 (25.0)	2	30
Salmonella spp.	1 (0.0)	1	4 (50.0)	1 (0.0)	0	7	14
Proteus spp.	1 (100.0)	4	3 (33.3)	3 (100.0)	0	0	11
Acinetobacter spp.	1 (0.0)	0	2 (50.0)	6 (16.7)	0	0	9
Citrobacter spp.	0	0	0	4 (75.0)	0	0	4
Alcaligenes faecalis	0	0	0	10 (80.0)	0	0	10
Hemophilus influenzae	0	0	0	0	0	2	2
Neisseria meningitidis	0	0	0	0	1 (100.0)	0	1
Other	0	0	0	2 (50.0)	0	0	2
Subtotal gram-negative	49	44	52	57	23	32	267
Total	88 (35.2)	74 (41.9)	116 (41.4)	100 (64.0)	70 (47.1)	54 (29.6)	502

[a] Group B streptococci.

Another characteristic feature of septicemia in developing countries is the high number of community-acquired septicemias due to nontyphoidal salmonellae. This most likely reflects the high prevalence of salmonellosis and cross-infection in the community. In a study from Jakarta, there was one case of nontyphoidal septicemia for every four cases of typhoid fever [3]. The ratio of community- and hospital-acquired septicemia also differs from USA/European findings [18]. Community-acquired septicemia accounted for 62%–63% of all cases of septicemia in Nigeria, Jamaica, and Kuwait [2, 12, 17]. This finding is largely attributable to a larger pediatric population, a lesser proportion of patients with underlying disease, and the endemicity of salmonellosis.

Streptococcus pneumoniae, nontyphoidal salmonellae, and *Hemophilus influenzae* were the most common pathogens of bacteremia in children in Johannesburg (Table 9.1). In contrast, both pneumococci and *H. influenzae*, which are the main

Table 9.6. Comparison of the etiology of neonatal septicemia in the USA and the tropics and subtropics. The range of pathogens encountered in the USA (New Haven) is similar to that for other developed countries. In Europe, the incidence of neonatal *Listeria* infection is somewhat higher than in the USA

Bacteria isolated (%)	Yale, New Haven Hospital, New Haven, USA [13]				Tropics and subtropics[a]
	1933–1943	1944–1957	1958–1965	1966–1978	1970–1981
Staphylococcus aureus	9	13	3	5	25
Staphylococcus epidermidis	0	0	1	1	9
Beta-hemolytic streptococci					
Group A	36	8	0	0	0.4
Group B	5	7	1	32	0.4
Enterococci	0	3	10	3	4
Streptococcus pneumoniae	11	5	3	1	2
Escherichia coli	25	37	45	32	20
Klebsiella Enterobacter spp.	0	0	11	12	16
Pseudomonas spp.	0	21	15	2	6
Salmonella spp.	0	2	0	0.5	3
Proteus spp.	0	0	1	2	2
Hemophilus influenzae	0	0	1	4	0.4
Listeria monocytogenes	2	2	0	0.5	0
Mixed	7	2	0	5	3
Other	5	2	8	3	11
Total cases *(n)*	44	62	73	239	502

[a] Cumulated data from Table 9.5.

pathogens of community-acquired septicemia in infants and young children in West Europe and the USA, were isolated only rarely in children in Nigeria (Table 9.2). In a retrospective study of 75 children with pneumococcal bacteremia, 68% were younger than 2 years [4]. Mortality was higher than the overall case fatality rate of 26.7% in children who were severely malnourished (42.8%) and in children with associated meningitis (54.5%) [4].

Studies of the etiology of neonatal septicemia in the tropics and subtropics are summarized in Table 9.5.

Staphylococcus aureus accounts for approximately 25% of cases, whereas group B streptococcal septicemia is reported only infrequently from the tropics [23, 29], even though a high carriage rate of group B streptococci was found in women of child-bearing age [22]. This is in contrast to the pattern in the USA and Europe, where group B streptococci have emerged as a major pathogen of neonatal septicemia since the early 1970s (Table 9.6). The etiology of neonatal septicemia as observed in Kingston, Jamaica, in 1982–1983 appears to represent an intermediate position. Of 27 cases of neonatal septicemia included in a study recently published [17], six were caused by *Staphylococcus aureus* and nine by group B streptococci. Another well-documented cause of neonatal septicemia in the USA and Europe, *Listeria monocytogenes,* seems to be difficult to diagnose under tropical conditions [24].

There are certain waterborne or soil-associated bacteria which infrequently cause septicemia in the tropics and subtropics, but are rare or absent in temperate climates, namely *Chromobacterium violaceum, Aeromonas hydrophila, Vibrio vulnificus,* and *Pseudomonas pseudomallei.* Portals of entry are minor traumas (such as abrasions) or the gastrointestinal tract (commonly associated with fresh water contact) [15].

Diagnosis and Investigations

Septicemia should be suspected in patients with swinging pyrexia, leukocytosis, and no localizing signs. In that situation bacteria may enter the bloodstream from a metastatic focus, while the primary portal of entry, an infected wound for example, has already resolved. The presence of a localized infection does not exclude a secondary invasion of the bloodstream by the infective agent, but gives important clues as to the etiology (see section "Nondirected Therapy" below). The definitive diagnosis of septicemia can only be made from blood cultures.

Blood cultures should be obtained, if possible, before antibiotic treatment is started. They should be taken during the initial rigors and spikes of temperature; fever and leukocytosis are inconstant findings in septicemia of premature babies and neonates, the elderly, and in severe protein-energy malnutrition. In these conditions bacteremia is often afebrile and blood cultures should be taken once the diagnosis is suspected.

Blood cultures should be taken by venipuncture, after careful skin disinfection with tincture of iodine or 70% isopropyl alcohol, and inoculated aseptically into a set of two bottles containing appropriate liquid media for aerobic and anaerobic culture. Inoculation of liquid blood cultures at the bedside is important, since sensitive organisms may die in blood during transport.

Aspiration of the sites of septic emboli (e.g., in meningococcemia) and, in septicemias secondary to infection elsewhere, bacteriological culture of pus, sputum, urine, and CSF may also yield the causative organism. Even in the absence of facilities for culture, microscopy of specimens may still be helpful. A Gram's stain of the pus, purulent CSF, and sputum can be done at the bedside, thus differentiating between the main pathogens of septicemia, namely, staphylococci, streptococci, pneumococci, and the gram-negative bacilli. An infection of the urinary tract as source of the septicemia can also be excluded by microscopy.

Under optimal conditions, antibiotic treatment is guided by the results of the antibiotic sensitivity testing. If there is no improvement or in case of relapse, further blood cultures are obtained to detect reinfection, superinfection, or further components of a mixed infection.

Treatment

General Rules

Clearance of the *initial septic focus* by drainage of pus or by operation may be essential for successful treatment. In septicemia, *bactericidal antibiotics* (penicillins, cephalosporins, and aminoglycosides) are preferred to bacteriostatic antibiotics

Table 9.7. Daily dose levels, dose intervals, and preferred routes of administration in septicemia

Antibiotic	Adults	Children[a]	Administration
Benzylpenicillin	20-40 MU	0.5 MU/kg (infants: 1 MU/kg)	I.v. injection or short i.v. infusion every 6 h, or i.m.
Ampicillin Amoxicillin	6-10 g (max. 20 g)	150-200 mg/kg (max. 400 mg/kg)	Short i.v. infusion or slow i.v. injection every 6-8 h, or i.m.
Cloxacillin Oxacillin	6-10 g	120-200 mg/kg	Short i.v. infusion or slow i.v. injection every 6-8 h, or i.m.
Carbenicillin	20-30 g	300 mg/kg	Short i.v. infusion or slow i.v. injection every 6-8 h.
Ticarcillin	15-20 g	200-300 mg/kg	Short i.v. infusion or slow i.v. injection every 6-8 h.
Mezlocillin Azlocillin Piperacillin	6-15 g (max. 20 g)	200-300 mg/kg	Short i.v. infusion or slow i.v. injection every 6-8 h, or i.m.
Streptomycin	1 g	20-30 mg/kg	I.m. (or i.v. infusion) every 12-24 h
Kanamycin	1 g	10-15 mg/kg	I.m. (or short i.v. infusion 20-30 min) every 12 h; children: the high dose every 8 h
Amikacin	1-2 g	10-15 mg/kg	I.m. or short i.v. infusion every 12 h
Gentamicin Tobramycin Netilmicin Sisomicin	240-320 mg	3-5 mg/kg	Every 8-12 h i.m., slow i.v. injection, or short i.v. infusion
Cephalothin	6-12 g	100 mg/kg (max. 200 mg/kg)	Short i.v. infusion or slow i.v. injection every 6-8 h
Cephaloridine	3-4 g	25-50 mg/kg (max. 100 mg/kg)	Short i.v. infusion or slow i.v. injection every 6-8 h, or i.m. (nephrotoxic)
Cefazolin Cefamandole Cefuroxime Cefotiam Cefoxitin[b]	4-6 g	100-150 mg/kg	Short i.v. infusion or slow i.v. injection, every 6-8 h, or i.m.
Cefotaxime Ceftazidime[c]	4-6 g	100-150 mg/kg	Short i.v. infusion or slow i.v. injection every 6-8 h, or i.m.
Ceftriaxon	2 g	100 mg/kg	Short i.v. infusion or slow i.v. injection, every 12-24 h, or i.m.
Chloramphenicol Thiamphenicol	3 (-4) g	50-80 mg/kg (max. 100 mg/kg)	P.o. or rapid i.v. injection every 6-8 h
Tetracycline Oxytetracycline Rolitetracycline	0.5 g	10 mg/kg	Slow i.v. injection every 6-8 h
Doxycycline	0.2 g	4 mg/kg	Slow i.v. injection, or p.o. every 12-24 h

Table 9.7 (continued)

Antibiotic	Adults	Children[a]	Administration
Co-trimoxazole	1.92–2.88 g	6 months to 5 years: 0.48 g 6–12 years: 0.69 g (6 mg/kg trimethoprim + 30 mg/kg sulfamethoxazole)	P.o. every 12 h; i.m. injection or long i.v. infusion (1 h) possible, but there are many side effects. Vials for i.m. use and i.v. use contain different amounts of substance
Erythromycin	2 g (max. 4 g)	30–60 mg/kg (max. 80 mg/kg)	P.o. as succinate, i.v. as lactobionate or glucoheptonate every 6 h
Vancomycin	1–2 g	40 mg/kg	Slow i.v. infusion every 12 h
Clindamycin	1.2–1.8 g (max. 2.4 g)	20–40 mg/kg	I.v. infusion or i.m. every 6 h
Lincomycin	1.2–1.8 g	20–40 mg/kg	I.v. infusion or i.m. every 8–12 h
Fusidic acid	(1.5–)3 g	20–40 mg/kg (max. 50 mg/kg)	P.o. or slow i.v. infusion every 8 h (no i.v. or i.m. injection!)
Metronidazole	1.5–2 g	20–30 mg/kg	Short i.v. infusion every 6–8 h
Ciprofloxacin	400 mg	—[d]	Short i.v. infusion every 12 h

This table should be used only as a reference for dosage. The antibiotics are *not* listed in order of choice (see Table 9.8 for choice of antibiotic).
MU, megaunits.
[a] Except neonates.
[b] And other first- and second-generation cephalosporins.
[c] And other third-generation cephalosporins.
[d] Contraindicated in children.

(tetracyclines, erythromycin, and chloramphenicol), although the former cannot completely prevent relapses.

Antibiotics must be given in *large doses* (Table 9.7) for an adequate length of time to prevent relapse. *Parenteral administration,* (preferably i.v.) is important, since, in general, serum levels are higher with i.v. than with oral administration. Exceptions are oral chloramphenicol, doxycycline, and metronidazole. In cases where the pathogen is not (yet) known and because of the synergistic effects, *combinations of antibiotics* are preferred in order to cover a wider range of pathogens.

The classical rule of not combining bactericidal and bacteriostatic antibiotics has been put in doubt recently, because observed in vitro antagonisms do not seem to hold true under clinical conditions. Nevertheless, combinations of penicillins or cephalosporins with tetracyclines, erythromycin, or chloramphenicol are not recommended.

Since septicemia may impair renal function and is treated with large doses of antibiotics, patients must be closely monitored for side effects. Impaired renal function may necessitate adjustment of the dose. Side effects of aminoglycosides may be oto- und nephrotoxicity; of penicillins, neurotoxicity. Convulsions can also arise from the increased permeability of the blood-brain barrier in meningitis.

High doses of carbenicillin may lead to sodium overload and an associated fall in serum potassium.

Supportive Treatment. Treatment may be necessary for shock, restoration and maintenance of electrolyte balance and hydration, or correction of acidosis. Surgical measures may also be required.

Reasons for Treatment Failure. Inappropriate choice of antibiotics, inadequate dosage, superinfection with a different organism, development of resistance, persistence of susceptible organisms in sites into which antibiotics cannot penetrate (such as collection of pus in abscesses), and protracted septic shock are possible reasons for treatment failure.

Directed Therapy

The choice of antibiotic should be guided by the causative organism, its sensitivity pattern, and the clinical picture (Table 9.8). Factors such as tissue penetration into bone, urine, or CSF should also be taken into account.

In Europe, beta-lactams (penicillins and cephalosporins) are the main agents used because of their high clinical efficacy; even large doses carry only a minimal risk of irreversible side effects, compared with aminoglycosides. Large doses of benzylpenicillin are, worldwide, still the antibiotic of choice for all septicemias due to streptococci, pneumococci, and meningococci.

Nondirected Therapy

As long as the causative organism is not known, the choice of antibiotic is necessarily empirical; "calculated chemotherapy" takes account of the range of pathogens most likely to be encountered in the respective clinical situation.

Table 9.8. Directed therapy of septicemia

Pathogens	International standard therapy		Limited resources	
	Drug of choice	Alternatives	Drug of choice	Alternatives
Streptococci and pneumococci	Benzylpenicillin	Cefazolin, clindamycin, vancomycin	Benzylpenicillin	Chloramphenicol, ampicillin, tetracyclines[a]
Enterococci	Ampicillin + gentamicin	Vancomycin, mezlocillin, imipenem	Ampicillin (+ streptomycin or gentamicin)	High-dose benzylpenicillin, tetracyclines (+ streptomycin or gentamicin)
Staphylococcus aureus				
Penicillin-sensitive	Benzylpenicillin (+ fusidic acid)	Cefazolin, vancomycin	Benzylpenicillin	Chloramphenicol[a], erythromycin[a]
Penicillin-resistant	Oxacillin, cefazolin (+ fusidic acid)	Clindamycin, vancomycin	Cloxacillin	Chloramphenicol[a], erythromycin[a]
Methicillin-resistant	Vancomycin	Clindamycin, rifampicin	Clindamycin	Chloramphenicol[a], erythromycin[a], co-trimoxazole[a], rifampicin
Meningococci	Benzylpenicillin	Chloramphenicol	Benzylpenicillin	Chloramphenicol

Table 9.8 (continued)

Pathogens	International standard therapy		Limited resources	
	Drug of choice	Alternatives	Drug of choice	Alternatives
Hemophilus influenzae	Ampicillin[a], cefuroxime, cefamandole, cefotaxime	Quinolones	Ampicillin[a], chloramphenicol	Co-trimoxazole, tetracyclines
Listeria monocytogenes	Ampicillin	Acylaminopenicillins	Ampicillin, benzylpenicillin	Chloramphenicol, tetracyclines, erythromycin
Escherichia coli	Cefotaxime, gentamicin, amikacin	Ampicillin[a], mezlocillin, piperacillin	Gentamicin, amikacin, first-generation cephalosporins	Ampicillin[a], tetracyclines[a], quinolones
Klebsiella/ Enterobacter	Cefotaxime + gentamicin or amikacin	Mezlocillin[a], quinolones	Chloramphenicol[a], first-generation cephalosporins (+ gentamicin)	Tetracyclines[a], co-trimoxazole[a], quinolones
Proteus mirabilis	Ampicillin, mezlocillin, piperacillin	Parenteral cephalosporins	Ampicillin	Chloramphenicol, co-trimoxazole, quinolones
Proteus, indole-positive	Cefotaxime, cefoxitin + gentamicin	Mezlocillin[a], piperacillin[a], quinolones	Chloramphenicol[a] + gentamicin	Co-trimoxazole, quinolones
Pseudomonas aeruginosa	Azlocillin or piperacillin + tobramycin	Cefsulodin, ceftazidime, imipenem, ciprofloxacin, aztreonam	Azlocillin or piperacillin + gentamicin, quinolones	Polymyxins, mezlocillin, carbenicillin, ticarcillin, quinolones
Salmonella typhi	Co-trimoxazole, amoxicillin	Chloramphenicol, quinolones	Chloramphenicol	Co-trimoxazole, amoxicillin
Salmonella spp.	Ampicillin[a] + gentamicin	Mezlocillin, cefotaxime, quinolones	Chloramphenicol[a], ampicillin[a], gentamicin	Co-trimoxazole[a]
Serratia marcescens	Cefotaxime + gentamicin or amikacin	Mezlocillin, piperacillin, ceftazidime	Gentamicin, amikacin	Co-trimoxazole[a], quinolones
Clostridium perfringens group	Benzylpenicillin	Metronidazole, cephalosporins	Benzylpenicillin	Chloramphenicol, metronidazole, tetracyclines
B. melaninogenicus	Benzylpenicillin	Tetracyclines, metronidazole	Benzylpenicillin	Tetracyclines, metronidazole
Bacteroides fragilis	Clindamycin, metronidazole, cefoxitin	Latamoxef, imipenem	Metronidazole, chloramphenicol	Tetracyclines, rifampicin
Gonococci	Cefuroxime, cefotaxime, benzylpenicillin	Spectinomycin	Thiamphenicol, spectinomycin, benzylpenicillin	Co-trimoxazole[a], kanamycin, cephalosporins, quinolones
Pasteurella multocida	Benzylpenicillin	Tetracyclines	Benzylpenicillin, tetracyclines	Many agents active
Campylobacter fetus	Tetracyclines, erythromycin	Chloramphenicol	Tetracyclines, erythromycin	Chloramphenicol
Acinetobacter spp.	Ceftazidime	Ampicillin, mezlocillin, gentamicin	Gentamicin	Co-trimoxazole
Chromobacterium violaceum	Chloramphenicol + gentamicin	Tetracyclines, erythromycin, streptomycin	Chloramphenicol + gentamicin	Tetracyclines, erythromycin, streptomycin
Aeromonas hydrophila	Tetracyclines, ampicillin	Co-trimoxazole, parenteral cephalosporins	Tetracyclines, ampicillin	Co-trimoxazole

For fungal septicemia, see p. 311, 316; melioidosis, see p. 282.

[a] Resistance is possible.

The traditional use of fixed proprietary combinations, such as penicillin + streptomycin and ampicillin + cloxacillin as nondirected therapy must be viewed critically since in many instances, other antibiotics are more appropriate. In these combinations, the dose of penicillin is usually too low for treatment of septicemia. Streptomycin should primarily be used as an antituberculous drug, particularly in view of its suboptimal efficacy and the reported rates of resistance. The combinations of ampicillin and cloxacillin and of amoxicillin and clavulanic acid promote the selection of resistant gram-negative enterobacteria and their activity their inferior to that of cephalosporins.

In Europe, modern cephalosporins (cefotaxime, cefoxitin, cefotiam, cefuroxime, ceftazidime) are used for nondirected treatment of septicemia prior to the isolation of the causative organism. A second agent (acylaminopenicillins, metronidazole, or aminoglycosides) is usually added to extend the range of pathogens covered and/or to achieve synergistic effects. This concept of broad-spectrum therapy covering nearly all bacterial pathogens of septicemia is directed at a different pattern of septicemia - a high incidence of hospital-acquired septicemia, mostly in elderly patients with underlying, immunosuppressive diseases. All these combinations are less effective against *Staphylococcus aureus* than cloxacillin or first-generation cephalosporins (e.g., cefazolin), not to mention their costs, which will often be beyond the means of developing countries.

In the tropics, chloramphenicol still has to be given if the presumptive diagnosis of *community-acquired septicemia* (acute fever, leukocytosis, no obvious localizing signs) has been made. An alternative is co-trimoxazole i.v., which also covers gram-positive cocci and septicemic salmonellas. Tetracyclines or doxycycline may be used if rickettsiosis (e.g., scrub typhus in Southeast Asia) has to be considered in the differential diagnosis. (Since varying percentages of strains of *Staphylococcus aureus* are resistant to chloramphenicol, co-trimoxazole, or tetracycline, staphylococcal septicemia may not respond!) Quinolones may prove to be an effective alternative in community-acquired septicemia. The in vitro data show good activity against gram-positive cocci, including *S. aureus,* and gram-negative enteric bacilli, including *Salmonella* spp.

Septicemia After Minor Skin Sepsis (with Lymphangitis). This is usually caused by staphylococci or *Streptococcus pyogenes.* The *international standard therapy* consists of flucloxacillin or first-generation cephalosporins (cefazolin) i.v. Under conditions of *limited resources,* cloxacillin, chloramphenicol, or erythromycin are alternatives, albeit less effective.

*Septic Abortion and Puerperal Sepsis.*The main pathogens are aerobic and anaerobic streptococci (including *Streptococcus pyogenes*), anaerobic bacteria (such as *Bacteroides* spp.), and clostridia, enterobacteria, and staphylococci - often in mixed infection. The *international standard therapy* consists of second- to third-generation cephalosporins (cefotaxime, cefoxitin, cefuroxime) or amoxicillin + metronidazole; an aminoglycoside is added in severely ill patients. *Limited resources:* A study from Ethiopia showed 20% of 45 patients with puerperal sepsis to be bacteremic [26]. Penicillin + chloramphenicol was found to be significantly more effective than the more conventional combination of penicillin + streptomycin [11].

A combination of penicillin and an aminoglycoside should be supplemented with metronidazole to cover the penicillin-resistant *Bacteroides* spp. If *Streptococcus pyogenes* or clostridia (gas gangrene) are isolated from blood culture or a cervical swab (Gram's stain and culture), benzylpenicillin in large doses is the treatment of choice. Retained portions of the placenta must be removed by curettage.

Postoperative Septicemia Arising from Infected Wounds. This is frequently caused by resistant staphylococci. Mixed infection with enterobacteria is not unusual. The *international standard therapy* consists of first-generation cephalosporins (cefazolin) with good activity against staphylococci. Flucloxacillin or a combination of first-generation cephalosporins with an aminoglycoside (gentamicin) is given to severely ill patients. Under conditions of *limited resources,* cloxacillin, chloramphenicol, or erythromycin are alternatives, albeit less effective.

Septicemias After Intestinal or Gynecological Operations. The range of pathogens encountered includes the enterobacteria and the anaerobic bowel flora, namely *Bacteroides* spp. and anaerobic streptococci, typically in mixed infection. The *international standard therapy* consists of cefotaxime + metronidazole, cefoxitin + azlocillin, or ampicillin + gentamicin + metronidazole. Under conditions of *limited resources,* tetracycline i.v., penicillin + streptomycin + metronidazole, chloramphenicol + gentamicin, or co-trimoxazole + metronidazole should be administered.

Septicemias After Urological Operations. These are commonly caused by resistant enterobacteria *(Proteus, Klebsiella, Enterobacter)* and *Pseudomonas aeruginosa.* The *international standard therapy* consists initially of a combination of a third-generation cephalosporin or a modern penicillin with an aminoglycoside. This is changed, if necessary, according to the results of antibiotic sensitivity testing. Under conditions of *limited resources,* co-trimoxazole, ampicillin + gentamicin, or quinolones should be prescribed; amikacin may be necessary in cases of gentamicin resistance.

Septicemia Complicating Infections of the Normal Urinary Tract. These are usually caused by *Escherichia coli.* Other enterobacteria are less common.The *international standard therapy* consists of third-generation cephalosporins, an acylaminopenicillin, quinolones, or an aminoglycoside. Under conditions of *limited resources,* tetracycline, co-trimoxazole, chloramphenicol, ampicillin, gentamicin or quinolones should be given.

Biliary Septicemia Associated with Cholangitis or Cholecystitis. This is rare in the tropics, where a major predisposing factor, cholelithiasis, is almost nonexistent. An exception is Southeast Asia, where clonorchiasis is common. Mixed infection due to *E. coli,* other enterobacteria, and microaerophilic and anaerobic streptococci is typical. The relief of mechanical obstruction causing cholestasis is necessary. The *international standard therapy* consists of third-generation cephlosporins or an acylaminopenicillin, supplemented by an aminoglycoside where necessary. Under conditions of *limited resources,* ampicillin or cephalosporins should be administered. Tetracyclinces are no longer recommended for severe biliary infections due to loss of activity in bile. If amebic abscess cannot be excluded, addition of metronidazole is advisable.

Infusion Bacteremia. Severe febrile reactions and even shock can result from contamination of i.v. infusion fluids or blood stored for transfusion. Both faulty sterilization procedures and unsterile handling of i.v. equipment in current use have been reported [16]; *Klebsiella* and *Bacillus* species were the main pathogens involved, but many other bacteria and fungi are possible. Antibiotic treatment with an aminoglycoside or chloramphenicol may be indicated, although most cases resolve spontaneously.

Septicemias Originating from Acute Tonsilitis or Tonsillar Abscess. Bacteroides spp., staphylococci, and streptococci are the main pathogens. *Enterobacteria* are rarely responsible. The *international standard therapy* consists of cefoxitin or clindamycin; benzylpenicillin or flucloxacillin are possible alternatives. Under conditions of *limited resources,* benzylpenicillin, chloramphenicol, or tetracycline should be given.

Septicemias Arising in Patients with Bone-Marrow Suppression (e.g., Leukemia). A wide range of pathogens, including opportunistic pathogens, may be encountered; *E. coli, Proteus, Klebsiella, Pseudomonas,* and staphylococci are, however, the pathogens most commonly isolated. The *international standard therapy* consists of broad-spectrum bactericidal combinations like cefotaxime + azlocillin + gentamicin. Under conditions of *limited resources,* co-trimoxazole covers the common pathogens, with the exception of *Pseudomonas.* Alternatively, ampicillin + cloxacillin may be combined with an aminoglycoside, preferably gentamicin or amikacin. Quinolones may be of value.

Antibiotics may have to be continued for several weeks in case of long-lasting bone marrow suppression, because of the high risk of relapse or reinfection.

Neonatal Septicemia. The clinical presentation may be very vague; the temperature may be normal or subnormal instead of high. In developing countries, septicemia is one of the main causes of neonatal jaundice, especially if the jaundice appears later than 24 h after birth and if the child is ill. Signs of infection must be looked for in any unwell neonate, and treatment with a combination of antibiotics must be started as soon as the diagnosis is suspected. As septicemia is frequently associated with meningitis in the neonatal period, a lumbar puncture should always be performed (Gram's stain of CSF). The etiology of neonatal septicemia in the tropics (Table 9.5) is discussed elsewhere. The dose schedules used must be adapted to the physiologically impaired hepatic and renal function of the newborn (see Table 1.3 of chapter "Dosage of Antibiotics"). The *international standard therapy* consists of combinations of second- or third-generation cephalosporins with acylaminopenicillins, possibly supplemented by an aminoglycoside - e.g., cefuroxime or cefotaxime + piperacillin or azlocillin (+ gentamicin, tobramycin, or amikacin). Combination of one of these aminoglycosides with benzylpenicillin or ampicillin is a traditional alternative. Under conditions of *limited resources,* benzylpenicillin or ampicillin is combined with kanamycin or streptomycin, when gentamicin, the aminoglycoside to be preferred, is not available. Several investigators have recommended supplementation of the aminoglycoside by a penicillinase-resistant penicillin (such as cloxacillin) in the initial treatment because of the high proportion of penicillin-resistant staphylococci [10, 21].

Alternative regimens such as ampicillin+cloxacillin (which is ineffective against *Klebsiella, Enterobacter,* and *Pseudomonas*) or monotherapy with chloramphenicol (which carries the risk of grey syndrome) are associated with comparatively high failure rates. If available, second- or third-generation cephalosporins should be preferred.

References

1. Adetuyibi A, Akisanya JB, Onadeko BO (1976) Analysis of the causes of death on the medical wards of the University College Hospital, Ibadan, over a 14-year period (1960-1973). Trans R Soc Trop Med Hyg 70: 466-473
2. Alausa KO, Montefiore D, Sogbetun A, Ashiru I, Onile BA, Sobayo E (1977) Septicaemia in the tropics. Scand J Infect Dis 9: 181-185
3. Anderson KE, Joseph SW, Nasution R, Sunoto, Butler T, Van Peenen PFD, Irving GS, Saroso JS, Watten RH (1976) Febrile illnesses resulting in hospital admission: a bacteriological and serological study in Jakarta, Indonesia. Am J Trop Med Hyg 25: 116-121
4. Berkowitz FE (1981) Pneumococcal bacteraemia - a study of 75 black children. Ann Trop Paediatr 1: 229-235
5. Berkowitz FE (1984) Bacteremia in hospitalized black south African children. Am J Dis Child 138: 551-556
6. Bhakoo ON, Agarwal KC, Narang A, Bhattacharjee S (1974) Prognosis and treatment of neonatal septicaemia - a clinicobacteriological study of 100 cases. Indian Pediatr 11: 519-528
7. Brown KGB (1975) Analysis of admissions to the adult medical wards at Queen Elizabeth Central Hospital, Blantyre, Malawi. East Afr Med J 52: 509-528
8. Bwibo NO (1971) The role of neonatal infection in neonatal and childhood mortality. J Trop Pediatr 19: 89-90
9. Charlton RW (1979) Observations on blood cultures. A 10 year survey of specimens from African patients. Cent Afr J Med 25: 221-223
10. Dawodu AH, Alausa OK (1980) Neonatal septicaemia in the tropics. Afr J Med Sci 9: 1-6
11. Duncan ME, Perine PL, Krause DW, Awoke S, Zaidi AA (1980) Pelvic inflammatory disease and puerperal sepsis in Ethiopia. II. Treatment. Am J Obstet Gynecol 138: 1059-1063
12. Elhag KM, Mustafa AK, Sethi SK (1985) Septicaemia in a teaching hospital in Kuwait. I: Incidence and aetiology. J Infect Dis 10: 17-24
13. Freedman RM, Ingram DL, Gross I, Ehrenkranz RA, Warshaw JB, Baltimore RS (1981) A half century of neonatal sepsis at Yale, 1928-1978. Am J Dis Child 135: 140-144
14. Gedebou M, Tassew A, Azene G (1984) Blood culture isolates from an Addis Abeba Hospital, frequency and antibiotic sensitivities. East Afr Med J 61: 190-196
15. Gratten M (1981) The laboratory diagnosis of opportunistic infections caused by uncommon bacteria in Papua New Guinea. Papua New Guinea Med J 24: 169-173
16. Koopman JS, Fighetti de Olave A (1980) An extensive outbreak of klebsiella pneumoniae bacteremias from in-use contamination of i. v. bottles. Bull Pan Am Health Organ 14: 185-192
17. Macfarlane DE, Narla VR (1985) Bacteraemia at the university hospital of the West Indies - a report of 222 cases. J Infect 10: 126-142
18. McGowan JE, Barnes MW, Finland M (1975) Bacteremia at Boston City Hospital: occurrence and mortality during 12 selected years (1935-1972), with special reference to hospital-acquired cases. J Infect Dis 132: 316-335
19. Mukonyora M, Mabiza E, Gould IM (1985) Staphylococcal bacteraemia in Zimbabwe 1983. J Infect 10: 233-239
20. Nazer H (1981) Neonatal septicaemia at the Jordan University Hospital. J Trop Pediatr 27: 199-204

21. Omene JA (1979) Neonatal septicaemia in Benin City. A review of 74 cases. Trop Geogr Med 31: 35-39
22. Onile BA (1980) Group B streptococcal carriage in Nigeria. Trans R Soc Trop Med Hyg 74: 367-370
23. Onile BA, Osinusi BO, Montefiore D, Sogbetun AO (1980) Neonatal septicaemia resulting from group B streptococci: a case report. Int J Obstet Gynecol 17: 393-395
24. Onyemelukwe GC, Lawande RV, Egler LJ, Mohammed I (1983) Listeria monocytogenes in Northern Nigeria. J Infect 6: 141-145
25. Patel KM, Lwanga SK (1971) A study of medical admissions to Mulago hospital, Kampala. East Afr Med J 48: 76-84
26. Perine PL, Duncan ME, Krause DW, Awoke S (1980) Pelvic inflammatory disease and puerperal sepsis in Ethiopia. I. Etiology. Am J Obstet Gynecol 138: 969-973
27. El Rifai MR (1982) A study of 214 neonates with infectino in the maternity and children's hospital of Riyadh, Saudi Arabia. Ann Trop Pediatr 2: 119-122
28. Shah PM, Schäfer V, Bruns F, Stille W, Knothe H (1984) Erregerspektrum von Septikämien. In: Wilms K, Stille W (eds) Infektionen bei Abwehrschwäche. FAC, Fortschritte der Antimikrobiellen-Antineoplastischen Chemotherapie, vol 3-6. Futuramed, München, pp 901-906
29. Sharma KB, Prakash K, Ravindran PC, Narang P (1975) Neonatal septicemia due to group B-beta hemolytic streptococci. Indian J Med Res 63: 781-786
30. Srinivasa DK, Danabalan M, Anand D (1974) Infant mortality trends in a rural health centre of Pondicherry. J Indian Med Assoc 62: 39-43
31. Stille W (1972) Septikämie - Problematik, Klinik und Therapie. Rheindruck, Boppard
32. Winfred I (1984) The incidence of neonatal infections in the nursery unit at the Ahmadu Bello University Teaching Hospital, Zaria, Nigeria. East Afr Med J 61: 197-202
33. Yánez MZ, Zamorano MG, Diaz CQ, Guerra PA (1981) Analisis bacteriana en el neonato en los últimos cinco anos en el Servicio de Pediatria del Hospital Regional Leonardo Guzmán, de Antofagasta. Rev Chil Pediatr 53: 213-217

10 Infective Endocarditis

Incidence and Prognosis

The incidence of infective endocarditis ranges between 0.5% and 3% of patients with cardiac disease in the few hospital-based series from Africa [5, 15]. In the Rwandian experience, infective endocarditis is relatively rare compared to the high prevalence of rheumatic heart disease [4].

The available data regarding the etiology of infective endocarditis are summarized in Tables 10.1 and 10.2. A feature to note in the African studies is the high frequency with which *Staphylococcus aureus* was isolated. Cases of endocarditis due to *S. aureus* commonly take a peracute course, carry a poor prognosis, and often are diagnosed at autopsy only. The propensity of this organism to attack normal heart values was confirmed in the African series [2, 5, 13].

The series from Malaysia with the predominance of streptococci, especially *Streptococcus viridans,* comes close to the classical descriptions of the disease as given, for example, in the series by Lerner and Weinstein from Boston [7, 12]. With the exception of 2 studies [12, 15], dental caries and prior dental procedures were not considered an important source of infection. In most instances, a source of infection was not readily apparent.

Organisms able to attack normal heart valves and frequently associated with an acute course of endocarditis are, apart from *S. aureus,* pneumococci, β-hemolytic streptococci, gonococci, and gram-negative enteric bacteria [14]. Mortality is high.

Subacute bacterial endocarditis (the classical "endocarditis lenta") is known for its association with streptococci, especially the viridans streptococci. It commonly affects heart valves already affected with rheumatic heart disease or congenital heart disease. With modern treatment, mortality can be reduced to 10%-15%. The time course of enterococcal endocarditis, which is observed more frequently in developed countries in recent decades, is more acute. Enterococcal endocarditis occurs mainly in elderly patients. Embolic complications are frequent and mortality is in the range of 30%.

In rare instances, subacute bacterial endocarditis is caused by certain fastidious, gram-negative organisms, namely *Hemophilus aphrophilus, H. influenzae, Cardiobacterium hominis,* and *Actinobacillus actinomycetemcomitans* [3], but also by *Coxiella burnetii,* and *Brucella.* Some of these cases respond to combined treatment with penicillin and aminoglycoside. Cefotaxime may also be useful. Endocarditis due to *Coxiella burnetii* and *Brucella* may respond to doxycycline.

As the prognosis depends among other factors upon the isolation of the pathogen, blood cultures must be taken prior to antibiotic treatment. In acute cases, two to three sets of blood cultures from different sites suffice. In subacute cases four to six sets should be taken within a day.

Table 10.1. Underlying heart disease in infective endocarditis

Country [Reference] Period of review	Malaysia [12] 1968–1977	South Africa [8] 1974–1981	India [6] 1967–1979	Ivory Coast [15] 1978–1982	Uganda [13] 1968–1969	Nigeria [5] 1961–1970	Nigeria [2] 1958–1966	USA [7] 1956–1964
Cases reviewed *(n)*	101	13 (children only)	41 (children only)	40[a]	102	90	64 (autopsy review)	100
Cases diagnosed at autopsy *(n)*	-	-	16	-	45	41	64	-
Underlying heart disease	99	10	35	40	62	59	11	61
Rheumatic heart disease	69	9	15	32	50	57		40
Congenital heart disease	30	1	20	4	8	2	} 11	10
Endomyocardial fibrosis	-	-	-	3	3	-	-	-
Aortic calcific sclerosis	-	-	-	-	1	-	-	-
Normal heart	1	3	6	-	40	31	53	39
Prosthetic heart valve	1	-	-	9	-	-	-	-

Table 10.2. Etiology and mortality of infective endocarditis

Country [Reference] Period of review	Malaysia [12] 1968–1977	South Africa [8] 1974–1981	India [6] 1967–1979	Ivory Coast [15] 1978–1982	Uganda [13] 1968–1969	Nigeria [5] 1961–1970	Nigeria [2] 1958–1966	USA [7] 1956–1964
Cases reviewed *(n)*	101	13 (children only)	41 (children only)	40[a]	102	90	64 (autopsy review)	100
Cases diagnosed at autopsy *(n)*	–	–	16	–	45	41	64	–
Bacteria isolated *(n)*								
Staphylococcus aureus	14	5	4	5	16	9	7	20
Staphylococcus epidermidis	3	2	–	7	4	12	2	3
Streptococcus viridans	43	3	5	2	9	2	4	27
Enterococci *(Streptococcus faecalis)*	4	1	2	2	2	4	–	8
Beta-hemolytic streptococci	–	1	2	–	–	2	3	3
Other streptococci	7	–	–	–	2	–	1	18
Streptococcus pneumoniae	1	–	–	–	2	3	6	–
Escherichia coli	3	–	1	–	2	} 5	–	–
Pseudomonas aeruginosa	–	–	–	1	1		–	–
Salmonella spp.	–	–	–	–	–	1	1	1
Serratia marcescens	–	1	–	–	–	–	–	–
Candida spp.	1	–	–	–	–	–	–	–
Hemophilus influenzae	–	–	–	–	–	–	–	1
Others	1	2	–	1	–	9	–	4
Mixed infections	–	2	2	–	2	5	1	1
No bacteria isolated	24	–	29	15	64	48	41	14
Mortality of cases diagnosed in life	22.7%	61.5%	ND	37.5%	19%	44.9%	NA	35.3%

ND, no data; NA, not applicable.
[a] Blood cultures were taken in 33 of 40 cases.

Because there is a continuous low-grade bacteremia in infective endocarditis, further blood cultures are rarely helpful if high-quality media are used. Negative blood cultures should be kept for at last 2 weeks to allow the isolation of slow-growing, fastidious organisms.

Treatment

General Principles

The following principles of treatment are generally accepted, even though treatment regimens conflicting with these principles have resulted in cure in some well-documented cases.

1. Bactericidal beta-lactam antibiotics (penicillins, cephalosporins) should be used, either singly or in combination with aminoglycosides. This is in contrast to other types of septicemia which can be treated with bacteriostatic agents like chloramphenicol. Vancomycin and clindamycin are reserved for special cases. Erythromycin, tetracyclines, and co-trimoxazole are mainly bacteriostatic, despite slight bactericidal effects in vitro, and are not generally recommended for the treatment of endocarditis.
2. Antibiotics should be administered parenterally and the dosage adjusted according to renal function.
3. Treatment should continue for an extended period of time, 4-6 weeks (usually until the erythrocyte sedimentation rate (ESR) returns to normal).
4. Antibiotic sensitivity testing should be done. Methods more sophisticated than a simple disc diffusion test are recommended such as determination of the minimal inhibitory concentration (MIC), minimal bactericidal concentration (MBC), or the bactericidal titer of the patient's serum against his own isolate.
   Blood cultures during the first few days of treatment facilitate recognition of a persistent bacteremia. Blood cultures usually become sterile within 24-48 h of treatment with the exception of *S. aureus*, where the bacteremia may last for several days despite appropriate treatment. Persistent bacteremia in the presence of apparently adequate antibiotics suggests occult metastatic abscesses or mycotic aneurysms, but also inadequate bactericidal activity (mainly in enterococcal endocarditis).
5. Anticoagulants are contraindicated in usual endocarditis (risk of embolization and severe bleeding complications). The use of anticoagulants in prosthetic valve endocarditis is controversial. Corticosteroids are also contraindicated (risk of valve perforation).

Appropriate treatment will lead to defervescence and cessation of embolization within a few days. Nevertheless, prolonged treatment for 3-4-6 weeks (usually until the ESR returns to normal) is necessary. Absence of fever after discontinuation of antibiotic therapy indicates successful treatment, but does not rule out late relapses. There is a lifelong risk of new episodes of endocarditis (new infections of the damaged valve).

Recurrent fever after initial defervescence, while the patient is still on antibiotics, may suggest superinfection (obtain additional blood cultures), an occult metastatic abscess, or, most frequently, drug fever. Drug fever and rashes occur in a significant proportion of patients on long-term high-dose penicillin. Many clinicians prefer to manage these allergic reactions with histamines or, if necessary, corticosteroids, which carries the risk of shock, granulocytopenia, or interstitial nephritis. Desensitization with increasing doses may also be considered. The change to another effective treatment regimen is recommended whenever possible. Embolization and progressive heart failure (perforation of a valve) may occur despite appropriate antibiotic therapy. Both these and failure of eradication of a resistant organism are indications for valve replacement.

Streptococcal Endocarditis

Streptococcal endocarditis due to viridans and nonhemolytic streptococci (the classical subacute bacterial endocarditis) can be treated with large doses of benzylpenicillin alone. An accepted regimen is 20 (-30) megaunits (MU) benzylpenicillin i.v. daily, divided into two to three doses, for at least 28 days; 5-10 MU benzylpenicillin every 12 h is a compromise under conditions of limited resources. In children, it is recommended to use 0.5 MU/kg body weight i.v. daily, divided into four doses.

Because of the observed in vitro synergism of penicillins with aminoglycosides against streptococci and because of nonenterococcal streptococci with increased penicillin resistance (MIC >0.1 μg/ml), the concomitant administration of an aminoglycoside is generally recommended. The benefits of aminoglycoside therapy must be weighed against its risks of oto- and nephrotoxicity, which are increased in case of prior renal and vestibular diseases and in elderly patients. In cases where the determination of the MIC of the isolate is not possible and where patients tend to abscond when feeling better, it appears safer to give an aminoglycoside concomitantly at least for the first 14 days of treatment.

In Europe, streptomycin is replaced by gentamicin for reasons of toxicity and activity. Streptomycin carries a high risk of vestibular toxicity (check cochlear function and hearing regularly), but is less nephrotoxic than gentamicin (Table 10.3).

Table 10.3. Dosage of aminoglycosides in streptococcal endocarditis

Aminoglycoside	Daily dosage	Frequency	Route
Adults (normal renal function)			
Streptomycin	1 g	2 doses	i.m.
Gentamicin	240 mg	2 doses	i.m., or slow i.v. injection
Children			
Streptomycin	10(-15) mg/kg b.w.	2 doses	i.m.
Gentamicin	3-5 mg/kg b.w.	2 doses	i.m., or slow i.v. injection

When monitoring blood concentrations, the peak levels should be within 15–25 μg/ml for gentamicin. Trough levels of gentamicin (taken immediately prior to the next dose) seem to be more important with regard to the nephrotoxicity of this agent; they should be less than 2 μg/ml and preferably less than 1 μg/ml.

Two weeks of an i.m. combined penicillin/streptomycin regimen have been shown to be effective in streptococcal endocarditis. Procaine penicillin (1.2 MU every 6 h) and streptomycin (0.5 g every 12 h) are given for 2 weeks. Early relapses have not been reported. This regimen is not recommended for patients with an illness longer than 3 months in duration, complications, abnormal renal function, or preexisting vestibular disease.

Benzathine penicillin leads to very low serum levels and is considered inappropriate for the treatment of bacterial endocarditis.

Penicillin Allergy

In case of *penicillin allergy,* a cephalosporin (risk of cross-allergy less than 10%) such as cefazolin, 2 g every 8 h i.v., can be used. Parenteral lincomycin, clindamycin, vancomycin or imipenem are expensive alternatives.

Endocarditis Due to Pneumococci or Group A Streptococci

The same treatment procedure as for streptococcal endocarditis should be used for endocarditis due to pneumococci or beta-hemolytic streptococci. The course is often complicated by metastatic abscesses, meningitis and valvular destruction. The prognosis is poor.

Enterococcal Endocarditis

In enterococcal endocarditis, treatment with a penicillin/aminoglycoside combination is necessary to achieve bactericidal synergy against these resistant pathogens. One usually combines gentamicin with either benzylpenicillin, ampicillin, or amoxicillin. A combination with streptomycin instead of gentamicin is also effective (as was shown by earlier clinical experience), but is regarded inferior because of the existence of strains with high-level streptomycin resistance (Table 10.4).

The far more expensive acylaminopenicillins (e.g., mezlocillin) are of no advantage compared to ampicillin or amoxicillin. Prolonged treatment for 6 weeks is recommended. Abbreviated courses of 4 weeks have been successful.

Penicillin Allergy

In case of penicillin allergy, vancomycin (2 g daily) is the drug of choice. If vancomycin is not available, desensitization or a therapeutic trial with large doses of erythromycin plus rifampicin or co-trimoxazole seems justified. Palliative long-term therapy with doxycycline may also be considered. Cephalosporins and peni-

Table 10.4. Combination penicillin/aminoglycoside treatment of enterococcal endocarditis

Antibiotic	Daily Dosage	Frequency	Route
Adults (normal renal function)			
Benzylpenicillin	20-40 MU	3-4 doses	i.v.
Ampicillin	10-15 g	3-4 doses	i.v.
Amoxicillin	8-12 g	3-4 doses	i.v.
Streptomycin	1.0 g	2 doses	i.m.
Gentamicin	240 mg (3 mg/kg b.w.)	2 doses	i.m., or slow i.v. injection
Children			
Benzylpenicillin	0.5 MU/kg b.w.	4 doses	i.v.
Ampicillin	250-300 mg/kg b.w.	3-4 doses	i.v.
Amoxicillin	100 mg/kg b.w.	3-4 doses	i.v.
Streptomycin	10-15 mg/kg b.w.	2 doses	i.m.
Gentamicin	3-5 mg/kg b.w.	2 doses	i.m.,or slow i.v. injection

cillinase-resistant penicillins are clinically ineffective in enterococcal endocarditis, as is monotherapy with aminoglycosides.

Oral Treatment

Oral treatment of endocarditis is not an approved form of therapy and should be reserved for cases where i.v. access is impossible [1, 7]. Successful treatment of both streptococcal and enterococcal endocarditis with oral amoxicillin (1 g every 3-4 h) and gentamicin i.m. has been reported [9]. Amoxicillin is preferred to phenoxymethylpenicillin or ampicillin because of its high and reliable rate of intestinal absorption. Supervised intake of amoxicillin is advisable. Cases of *Streptococcus viridans* endocarditis have been treated successfully with phenoxymethylpenicillin (0.5-1 g every 4 h) with or without probenecid. In children, subacute bacterial endocarditis has been cured with large doses of oral ampicillin *(Streptococcus viridans)* and cloxacillin *(Staphylococcus aureus)* [10]. In some of these cases, antibiotics were initially administered parenterally. With phenoxymethylpenicillin and cloxacillin the highest serum levels are achieved by administration 1 h before meals.

Staphylococcal Endocarditis

Staphylococcus Aureus

Endocarditis due to *S. aureus* usually presents as acute ulcerative (septic) endocarditis. For initial treatment, use parenteral cloxacillin, 6-10 g daily divided into three to four doses. Fusidic acid (1 g every 8 h p.o.) or rifampicin (0.6 g once daily p.o.) may be added in severe cases.The addition of gentamicin is often recommended, but appears to be of questionable efficacy. Vancomycin is an alternative treatment for all cases of staphylococcal endocarditis.

If the isolate proves to be sensitive to penicillin, treat with i.v. benzylpenicillin (20-40 MU divided into four doses).

Penicillin Allergy

In cases of penicillin allergy, i.v. cephalosporins such as cefazolin (2 g every 8 h) may be given (with some risk of cross-allergy).

Cloxacillin (Methicillin)-Resistant S. Aureus

Parenteral lincomycin (0.6 g every 8-12 h) or vancomycin (1 g every 12 h) is recommended. Rifampicin may be added in special cases. Treatment of endocarditis due to methicillin-resistant *S. aureus* with parenteral co-trimoxazole has been successful, but is not regarded as an accepted form of therapy.

Following clinical improvement, usually after 2 weeks, treatment may be continued with oral cloxacillin (1 g every 6 h) or phenoxymethylpenicillin (1 g every 6 h) depending on sensitivity - for altogether 6 weeks. An extended course of antibiotics is always necessary in staphylococcal endocarditis.

Staphylococcus Epidermidis

Endocarditis due to *Staphylococcus epidermidis* is characteristically seen in patients with prosthetic heart valves. As in other foreign body infections, elimination of bacteria is difficult and relapses are common.

Antibiotic therapy is the same as for *S. aureus,* but valve replacement is frequently needed for definitive treatment.

Endocarditis Due to Gram-Negative Bacteria

Endocarditis due to gram-negative bacteria carries a poor prognosis - even with optimal management (combination therapy) with a third-generation cephalosporin (e.g., cefotaxime) and an aminoglycoside and early valve replacement, if possible. Parenteral co-trimoxazole may be tried, if an effective cephalosporin is not available.

Culture-Negative Endocarditis

Culture-negative endocarditis should be treated in the same way as enterococcal endocarditis (40 MU benzylpenicillin + 240 mg gentamicin daily).

If, within 48-72 h, there is *no clinical response* in subacute bacterial endocarditis to 40 MU benzylpenicillin + 240 mg gentamicin daily, large doses of doxycycline or another tetracycline may be given. Alternatively, treatment with a third-generation cephalosporin (e.g., cefotaxime) and an aminoglycoside may be attempted.

Blood Culture Facilities Not Available

Where no blood culture facilities are available, the treatment should be performed according to the clinical pattern. Acute septic endocarditis should be treated as staphylococcal endocarditis and subacute bacterial endocarditis as enterococcal endocarditis. This also applies for *initial treatment before culture results are available.*

Prevention of Endocarditis

The recommendations given by the American Heart Association [11] may be difficult to practise under limited conditions and may have to be modified accordingly.

Patients with congenital heart disease, rheumatic heart disease, or a previous episode of infective endocarditis are at risk of developing endocarditis when undergoing procedures associated with bacteremia. These are, in particular, dental and upper respiratory tract surgical procedures and surgery and instrumentation of the gastrointestinal tract and genitourinary tract.

Head and Chest

Prophylaxis in dental extraction, adenectomy, tonsillectomy, and bronchoscopy is directed against viridans streptococci. The treatment involves:

1. Phenoxymethylpenicillin, 2 g orally 1-2 h before the procedure, followed by a 1-g dose 6 h later; or
2. 2 MU benzylpenicillin or procaine penicillin parenterally 1-2 h before the procedure, followed by 1 MU procaine penicillin or benzylpenicillin or by 1 g phenoxymethylpenicillin orally 6 h later.
3. For penicillin-allergic patients, 1.5 g erythromycin orally 1-2 h before the procedure, followed by a second dose of 500 mg 6 h later.

Patients who have been on penicillin for secondary prevention of rheumatic fever should receive erythromycin, because viridans streptococci relatively resistant to penicillin may have been selected.

Gastrointestinal Tract and Urinary Tract

Prophylaxis in gastrointestinal tract and genitourinary tract surgery and instrumentation is directed against enterococci. The treatment involves:

1. Ampicillin, 2 g i.m. or i.v. + gentamicin 120 mg ($\approx$1.5 mg/kg) i.m. or i.v. 30-60 min prior to the procedure. A second dose may be given 8 h later.
2. For penicillin-allergic patients, vancomycin, 1 g i.v. *slowly* over 1 h + gentamicin 120 mg ($\approx$1.5 mg/kg) i.m. or i.v., 1 h before the procedure.

Pediatric Dosage

The dosages for prevention of endocarditis in children are shown in Table 10.5. In no case should the pediatric dose exceed the maximum adult dose.

Table 10.5. Antibiotic dosage for prevention of endocarditis in children [11]

Antibiotic	Dosage
Phenoxymethylpenicillin	> 27 kg body weight, full adult dose < 27 kg body weight, one-half adult dose
Benzylpenicillin Procaine penicillin	50000 MU/kg, 25000 MU/kg for second dose
Ampicillin	50 mg/kg per dose
Erythromycin	20 mg/kg; 10 mg/kg for second dose
Gentamicin	2 mg/kg per dose
Vancomycin	20 mg/kg per dose

References

1. Bisno A (1981) Treatment of infective endocarditis. Grune and Stratton, New York
2. Brockington IF, Edington GM (1972) Adult heart disease in western Nigeria: a clinicopathological synopsis. Am Heart J 83: 27-40
3. Davis M, Gratten M, Reé GH (1982) Infective endocarditis caused by Actinobacillus actinomycetemcomitans and Achromobacter xylosoxidans. Papua New Guinea Med J 25: 7-11
4. Dasnoy G (1982) Maladies cardiovasculaires y compris l'hypertension. In: Meheus A, Butera S, Eylenbosch W, Gatera G, Kivits M, Musafili I (eds) Sante et maladies au Rwanda. Universitaire Instelling, Antwerpen, pp 613-617
5. Falase AO, Jaiyesimi F, Iyun AO, Attah EB (1976) Infective endocarditis-experience in Nigeria. Trop Geogr Med 28: 9-15
6. Kohli RS, Anand IS, Nagrani B, Bidwai PS, Walia BNS, Mchta S, Kumar L, Wahi PL (1983) Infective endocarditis in children. Indian Pediatr 20: 439-443
7. Lerner PI, Weinstein L (1966) Infective endocarditis in the antibiotic era. N Engl J Med 274: 199-206, 259-266, 323-331, 388-393
8. Moethilal R, Coovadia HM (1982) Infective endocarditis in thirteen children, a retrospective study (1974-1981). Ann Trop Paediatr 2: 57-62
9. Neu HC (1979) Diagnosis and treatment: drugs five years later. Amoxicillin Ann Intern Med 90: 356-360
10. Phillips D, Watson GH (1977) Oral treatment of subacute bacterial endocarditis in children. Arch Dis Child 52: 235-237
11. Shulman S, Amren DP, Bisno AL, Dajani AS, Durack DT, Gerber MA, Kaplan EL, Millard HD, Sanders WE, Schwartz RH, Watanakunakorn C (1984) Prevention of bacterial endocarditis. A statement for health professionals by the Committee of rheumatic fever and infective endocarditis of the council on carciovascular diseases in the young of the American heart association. Circulation 70: 1123A-1127A
12. Singham KT, Annuar M, Puthucheary SD (1981) Infective endocarditis 1968-1977: an Asian experience. ASEAN J Clin Sci 2: 89-92
13. Somers K, Patel AK, Steiner I, D'Arbela PG, Hutt MS (1972) Infective endocarditis. An African experience. Br Heart J 34: 1107-1112
14. Sykes RM (1984) Salmonella endocarditis in a Nigerian child. East Afr Med J 61: 326-327
15. Touze JE, Erka A, Ipou KN, Lucht F, Bertrand E (1983) Aspects particuliers des endocarditis infectieuses en Cote d'Ivoire. A propos de 40 cas. Med Mal Infect 13: 745-749

11 Purulent Pericarditis

In comparison with the decline in incidence of infective pericarditis observed in developed countries during the past decades, both purulent and tuberculous pericarditis are far more common in developing countries. The overall incidence is, however, low. Infective pericarditis accounted for less than 1% of total pediatric hospital admissions in one series [4].

Data on the etiology of purulent pericarditis in the tropics are given in Table 11.1. *Staphylococcus aureus* [1] is the pathogen most frequently isolated, followed by *Streptococcus pneumoniae*. By contrast, in pediatric patients in the USA, *Hemophilus influenzae* is second in frequency to *S. aureus* as cause of purulent pericarditis. Spread of infection to the pericardium may arise either hematogenously from a primary focus such as osteomyelitis, septic arthritis, or pyomyositis, or by

Table 11.1. Etiology of purulent pericarditis

Country [Reference]	Uganda [3]	Papua New Guinea [2]	Nigeria [4]	South Africa [7]	USA [5]	
Period	1968-1972	1979-1982	1967-1976	1971-1975	1912-1943	1944-1975
Total cases (n)	66	12	22[a, b]	28[a]	114[c]	53[d]
Cases diagnosed at autopsy only (n)	33	0	11	2	85[e]	45[e]
Pathogens isolated						
Staphylococcus aureus	19	6	11	22	23	15
Streptococcus pneumoniae	7	2	3	4	62	6
Streptococcus spp.	0	1	0	1	12	9
Hemophilus influenzae	0	0	0	1	0	0
Escherichia coli	0	0	4	0	3 (*E. coli*, *Klebsiella*, *Proteus*, *Pseudomonas* combined)	22 (*E. coli*, *Klebsiella*, *Proteus*, *Pseudomonas* combined)
Klebsiella spp.	3	2	2	0		
Proteus spp.	0	0	0	0		
Pseudomonas spp.	2	0	3	0		
Salmonella/Shigella spp.	0	0	1	0	1	3
Neisseria meningitidis	0	0	0	0	3	2
No organism isolated	35	1	0	0	10	2
Mortality of cases diagnosed in life	ND	50	36	31	NA[e]	NA[e]

ND, no data; NA, not applicable.
[a] Children only. [b] 24 isolates in 22 cases. [c] 122 isolates in 114 cases. [d] 55 isolates in 53 cases.
[e] Autopsy review.

direct extension from a pleuropulmonary infection like pneumonia, empyema, or lung abscess. The incidence of purulent pericarditis secondary to pneumonia was 0.64% in 943 children hospitalized with pneumonia in Zimbabwe [6].

Treatment

The treatment of infective pericarditis consists in high-dose systemic antimicrobial chemotherapy for 4–6 weeks. When the pathogen isolated is *Staphylococcus aureus,* treat with cloxacillin. For *Streptococcus pneumoniae,* other streptococci, and penicillin-sensitive *S. aureus,* benzylpenicillin should be prescribed. As long as no pathogen has been isolated the drug to be administered is cloxacillin.

In cases where tuberculous and purulent pericarditis cannot be differentiated, nondirected therapy of tuberculosis with a combination including rifampicin is advisable, since rifampicin covers the common pathogens of purulent pericarditis as well. Repeated needle aspiration, drainage, or surgery may be necessary to prevent cardiac tamponade and sequelae, such as constrictive pericarditis.

References

1. Bentley SJ, Andy JJ, Adeyomo A (1979) Staphylococcal pericarditis with pyopericardium. Trop Geogr Med 31: 525–529
2. Corachan M, Poore P, Hadley GP, Tanner A (1983) Purulent pericarditis in Papua New Guinea: report of 12 cases and review of the literature in a tropical environment. Trans R Soc Trop Med Hyg 77: 314–343
3. D'Arbela, Patel AK, Grigg GL, Somers K (1972) Pericarditis, with particular emphasis on pyogenic pericarditis: a Uganda experience. East Afr Med J 49: 803–816
4. Jaiyesimi F, Abioye AA, Antia AU (1979) Infective pericarditis in Nigerian children. Arch Dis Child 54: 384–390
5. Klacsman PG, Bulkley BH, Hutchins GM (1977) The changed spectrum of purulent pericarditis. An 86 year autopsy experience in 200 patients. Am J Med 63: 666–673
6. Sinzobahamvya N, Ikeogu MOC (1987) Purulent pericarditis. Arch Dis Child 62: 696–699
7. Weir EK, Joffe HS (1977) Purulent pericarditis in children: an analysis of 28 cases. Thorax 32: 438–443

12 Infections of the Central Nervous System

Bacterial Meningitis

Incidence and Etiology

Neisseria meningitidis, Streptococcus pneumoniae, and *Hemophilus influenzae* type B are, worldwide, the main pathogens of acute bacterial meningitis. These three organisms account for 65%-95% of all cases of pyogenic meningitis reported in series from several developing countries (Table 12.1). It has been estimated that in many developing countries the morbidity and mortality from acute bacterial meningitis approaches that from poliomyelitis [13].

At Dakar, Senegal, the overall incidence of pyogenic meningitis was approximately 50 per 100000 population per year and the mortality from pyogenic meningitis was 20 per 100000 population per year during the decade 1970-1979 [6]. Of all cases, 53.9% occurred in children below the age of 2 years. Based on population data, age-specific incidence rates were calculated (Table 12.2). All pneumococcal, meningococcal, and *H. influenzae* meningitis, as well as meningitis due to enterobacteria, are commonest in the first year of life. Thereafter, the incidence rates of *H. influenzae* and enterobacterial meningitis fall very rapidly, while those of meningococcal and pneumococcal meningitis decrease more slowly. Of interest is the recurrence of both pneumococcal and enterobacterial meningitis with old age. Also of note is the frequency of salmonella meningitis, which, in several series, is the fourth or fifth commonest pathogen (Table 12.1).

Escherichia coli and other enterobacteria, including salmonellae, *Pseudomonas aeruginosa,* and gram-positive cocci, such as *Staphylococcus aureus* or *Streptococcus pneumoniae,* are the organisms commonly encountered in neonatal meningitis. Infections with group B streptococci and *Listeria monocytogenes* are apparently, as in neonatal septicemia, rare, compared to their incidence in Europe and the USA. However, group B and D streptococci accounted for 14 of 38 cases of bacteriologically proven meningitis in neonates and infants under 8 weeks of age in a recent study from Cameroon [2].

Although large epidemics of meningococcal disease have exempted no continent, it is only in an area of sub-Saharan Africa that epidemics occur regularly every 5 to 10 years having caused hundreds of thousands of deaths since the turn of the century. The "meningitis belt" lies between the 300-mm and 1100-mm rainfall lines and extends from the Sudan and Ethiopia in the east to Mali and Guinea in the west. Climatic factors such as the harmattan (a dusty desert wind) are, in part, held responsible for this peculiar epidemiology, possibly due to exsiccation of mucous membranes. Thus, many epidemics start towards the end of the dry season, when it is hot, dry, and dusty, and finish shortly after the onset of the rainy season (Fig. 12.1).

Table 12.1. Etiology of acute bacterial meningitis

Country	Brazil		India	Papua New Guinea	Egypt	Senegal	Nigeria	Zambia	Kenya	Zaire	Ethiopia	Cameroon	Papua New Guinea	Nigeria
[Reference]	[1]		[31]	[26]	[16]	[6]	[25]	[9]	[27]	[30]	[17]	[2]	[11]	[21]
Period of study	1958–1967	1968–1972	1966–1967	1973–1978	1977–1978	1970–1979	1976–1977	1978–1981	1970–1981	1958–1977	1975–1976	1982–1983	1980–1984	1974–1982
Age group	All ages	All ages	All ages	Adults	All ages	All ages	All ages	All ages	All ages	1 month -13 years	1 month -14 years	Neonates -14 years	2 days -10 years	Neonates
Total cases (*n*)	6957	8110	146	228	1627	3422	2130[b]	ND	26851[b]	471	ND	215	ND	53
Cases bacteriologically proven (*n*)	3428[a]	4392[a]	84	84	350	2515	130	445[d]	1631	252	120	174	155	45
Etiology of bacteriologically proven cases (%)														
Streptococcus pneumoniae	31.3	14.2	53.6	59.5	40.6	39.1	53.8	34.0	25.2	32.9	21.7	42.5	43.0	8.9
Neisseria meningitidis	30.8	61.2[c]	28.6	35.7	25.4	14.6	7.7	23.4[d]	36.2[e]	1.6	25.0	3.4	5.0	0
Hemophilus influenzae	21.7	8.6	13.1	3.6	12.0	26.7	23.1	8.3	10.4	45.6	41.7	30.5	49.5	0
Others, including	16.2	16.4	4.8	1.2	22.0	19.7	15.4	34.3	28.2	19.8	11.7	23.6	2.5	91.1
Escherichia coli	1.2	0.6	0	0	0	2.0	2.3	1.5	4.5	0.4	0.8		0	20.0
Proteus spp.	1.1	0.4	0	0	0	1.3	0	1.3	0.7	4.0	0		0	4.4
Klebsiella spp.	ND	ND	0	0	0	2.1	3.1	12.0	4.0	3.6	0.8	10.9	0.5	8.9
Salmonella spp.	2.3	3.1	0	1.2	0	3.6	1.5	2.2	5.3	0.8	3.3		0	4.4
Pseudomonas aeruginosa	0.2	0.3	0	0	0	1.0	1.5	4.3	1.1	2.4	0		0	6.7
Staphylococcus aureus	2.8	0.3	0	0	0	2.3	3.1	6.5	2.3	2.4	0	1.1	0	28.9
Streptococcus spp.	1.1	0.5	0	0	0	2.6	3.1	5.7	3.0	4.4	1.7	11.5[f]	2.0	4.4

ND, no data.
[a] Cases of tuberculous meningitis were excluded.
[b] Total liquor specimens examined *(n)*.
[c] A group C meningococcal epidemic was observed from July 1971 onwards.
[d] Additional 54 cases of meningococcal meningitis in 1978 were considered to be due to an outbreak and not included in the statistics.
[e] Since 1975 there was a 10-fold increase in cases of meningococcal meningitis.
[f] Group B and D streptococci; 14 of a total of 20 cases occurred in neonates and infants under 8 weeks of age.

Table 12.2. Incidence of pyogenic meningitis according to age and etiology in Dakar, Senegal. From [6]

Age group (years)	Effective population (1000 s)	Incidence (per 100000 population per year) of cases with individual etiology				
		All cases of pyogenic meningitis	*Streptococcus pneumoniae*	*Neisseria meningitidis*	*Hemophilus influenzae*	Entero-bacteria
<1	37	400.0	94.6	8.9	132.0	58.6
1-4	131	53.9	16.6	4.4	12.5	3.3
5-9	136	19.7	4.5	6.7	0.4	0.8
10-14	112	15.9	3.7	5.4	0.4	1
15-19	103	15.4	2.4	6.2	0	1.1
20-29	169	13.5	4.0	1.9	0.1	0.4
30-39	100	12.1	5.2	0.8	0	0.7
40-49	73	12.5	7.8	0.8	0	0.3
50-59	41	16.6	11.0	2.0	0	0.5
60-69	21	23.4	15.2	0.9	0	1.9
⩾70	13	51.4	13.9	0.6	0	3.8

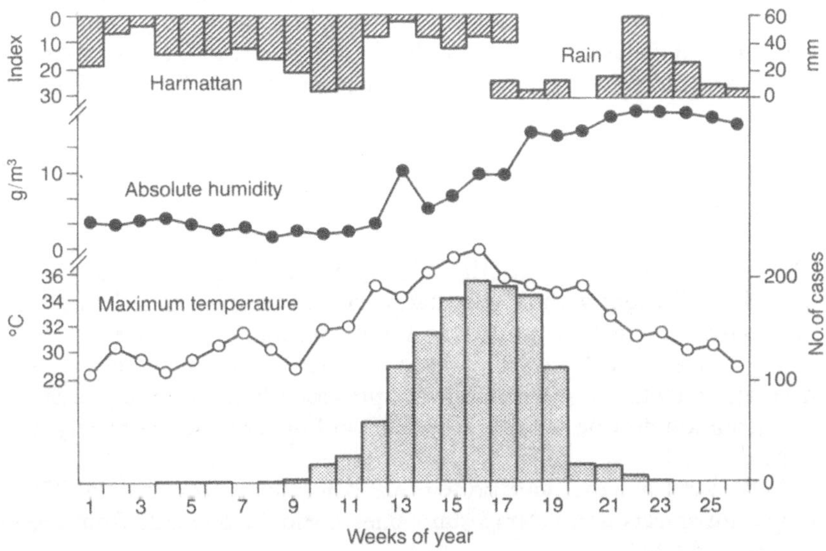

Fig. 12.1. Time-course of climatic factors in relation to the weekly number of patients with meningococcal disease admitted to Ahmadu Bello University Teaching Hospital, Zaria, Northern Nigeria, during an epidemic in 1977. From [15]

Most of the African epidemics are caused by serogroup A, although epidemics due to serogroups B and C meningococci have been reported [5, 20, 46]. Cases of meningococcal disease occurring in an endemic situation are usually due to serogroups B or C. The age distribution during epidemics is commonly shifted to the right, the age group from 5 to 15 years being most affected.

Table 12.3. Mortality due to pneumococcal, meningococcal, and *H. influenzae* meningitis[a]

Country [Reference]	Period of study	*Streptococcus pneumoniae*		*H. influenzae*		*Neisseria meningitidis*	
		No. of cases	Mortality (%)	No. of cases	Mortality (%)	No. of cases	Mortality (%)
Brazil [1]	1958-1972	1694	46.8	1094	34.3	3741	15.3
India [36]	1966-1967	45	46.7	11	63.6	24	4.2
Egypt [16]	1977-1978	142	44.4	42	57.1	89	21.3
Senegal [6]	1970-1979	983	59.5	671	33.5	366	13.9
Nigeria [25]	1976-1977	70	28.6	30	20.0	10	20.0
Zaire [30]	1958-1977	83	28.9	115	13.0	4	0
Ethiopia [15]	1975-1976	26	38.5	50	18.8	30	6.7
Cameroon [2]	1982-1983	74	39.2	53	24.5	6	33.3

[a] See Table 12.1 for age groups studied.

Of the three main pathogens, meningitis due to *Streptococcus pneumoniae* is usually associated with the highest mortality, whereas the lowest mortality is observed with meningococcal meningitis (Table 12.3). The prognosis of meningococcal septicemia without meningitis, however, is considerably poorer. The mortality from acute meningococcemia was 40.6% in contrast to 6.1% in cases of meningitis during an epidemic of group A meningococcal disease in Zaria, Northern Nigeria [14, 15].

Diagnosis

The diagnosis of acute bacterial meningitis should be confirmed by examination of CSF. Cell count with differentiation of cells and a stained smear are the most important investigations under conditions of limited resources.

Suggestive clinical features - fever, headache, vomiting, alteration of consciousness, and meningism - should always prompt a lumbar puncture. Signs of meningeal irritation may be missing in infants and in very old or severely obtunded patients.

Cell count, glucose, and protein are determined from native CSF. Methylene blue stain smears and Gram's stain smears should be made from the sediment of centrifuged CSF.

Turbid liquor with a cell count of usually 1000 to 5000/mm^3 of predominantly polymorphonuclear leukocytes is typical of acute bacterial meningitis. In partially treated meningitis, the liquor formula may be similar to that of aseptic meningitis. Viral meningoencephalitis, tuberculous, fungal, leptospiral, and spirochetal meningitis, or sympathetic meningitis due to otitis, sinusitis, mastoiditis, or a brain abscess then enter into the differential diagnosis. Repeated lumbar punctures may be necessary to clarify the diagnosis. Measurement of lactate dehydrogenase (LDH), lactic acid, or C-reactive protein (CRP) are additional laboratory procedures to distinguish between bacterial and aseptic meningitis (the counts will be high in bacterial meningitis).

Microscopy alone will determine the cause of untreated bacterial meningitis in about 50% (up to 90%) of patients. Pneumococci and *Hemophilus influenzae* are more readily recognized than meningococci. Both are usually present in larger numbers and lie extracellularly. Pneumococci are gram-positive, lanceolate diplococci often surrounded by mucus. *H. influenzae* is a tender, pleomorphic, gram-negative rod, which, on some occasions, resembles coccobacilli, and on others, appears as a long, fine cacillus. Usually only few meningococci are seen in the CSF. Meningococci are gram-negative kidney-shaped diplococci usually found intracellularly in the cytoplasm of polymorphonuclear leukocytes. Distinction of small numbers of intracellular meningococci from nuclear material can be difficult.

Some experience is needed in the interpretation of Gram's stain smears of CSF. Precipitated dye or cell debris may resemble pneumococci. Confusion of *H. influenzae* with pneumococci is another common error. Coccobacilli of *H. influenzae* may resemble gram-positive cocci by staining more heavily at their poles, whereas older pneumococci may lose their capacity to take on a gram-positive stain.

Methylene blue stained smears of CSF permit the differentiation of cells and are advantageous because of the marked coloration of intracellular meningococci and the absence of dye spots.

Confirmation of microscopical results by culture is advisable. Culture is a prerequisite for identification of rare causes of bacterial meningitis where microscopy may only differentiate between gram-positive and gram-negative cocci or bacilli. Culture and sensitivity testing is required in particular for adequate treatment of meningitis due to enterobacteria and *Staphylococcus aureus*.

Ziehl-Neelsen and india ink smears should be examined when tuberculosis or cryptococcal meningitis are considered in the differential diagnosis of subacute or chronic meningitis.

Blood cultures should be obtained whenever possible, because bacterial meningitis is frequently associated with bacteremia. The association of neonatal meningitis with septicemia is well known. Positive blood cultures may be expected in 30%-50% of cases, when culture of CSF remains negative.

X-rays of chest, skull, and sinuses are advisable in order to detect pneumonia, skull fracture, or sinusitis. Cardiac murmur may be the sign of underlying septic endocarditis. Peripheral leukocytosis is common in bacterial meningitis, but leukopenia is a bad prognostic sign.

Treatment

General Principles

CSF penetration of antibiotics is a key determinant of successful treatment. The percentage of the serum concentration achieved in the CSF of normal, healthy persons is less than 1% for penicillins and cephalosporins, about 5% for aminoglycosides, about 10% for tetracyclines, 30%-50% for chloramphenicol, and 50% for sulfonamides. CSF levels of penicillins, cephalosporins, and aminoglycosides are increased to ~10% of the serum concentration in bacterial meningitis, since the diffusion barrier imposed by normal meninges is impaired. CFS levels must be

Table 12.4. Daily dosage in parenteral treatment of meningitis

Antibiotic	Dosage		Dose interval (h)
	Children	Adults	
Benzylpenicillin	(Max. dose 12.0 MU) 0.5 MU/kg (newborn) 1.0 MU/kg (infant) 0.5 MU/kg (older children)	10-20 MU	4-6
Chloramphenicol	25 mg/kg (newborn, 1st and 2nd week) 50 mg/kg (3rd-4th week) 80-100 mg/kg (infants) 50-100 mg/kg (older children)	3 g (max. 4 g)	6-8
Ampicillin/ Cloxacillin and other penicillinase-resistant penicillins	100 mg/kg (newborn) 300 mg/kg (infants up to 3 months) At least 2 g (children ¼-2 years) At least 4 g (children 2-6 years) At least 6 g (children older than 6 years)	6-10 g (max. 20 g)	4-6
Azlocillin/ Piperacillin	100 mg/kg (newborn) 300 mg/kg (older children)	15-20 g	6
Cefotaxime/ Cefuroxime	100 mg/kg (newborn) 150 mg/kg (older children)	6-8 g	6-8
Latamoxef	60 mg/kg (older children) 100 mg/kg (infants)	4-6 g	6-8
Ceftriaxone	60 mg/kg (newborn) 100 mg/kg (infants)	4 g	12
Kanamycin	(Max. dose 1 g) 15-20 mg/kg (newborn) 20-30 mg/kg (children)	1 g	12
Gentamicin/ Tobramycin	5 mg/kg (newborn) 7.5 mg/kg (infants) 5-6 mg/kg (children)	240-300 mg	12

I. v. injection or short i. v. infusion produces effective CSF levels more rapidly than i. m. injection; chloramphenicol may also be given by mouth.

related to the antibacterial activity of an agent. The cephalosporins cefotaxime, cefuroxime, latamoxef, and ceftriaxone have been shown to be reliably effective in bacterial meningitis. First-generation cephalosporins are generally ineffective.

Parenteral administration and high dosage (Table 12.4) are essential to obtain effective antibiotic concentrations in the CSF. According to the present European opinion, 10-20 megaunits (MU) benzylpenicillin daily are required to treat meningococcal or pneumococcal meningitis. Lower doses have been used in the past; higher doses may predispose to seizures.

Since the percentage of antibiotic to diffuse into the CSF decreases as healing proceeds, parenteral treatment at high dosage must be continued. Oral chloramphenicol which diffuses well into the CSF is a cheaper and more practical alternative in the healing period. Oral sulfonamides and co-trimoxazole may be given alternatively, but their use requires sensitivity testing of the isolate. If initial therapy is performed with chloramphenicol, parenteral administration may be necessary, for the patient may be unable to take oral medication.

Intrathecal administration of antibiotics is nowadays generally not recommended. Exceptions are cases of meningitis due to highly resistant enterobacteria and *Pseudomonas aeruginosa*, which are extremely rare in previously healthy adults. Intrathecal administration of gentamicin in addition to systemic beta-lactam + gentamicin (adequate tissue levels are equally important) may be required owing to gentamicin's poor CSF penetration.

When giving gentamicin, a special formulation without additives is required. It should be injected slowly (1 ml/min), and the risk of side effects is high. The dosage for once daily intrathecal administration is 0.5–1 mg for children and 1 mg for adults.

However, the effect of additional administration of gentamicin by the intralumbar route, or the intraventricular route in infants with an open fontanelle, is controversial. The former recommendation of intrathecal therapy with gentamicin, mainly for neonatal meningitis, has now been largely superseded by treatment with parenteral cefotaxime.

Acute bacterial meningitis is a medical emergency, and treatment must be instituted as soon as possible. Diagnostic lumbar puncture and institution of therapy according to the Gram's stain of CSF within 30 min is the procedure usually advised.

When adequate facilities are unavailable, immediate treatment with chloramphenicol (all ages except infants under 2 months) or benzylpenicillin (adults and children over the age of 3 years) is indicated. Immediate treatment may be lifesaving in fulminant bacterial meningitis and in the case of acute meningococcemia (spreading rash and shock).

Chloramphenicol alone i.m. was found to be as effective as chloramphenicol + penicillin i.v. in a recent study of 367 children with bacterial meningitis carried out in Papua New Guinea [38].

A second lumbar puncture is often advised 24–36 h after the institution of therapy. By that time, staining of CSF smears should reveal no organisms or only in greatly reduced number. Normalization of the cell count as part of the inflammatory process takes several days, with a shift from predominantly polymorphonuclear leukocytes to lymphocytes and monocytes. Further lumbar punctures are indicated in cases not responding to treatment (persistent fever) and should be undertaken to monitor the results of therapy in cases due to enterobacteria, *Pseudomonas aeruginosa*, and *Staphylococcus aureus*.

Treatment Failure

Persistent fever, continuing presence of bacteria in the CSF, or increase of cells in the CSF may indicate failure of treatment. Resistance of the organism, diagnostic error, wrong dosage, low levels of antibiotic (occasional poor absorption from intramuscular injection), and superinfection have to be considered when therapy is reevaluated. Other causes of treatment failure are persistent foci of infection, such as brain abscess, subdural empyema, sinusitis, mastoiditis, pneumonia, purulent arthritis, or metastatic abscesses from associated endocarditis.

Intraventricular accumulation of pus (quite frequent in severe pneumococcal meningitis) is a common cause of persistent fever despite optimal chemotherapy.

Allergic reactions, such as drug fever or a serum sickness-like reaction, may be the cause of recurrent fever. A serum sickness-like reaction, often with joint effusions, may be seen in about 5% of cases of meningococcal meningitis, usually 5-15 days after the onset of the illness. Therapy other than analgesics is not indicated.

Corticosteroids do not improve the clinical outcome in bacterial meningitis. The only indications for treatment with steroids are fulminant meningococcal disease with shock (Waterhouse-Friderichsen syndrome) or cerebral edema.

Supportive Measures

Measures of supportive treatment, such as intensive care with mechanical ventilation of comatose patients, depend on the facilities available. Treatment of shock and hyperpyrexia, prevention of aspiration, maintenance of fluid and electrolyte balance (the syndrome of inadequate ADH secretion is particularly common in pneumococcal meningitis) and prophylaxis of stress ulcers are further aspects of therapy.

Cerebral edema is treated with diuretics and steroids. Repeated lumbar punctures for relief of high CSF pressure carry a risk of cerebellar herniation. Relative overdosage of penicillin (neurotoxicity!) may predispose to seizures (control by diazepam or barbiturates); 20 MU in adults and 12 MU in children are the maximum daily doses and should not be exceeded in the acute stage because of the increased permeability of the blood-CSF barrier. Subdural effusions should be drained regularly.

Initial Therapy (Cause Unknown)

Use of cefotaxime (or cefuroxime or ceftriaxone) is now often recommended for patients of all ages, except infants under the age of 2 months, when the causative organism is not known. These agents are highly active against *N. meningitis, Streptococcus pneumoniae,* and *H. influenzae* as well as against most of the rare causes of bacterial meningitis.

Recommendations concerning the use of penicillin and chloramphenicol are certainly not outdated under conditions of limited resources.

Adults and children older than 3 years: benzylpenicillin active against *N. meningitis* and *Streptococcus pneumoniae* is usually sufficient. Chloramphenicol covers a wider range of pathogens, but carries the small risk of aplastic anemia. Chloramphenicol is the alternative in patients allergic to penicillin and should be preferred in areas like Papua New Guinea, where penicillin-insensitive pneumococci occur with noticeable frequency.

Children up to the age of 3 years: Chloramphenicol should be used to cover *H. influenzae.*

Combination therapy is required in areas where unusual resistance occurs; for instance, in Cameroon a combination of ampicillin and chloramphenicol is used for the initial treatment of childhood bacterial meningitis, since 9% of strains of *H. influenzae* isolated from patients with bacterial meningitis were found to be resistant to ampicillin because of beta-lactamase production and, vice versa, 4% of the pneumococci were resistant to chloramphenicol [2].

Neonates and infants up to the age of 2 months: The wide range of organisms that cause neonatal meningitis requires combination therapy, at least initially, as long as the causative organism is not known. Drug combinations like cefotaxime (enterobacteria), azlocillin/piperacillin (*Listeria monocytogenes,* enterococci, group B streptococci, and *Pseudomonas aeruginosa*), and gentamicin (synergistic activity against enterobacteria) are the modern recommendations in Europe. The earlier recommendation of ampicillin + gentamicin, which covers sensitive enterobacteria, *L. monocytogenes,* pneumococci, and streptococci (including enterococci), is still widely used in the United States. Kanamycin is an inferior alternative to gentamicin for reasons of toxicity and resistance.

The combination used may be adapted according to the relative frequency of pathogens encountered in a neonatal unit. Thus, in a Nigerian series where *Staphylococcus aureus* was the single most common organism (see Table 12.1), initial standard therapy consisted of cloxacillin + gentamicin [21].

Dosage must be carefully calculated (Table 12.4), when chloramphenicol (sensitive enterobacteria and staphylococci) is used for initial therapy of neonatal meningitis. Overdosage of chloramphenicol predisposes to the grey syndrome in neonates.

Warning. Other agents, such as co-trimoxazole or sulfonamides (sensitivity testing of the *isolate* required), or erythromycin, tetracyclines, or first-generation cephalosporins (of doubtful or no efficacy) should not be used in the initial therapy of bacterial meningitis. This warning also applies to the use of new agents (e.g., quinolones). New antimicrobials should be used for this condition only after clinical trials have documented their clinical efficacy.

Directed Therapy (Cause is Known)

Meningococcal Meningitis. Penicillin is usually considered the drug of choice because of the risk of aplastic anemia associated with chloramphenicol. Chloramphenicol is equally effective and has the advantage that it can be given by mouth after clinical improvement.

Recent African recommendations [40] are as follows:

Adults: benzylpenicillin 2-3 MU every 4-6 h i.v. or i.m. for 5-7 days, or, if patient allergic to penicillin, chloramphenicol 1 g every 6 h i.v.

Children: benzylpenicillin 250000 U/kg, daily for 5-7 days i.v. in 4-6 divided doses, or chloramphenicol 50-100 mg/kg daily i.v. in 4 divided doses for 5-7 days. *Note:* Benzylpenicillin is best given by i.v. injection.

Five days of treatment often suffices in meningococcal meningitis (Table 12.5). This was shown by the results of a randomized trial in Nigeria, where chloram-

Table 12.5. Antibiotic therapy for patients with acute meningococcemia or meningococcal meningitis (to be given for 5 days) [12]

Age group (years)	Antibiotic			
	Benzylpenicillin		Chloramphenicol	
	Dosage (every 6 h)	Route	Dosage (every 6 h)	Route
<1	0.5 MU	i.v. or i.m.	12.5 mg/kg	i.v., i.m., or p.o.
1-4	1 MU	i.v. or i.m.	25 mg/kg	i.v., i.m., or p.o.
5-14	2 MU	i.v. or i.m.	500 mg	i.v., i.m., or p.o.
15 or >	3 MU	i.v. or i.m.	750 mg	i.v., i.m., or p.o.

phenicol and penicillin were compared [45]. Clinical outcome was favourable in both groups. These results are in contrast to many recommendations from industrialized countries which suggest treatment for 7-14 days.

Meningococcal meningitis is nearly always caused by hematogenous spread. Some patients present with septicemia or pyogenic arthritis, rather than meningitis. Chemotherapy of acute meningococcemia (typically spreading purpuric rash in a hypotensive patient) is as for meningococcal meningitis. Hypotensive shock requires vigorous treatment. Treatment with heparin in disseminated intravascular coagulation, a common complication in acute meningococcemia, is controversial.

Chemoprophylaxis is indicated in close contacts, e.g., contacts who have slept in the same room as a patient (such as household contacts, school mates, or fellow military recruits). The three agents which are effective in eliminating nasopharyngeal carriage of meningococci (for reasons of resistance, costs, and side effects) are shown in Table 12.6. Neither minocycline nor rifampicin are practicable for mass prophylaxis during epidemics. Chemoprophylaxis with penicillin is not considered safe, but preliminary data suggest that therapeutic (!) courses of oral penicillin or procaine penicillin may be protective (for 7 and 5 days respectively).

Pneumococcal meningitis is either hematogenous in origin or the result of direct spread from the paranasal sinuses, the mastoid, or a skull fracture. Meningitis may occur with an interval of years after a skull fracture; recurrent meningitis, usually due to pneumococci, is common in such patients. An association of pneumococcal meningitis with lobar pneumonia is not uncommon in tropical Africa.

Benzylpenicillin in large doses (4-5 MU every 6 h for adults, 0.5 MU/kg/day in children, 1 MU/kg/day in infants) should be given i.v. by slow injection or slow infusion. Duration of treatment for a minimum of 12 days is advisible, since there is an appreciable risk of relapse in pneumococcal meningitis. When inaccessible infections of the paranasal sinuses or the mastoid have been the portal of entry, treatment on full dosage should be continued for an even longer period (2-3 weeks).

Chloramphenicol is the most effective alternative available for patients allergic to penicillin. Chloramphenicol is also appropriate where meningitis caused by penicillin-insensitive pneumococci is known to occur (e.g., Papua New Guinea). Several cases of systemic disease due to multiply resistant pneumococci have been reported from South Africa. Rifampicin or vancomycin are the only effective agents in many of these cases.

Table 12.6. Chemoprophylaxis of meningococcal meningitis

	Infants <1 year	Children 1-12 years	Adults	Duration	Limiting factors
Minocycline	2 mg/kg twice daily	2 mg/kg twice daily	100 mg twice daily	2-5 days	High incidence of side effects, giddiness, and vomiting
Rifampicin	5 mg/kg twice daily	10 mg/kg twice daily	600 mg twice daily	2 days	Expensive, rifampicin-resistant meningococci can occur; risk of promotion of resistance in areas where tuberculosis and leprosy are prevalent; side effects
Sulfonamides (sulfadimidine, sulfadiazin)	500 mg daily	500 mg twice daily	1 g twice daily	2 days	Sulfonamide resistance is widespread; use only for sulfonamide-sensitive strains

Hemophilus influenzae meningitis is largely a disease of infants and young children. It carries a prognosis intermediate between pneumococcal and meningococcal meningitis. Development of the illness is less acute than in pneumococcal or meningococcal meningitis. The child often has vague, nonspecific symptoms for several days before onset of meningitis. The mastoid and the paranasal sinuses may be the portal of entry in a proportion of cases.

Parenteral chloramphenicol (80-100 mg/kg/day divided into three or four doses) should be given initially; after definitive improvement and when the child accepts oral medication, 50-80 mg/kg/day may be administered by mouth for a total of 5-7 days.

The use of ampicillin for *H. influenzae* meningitis is now superseded by the increasing incidence of ampicillin-resistant strains. Chloramphenicol-resistant strains have, fortunately, as yet remained rare. Modern therapy in Europe or the United States consists in treatment with one of the cephalosporins of proven value in meningitis (for dosage, see Table 12.4), the reason being that these agents are active against both ampicillin- and chloramphenicol-resistant strains of *H. influenzae*.

When plump, gram-negative bacilli are seen by direct microscopy, therapy of *meningitis due to enterobacteria* with cefotaxime, cefuroxime, or ceftriaxone, or, alternatively, with chloramphenicol or ampicillin combined with gentamicin, should be instituted immediately. Therapy must be guided by antibiotic sensitivity testing because of the unpredictable pattern of resistance. Sensitivity testing can be performed directly with CSF in order to obtain preliminary results within 8-12 h.

Most cases of *Salmonella* meningitis in infants are seen following an episode of severe enteritis. Meningitis caused by other enterobacteria is rare except in the first 2 months of life. Meningitides occurring in later life are often hospital-acquired (in association with neurosurgery) or the consequence of trauma. Duration of therapy for 21 days is generally advised in meningitis caused by enterobacteria.

Escherichia coli meningitis accounts for a high proportion of cases of neonatal meningitis. Typical symptoms and signs of meningitis are often missing and a high degree of suspicion is needed for early diagnosis. The prognosis is generally grave.

Where modern treatment with cefotaxime (or one of the other cephalosporins effective in meningitis) is not available, chloramphenicol or ampicillin combined with gentamicin should be used (for dosage, see Table 12.4). Additional intrathecal administration of gentamicin in neonatal meningitis is no longer recommended by many authorities.

Klebsiella and *Enterobacter* meningitides carry an even poorer prognosis than *E. coli* meningitis. All penicillins, including ampicillin and the modern acylamino-penicillins are ineffective in *Klebsiella* meningitis (even if sensitive in vitro). Chloramphenicol or co-trimoxazole (in combination with gentamicin) must be relied upon, when cefotaxime is not available. *Enterobacter* meningitis, usually acquired in hospital, can often be treated only with co-trimoxazole or chloramphenicol (in combination with gentamicin), because many strains are resistant to all penicillins and cephalosporins. In addition to systemic therapy, gentamicin may be administered intrathecally once a day (for dosage, see Table 12.4), until the CSF becomes sterile.

Proteus meningitis is very rare. Treatment is guided by the antibiogram. Ampicillin, chloramphenicol, cefotaxime, or co-trimoxazole are used.

Salmonella meningitis may be a rare complication of typhoid fever. It is more commonly due to serotypes like *Salmonella typhimurium* as a complication of severe enteritis in infancy. Salmonella meningitis is traditionally treated best with chloramphenicol. Ampicillin or co-trimoxazole are inferior to chloramphenicol in salmonella meningitis by reason of their poorer CSF penetration and pharmacokinetics.

Meningitis due to multiply resistant salmonellae is difficult to treat, since these strains are often only sensitive to gentamicin (poor CSF penetration), quinolones (sufficient CSF penetration), or cephalosporins (expensive). Experience with cefotaxime or cefuroxime is limited, but their use appears to be preferable to monotherapy with gentamicin [37]. Quinolones (ciprofloxacin, pefloxacin) may be an alternative in future; there are first positive experiences.

Pseudomonas aeruginosa meningitis is mainly iatrogenic by direct inoculation at the time of diagnostic and therapeutic procedures, but may also be a complication of septicemia (e.g., in patients with extensive burns). The prognosis is poor even with optimal treatment.

Treatment with large doses of i.v. piperacillin or azlocillin in combination with gentamicin or tobramycin (dosage is as for gentamicin, see Table 12.4) is the modern recommendation. Carboxypenicillins like carbenicillin or ticarcillin were used before piperacillin or azlocillin became available. The duration of treatment should be at least 2–3 weeks. Treatment with other agents (e.g., polymyxins, sulfonamides, tetracycline) is not promising despite in vitro activity. Quinolones may be shown to be effective agents for this indication in future.

Staphylococcal meningitis may occur at any age. The rare cases are usually a complication of staphylococcal septicemia, septic endocarditis, or brain abscess. Skin infection is often the portal of entry in newborns. Shunts of hydrocephalus are common causes of recurrent staphylococcal septicemia or meningitis.

The optimal treatment of staphylococcal meningitis is controversial, partially because of the relatively poor CSF penetration of the penicillinase-resistant penicillins.

Initial combination therapy with i.v. benzylpenicillin and a penicillinase-resistant penicillin is, therefore, often suggested (for dosage, see Table 12.4). If the isolate proves to be sensitive, treatment with penicillin is sufficient. Meningitis due to resistant staphylococci is more difficult to treat. Penicillinase-resistant penicillins (e.g., flucloxacillin) or chloramphenicol may be used. Use of chloramphenicol requires sensitivity testing, since a variable proportion of staphylococci are chloramphenicol resistant.

First-generation cephalosporins which are recommended for treatment of staphylococcal septicemia in patients allergic to penicillin are ineffective in meningitis. Cefuroxime is the only cephalosporin which, when given in large doses (see Table 12.4), achieves effective CSF concentrations and is sufficiently active against *Staphylococcus aureus*. Vancomycin and also rifampicin are suggested for meningitis caused by methicillin (cloxacillin)-resistant staphylococci. The duration of therapy in full dosage should be at least 2 weeks.

Streptococcal meningitis is treated best with penicillin. Chloramphenicol or the cephalosporins of proven value are alternatives for patients allergic to penicillin. Meningitis caused by group B streptococci occurs mainly in the neonatal period.

Enterococcal meningitis is treated with i.v. ampicillin (in combination with systemic gentamicin).

Listeria meningitis is always hematogenous in origin. In Europe and the USA, *Listeria* meningitis is seen predominantly in newborns and, when occurring in later life, in patients with impaired immune defenses (e.g., malignancy, iatrogenic immunosuppression) and severe liver disease. Treatment with i.v. ampicillin is recommended. Chloramphenicol, tetracyclines or large doses of benzylpenicillin have also been effective. Cephalosporins are generally ineffective. Since relapses are frequent, treatment on full dosage should be continued for at least 3 weeks.

Meningitis caused by other bacteria should be treated as if for septicemia, but only with antimicrobials included in Table 12.4.

Fungal meningitis is caused mainly by *Candida albicans* and *Cryptococcus neoformans;* very occasionally, it is due to *Aspergillus fumigatus, Mucor,* or *Coccidioides immitis.*

A rising incidence of cryptococcal meningitis has been noted in Central Africa associated with the occurrence of AIDS.

Cryptococcal meningitis, always of hematogenous origin, may be diagnosed rapidly using direct microscopy (india ink smear of CSF), or serologically with a latex test. Antigen may be detectable in the CSF or serum. Culture (on Sabouraud's agar) may take up to 10 days.

Amphotericin B, 0.3 mg/kg/day, + flucytosine, 150 mg/kg/day, is a treatment regimen of well-documented efficacy in cryptococcal meningitis. Fluconazole (still investigational) is an alternative (400 mg/day p.o.). Combination treatment with amphotericin B + flucytosine is also recommended for *Candida* meningitis. In combination therapy, the dose of 0.3 mg/kg/day of amphotericin B is usually sufficient. Renal function must be closely monitored, since the dose of flucytosine should be modified according to the renal function (azotemia due to amphotericin

B) in order to avoid increased toxicity from flucytosine. Additional intrathecal instillation of amphotericin B is part of the treatment of coccidioidal meningitis and may be indicated in very severe cases of *Candida* meningitis. Amphotericin B is instilled in increasing doses up to 0.5 mg every 2-3 days together with concomitant injection of corticosteroids to decrease local reactions. Treatment should be continued for 4-8 weeks.

Single-Dose Treatment of Meningococcal Meningitis

Parenteral administration of antibiotics in four divided doses daily is difficult, if not impossible, during epidemics, when health facilities - shortage of staff and equipment - are confronted with large numbers of patients. The following form of treatment was practised at a teaching hospital during an epidemic of meningococcal meningitis in Nigeria:

> "Adult patients with meningitis were started on treatment with benzylpenicillin, four mega units intramuscularly six-hourly. This treatment was continued for five days for some patients whilst others were discharged from hospital after two or three days of parenteral treatment and then maintained on oral penicillin for a further two or three days as an out-patient. Children with meningitis were started on treatment with intramuscular chloramphenicol, 100 mg/kg/day, for 48 hours and then, if they were conscious, treatment was changed ot oral chloramphenicol 50 mg/kg/day for a further three days. Most patients, especially children who were often too confused to ask for oral fluids, received intravenous fluids; diazepam and analgesics were given when indicated." [14]

Table 12.7. Single-dose treatment of pyogenic meningitis with long-acting chloramphenicol in oily suspension (Tifomycine) - results of two studies from French West Africa, 1975 and 1977

Age group	Etiology	No. of patients	Uneventful recovery (*n*, % of total) Residual neurological defects/Arthritis	Treatment failure Deaths/Further antibiotic treatment necessary
Study I [35]	*Neisseria meningitidis*	39	31 (79%) 0/2	8 4/4
2-65 years[a]	Not determined	21	19 (90%) 0/0	2 1/1
Study II [36]	*Neisseria meningitidis*	49	38 (76%) 2/1	11 5/6
1-15 years[a]	*Streptococcus pneumoniae*	17	4 (24%) 0/0	13 5/8
	Not determined	60	53 (88%) 0/0	7 2/5

[a] Dosage of Tifomycine according to age (given i.m. divided between two sites).

Age (years)	1-2	3-6	7-10	11-14	15 and over
Dosage (g)	1	1.5	2	2.5	3

The above-described therapy is certainly beyond the means of many rural dispensaries. Prior to the emergence of sulfonamide resistance, single injections of long-acting sulfonamide provided a simple, cheap, and effective form of treatment for meningococcal meningitis [20]. Sulfonamide resistance of meningococci is still widespread in the countries of the "meningitis belt" in Africa.

Reports by French groups of good results with a single dose of repository chloramphenicol in oily suspension (Tifomycine) in meningococcal meningitis (Table 12.7) are confirmed by the results of a formal, randomized trial (Table 12.8, study I). Treatment with this unusual chloramphenicol preparation was as effective as a more conventional course of parenteral penicillin, in terms of both final outcome and clinical response [42]. Treatment with a mixture of short- and long-acting penicillins (Triplopen) was also effective, but the finding of CSF still growing meningococci 48–72 h after treatment in four patients, all of whom did well, militates against its use (Table 12.8, study II).

During two epidemics in Mali and Rwanda, single-dose long-acting chloramphenicol was used on a large scale for treatment of meningitis [3, 41]. The reported mortality of 10.6% and 4.8% (hospitalized patients only), respectively compares

Table 12.8. Single-dose treatment of confirmed meningococcal meningitis – results of two randomized trials in hospitalized patients

Treatment schedule	No. of patients	Uneventful recovery (*n*, % of total) / Residual neurological defects/Arthritis	Treatment failure / Deaths/Poor clinical response/Positive culture
Study I [42]			
Benzylpenicillin, 3 MU i.m. every 6 h for 3 days, followed by procaine penicillin, 2 MU i.m. every 12 h, for 2 days	66	59 (89%) / 5/1	7 / 3/4/0
Single-dose 3 g long-acting chloramphenicol in oily suspension (Tifomycine) i.m., half in each buttock	65	61 (94%) / 3/1	4 / 1/3/0
Study II [22]			
Benzylpenicillin, 3 MU i.m. or i.v. every 6 h for 3 days, followed by procaine penicillin, 2 MU i.m. every 12 for 2 days	64	53 (83%) / 6/1	11 / 4/6/1 (pericardial fluid)
Single-dose of four vials Triplopen i.m., half in each buttock (2 MU benzylpenicillin, 1 MU procaine penicillin, 2 MU benethamine penicillin)	59	46 (78%) / 5/1	13 / 2/7/4 (CSF)[a]

[a] 48 or 72 h after treatment was given.

well with figures of 3%–30% (average 10%) in earlier epidemics, when sulfonamides were used [20].

The use of single-dose chloramphenicol under field conditions (clinical diagnosis and treatment at rural dispensaries by auxiliary medical staff, nursing of patients by relatives) was studied during an epidemic of meningococcal meningitis in northern Nigeria [33]. A simplified dosage schedule was used: 1 g for patients under 5 years of age; 2 g for those of 5–10 years; and 3 g for patients older than 10 years. Of 86 patients, 79 made a full recovery, with residual neurological defects in three of them. There were five deaths, and two patients required further antibiotic treatment. The clinical course of 45 patients in whom the diagnosis was eventually proven serologically was similar to that of the other 41 patients in whom the diagnosis of meningococcal disease could not be established. However, only one of the five deaths occurred in the group of patients with serologically proven meningococcal infection.

Although it is reasonable to assume that during epidemics most cases of meningitis are due to *N. meningitidis,* when there are no facilities to establish the diagnosis in every patient, the risk exists that cases of different etiology are treated inadequately. Treatment with single-dose chloramphenicol is not recommended in cases of *H. influenzae* or pneumococcal meningitis. Careful observation of the clinical course is, therefore, required to detect these patients. Failure to respond to treatment within 48 h after injection should prompt further investigation and antibiotic therapy.

Control and Prevention

As can be seen from Table 12.10, routine childhood immunization is at present not recommended even for affluent societies. The usefulness of the available polysaccharide vaccines is limited by the fact that they are only poorly immunogenic in infants.

Finnish data show that immunization with the capsular polysaccharide *Hemophilus influenzae* type B (Hib) vaccine is efficacious only after the age of 16 to 20 months [32]. Based on the assumption of an age distribution of Hib meningitis, 50%–60% of cases could be prevented as observed in Finland (Table 12.9). Data from several developing countries, however, indicate that meningitis due to *H. influenzae* occurs at an earlier age in these countries (Table 12.9), so that vaccination at the age of 18 months would be too late. This conclusion is consistent with preliminary data from Upper Volta on serum levels of Hib antibodies in children aged 2–4 years [18]. A new conjugate vaccine (*Hemophilus influenzae* type B polysaccharide and diphtheria toxoid) appears to be more immunogenic and effective in infants in Finland [49].

When considering the high incidence of pneumococcal disease in many developing countries and the high mortality due to pneumococcal meningitis (up to 60%), in particular, effective immunization would certainly mean a major advance. Apart from poor antibody response in infants, matters are complicated by the fact that the 84 different serotypes known do not confer cross-protection. The composition of the presently available 23-valent vaccine is based on studies of the preva-

Table 12.9. Comparison of age distribution of *Hemophilus influenzae* meningitis in developing countries and in Finland

[Reference]	Country	Age group	Total no. of cases	Cases (%) occurring below the age of		
				1 year	18 months	2 years
[6]	Senegal	All ages	671	73	ND	89
[30]	Zaire	1 month–13 years	115	71	87	ND
[17]	Ethiopia	1 month–14 years	50	96[a]	ND	ND
[2]	Cameroon	Neonates–14 years	53	55	ND	79
[39]	Malaysia	0–4 years	39	82	92	95
[32]	Finland	All ages	583	27	47	61

ND, no data.
[a] 4% occurred in patients above the age of 1 year.

Table 12.10. Control measures against infections due to *Neisseria meningitidis, Streptococcus pneumoniae,* and *Hemophilus influenzae* type b in relation to the financial resources available. From [13]

Infection	Health budget		
	Low	Intermediate	High
N. meningitidis	Chemoprophylaxis with sulfonamides for at-risk groups if organism sensitive Vaccination for control of epidemics	Chemoprophylaxis with sulfonamides for at-risk groups if organism sensitive Vaccination of at-risk groups Vaccination for control of epidemics	Chemoprophylaxis with sulfonamides, minocycline, or rifampicin for at-risk groups Vaccination of at-risk groups Vaccination for control of epidemics Routine immunization of children (?)
S. pneumoniae		Vaccination of persons at very high risk, e.g., patients with sickle-cell disease	Vaccination of all at-risk groups including the elderly Chemoprophylaxis for at-risk children <2 years
H. influenzae			Rifampicin chemoprophylaxis for child contacts <2 years

lence of different serotypes in industrialized countries. Adaptation of the choice of serotypes contained in the vaccine according to the pattern of serotypes prevalent in other countries could, in theory, easily be undertaken. Serotype 5, for example, which accounts for a considerable proportion of cases of pneumococcal meningitis in West Africa [8], was not contained in the earlier vaccine, but it is now contained in the 23-valent vaccine.

At present, vaccination is recommended for persons at high risk - including patients with (homozygous) sickle-cell disease or splenectomy. Children with sickle-cell disease are at high risk for both pneumococcal meningitis and septicemia [29]. Vaccination should be carried out by the age of 2 years, when a satisfactory antibody response may be expected. Diagnosis of pneumococcal infection must never be dismissed in vaccinees, since systemic infections due to serotypes contained in the vaccine have occurred in vaccinated patients. Long-term penicillin prophylaxis in sickle-cell disease remains controversial, despite positive experience in the USA [48].

Meningococcal serogroups A, C, W135, and Y polysaccharide vaccines are now available. Group B capsular polysaccharide is only poorly immunogenic.

Vaccination is a very effective means for the control of epidemics [3, 13]. Several vaccination strategies, varying from mass vaccination to selective vaccination of the most susceptible part of the population (those aged 2-20 during an epidemic, and household contacts), have been proposed [4, 24].

Whereas vaccination confers high protection against clinical disease, its impact on elimination of nasopharyngeal carriage is poor. Persons at risk should, therefore, be given chemoprophylaxis in addition to vaccination, since secondary cases often occur within a few days of the onset of the illness in the primary case.

One scheme suggested for the management of a suspected outbreak of meningococcal disease is the following:

"1. Confirm the clinical diagnosis of meningitis in the index cases
2. Obtain CSF samples from as many patients as possible
3. Identify the causative organism
   a) Fresh CSF samples: culture and antigen detection
   b) Stored CSF samples: antigen detection
4. If a meningococcus is incriminated as the cause of an outbreak
   a) Determine its serogroup
   b) Determine its sulfonamide sensitivity
5. If a major outbreak of meningococcal infections seems likely
   a) Intensify surveillance schemes for the detection of new cases
   b) Establish temporary treatment centers
   c) Initiate mass chemoprophylaxis with sulfonamides if the organism is sulfonamide sensitive
   d) Initiate mass vaccination if the causative organism belongs to serogroup A or serogroup C" [13]

Brain Abscess

Available data on the incidence of brain abscesses are difficult to compare. Whereas only seven cases of brain abcess were encountered in Ibadan, Nigeria, during a 5-year period [28], 93 cases of cerebral abscess were seen in Colombo, Sri Lanka, during the same period of time [43]. Four hundred cases of brain abscess were treated over 20 years at a center in China [47] and 23 new cases were seen within 1 year in Durban, South Africa [19].

Chronic suppurative otitis media was the source of the infection in 66% of cases in the Chinese series [47]. Other frequent sources are infections of the paranasal sinuses, the scalp, and the skull. Head injury and infection secondary to neurosurgery are further causes of brain abscesses. Hematogenous spread occurs in cyanotic congenital heart disease, bacterial endocarditis, bronchiectasis, and lung abscess. Brain abscesses may also arise as a complication of bacterial meningitis.

The advent of computerized tomography has facilitated diagnosis, localization, and follow-up of all forms of intracranial abscesses, but carotid angiography or isotope scanning has to be used if this expensive technique is not available.

Etiology

Streptococci (aerobic, microaerophilic, and anaerobic species), *Streptococcus milleri,* enterobacteria (*Proteus* spp., *E. coli, Klebsiella* spp.), *Staphylococcus aureus,* and *Bacteroides* spp. are the main bacterial pathogens. Mixed aerobic and anaerobic infection is common. Successful isolation of pathogen depends on prompt microbiological processing of the specimen. Delay of 8-12 h until inoculation and incubation can result in "sterile" cultures [7]. *Streptococcus milleri,* a recently characterized aerobic streptococcal species with a distinct propensity to form abscesses, has been isolated in a considerable number of brain abscesses [7, 34]. *Salmonella* spp., *Nocardia asteroides, Entamoeba histolytica,* and fungi are rare causes of cerebral abscesses. Reactivation of an earlier *Toxoplasma gondii* infection is a frequent cause of single or multiple brain abscesses in patients with AIDS.

Treatment

Neurosurgery, aspiration, drainage, or excision, depending on the individual circumstances, and antimicrobial therapy in high dosage, as in bacterial meningitis, form the basis of treatment. Supportive therapy includes treatment of cerebral edema with furosemide, mannitol, or dexamethasone and of seizures with anticonvulsants.

Initial chemotherapy is usually empirical. Benzylpenicillin [16-24-(30) megaunits (MU) a day in adults, 1 MU/kg/day in infants, 0.5 MU/kg/day in older children] plus chloramphenicol (3-4 g a day in adults, 100 mg/kg/day in children), *or* benzylpenicillin (dosage see above) plus metronidazole (1.5-2.25 g a day in adults, 40 mg/kg/day in children) are reasonable combinations. The latter, however, does not cover enterobacteria at all, the former only incompletely (sensitivity testing of isolates is advisable).

High doses of benzylpenicillin given i.v. are recommended due to the poor penetration into brain tissue and abscess cavity. Benzylpenicillin is the drug of choice for anaerobic streptococci and penicillin-sensitive staphylococci. Staphylococcal infection is typically encountered in brain abscesses of traumatic or hematogenous origin. If staphylococci are suspected, treatment is as for penicillin-resistant staphylococci with chloramphenicol (resistance possible!), with clindamycin or lincomycin suggested as alternatives. Cloxacillin, in high dosage because of poor penetration (adults 10 g i.v. daily, children 300–400 mg/kg daily), is an alternative. Rifampicin, fosfomycin, and fusidic acid are antistaphylococcal agents with good penetration into abscess cavities, but monotherapy with anyone of them is not advisable (rapid emergence of secondary resistance).

Otogenic brain abscesses are typically caused by enterobacteria, often in mixed infection with anaerobic bacteria. The combination of benzylpenicillin + chloramphenicol is a reasonable choice for initial therapy. The use of ampicillin in initial therapy offers no substantial advantage, since *Klebsiella* and all indole-positive *Proteus* strains are resistant. In cases of non-responding gram-negative infections, third-generation cephalosporins have to be used (cefotaxime, latamoxef). The addition of metronidazole against anaerobic infection is advisable. Since aminoglycosides diffuse only poorly into the abscess cavity, monotherapy with these agents is unreliable.

The local instillation of antibiotics into brain abscesses is not recommended (risk of seizures).

Subdural Empyema

The incidence of subdural empyema as reported in several studies from developing countries varies from 1.2 cases per year in Nigeria and India over 5 cases per year in Ceylon to 7 and 8.5 cases per year in South Africa and Zimbabwe respectively [10, 19, 23, 28, 44]. The latter figures appear to indicate a higher incidence of subdural empyema in developing countries than in the USA and Europe [10].

The most common causes of subdural empyema are ENT diseases, in particular chronic otitis media. Trauma, surgery, osteomyelitis, bacterial meningitis, septicemia, and rupture of a cerebral abscess are less common causes.

The range of pathogens encountered in subdural empyema is similar to that seen in brain abscess.

Treatment

The treatment of subdural empyema consists of neurosurgery (aspiration through burr holes or drainage), chemotherapy (i.v. antibiotics in high dosage), and supportive measures. Chemotherapy is as for brain abscess: benzylpenicillin plus chloramphenicol is the regimen suggested for initial, empirical therapy. For dosage see section on treatment of brain abscess, above.

References

1. Bastos C, Taunay A, Galvao PA, Tiriba A, Saraiva PA, Castro I, Lomar AV (1973) Meningitis, consideracões gerais sobre 15067 casos internados no Hospital "Emilio Ribas" durante o qüindenio 1958-1972. Ocorrência, etiologia e letalidade. Rev Assoc Med Brasil 19: 451-456
2. Bernard-Bonnin A-C, Tetanye Ekoe (1985) Les meningites purulentes de l'enfant a Yaounde: aspects epidemilogiques et prognostiques. Ann Soc Belg Méd Trop 65: 59-68
3. Binkin N, Band J (1982) Epidemic of meningicoccal meningitis in Bamako, Mali: epidemiological features and analysis of vaccine efficacy. Lancet II: 315-318
4. Bosmans E, Vimont-Vicary P, Andre FE, Crooy PJ, Roelants P, Vandepitte J (1980) Protective efficacy of a bivalent (A+C) meningococcal vaccine during a cerebrospinal meningitis epidemic in Rwanda. Ann Soc Belg Méd Trop 60: 297-306
5. Broome CV, Rugh MA, Yada AA, LeVan Giat, Hien Giat, Zeltner JM, Sanborn WR, Fraser DW (1983) Epidemic group C meningococcal meningitis in Upper Volta, 1979. Bull WHO 61: 325-330
6. Cadoz M, Denis F, Diop Mar I (1981) Etude épidémiologique des cas de méningites hospitalisés à Dakar pendant la décennie 1970-1979. Bull WHO 59: 575-584
7. De Louvois J, Gortvai P, Huxley R (1977) Bacteriology of abscesses of the central nervous system: a multicentre prospective study. Br Med J II: 981-984
8. Denis FA, Greenwood BD, Rey JL, Prince-David M, Mboup S, Lloyd-Evans N, Williams K, Benbachir I, El Ndaghri N, Hansman D, Omanga V, Krubwa K, Duchassin M, Perrin J (1983) Etude multicentrique des sérotypes de pneumocoques en Afrique. Bull WHO 61: 661-669
9. Dube SD, Shenderov BA (1983) Incidence and pattern of bacterial meningitis in Lusaka. Cent Afr J Med 29: 100-103
10. Glasauer FE, Coots D, Levy LF, Auchterlonie WC (1978) Subdural empyema in Africans in Rhodesia. Neurosurgery 3: 385-391
11. Gratten M, Barker J, Shann F, Gerega G, Montgomery J, Kajoi M, Lupiwa T (1985) Non-type B Haemophilus influenzae meningitis (letter). Lancet I: 1343-1344
12. Greenwood BM (1984) Treatment and prevention of meningococcal disease. Trop Doct 14: 61-66
13. Greenwood BM (1984) Selective primary health care: strategies for control of disease in the developing world. XIII. Acute bacterial meningitis. Rev Infect Dis 6: 374-389
14. Greenwood BM, Bradley AK, Cleland PG, Haggie MHK, Hassan-King M, Lewis LS, Macfarlane JT, Taqi A, Whittle HC (1979) An epidemic of meningococcal infection at Zaria, Northern Nigeria. 1. General epidemiological features. Trans R Soc Trop Med Hyg 73: 557-562
15. Greenwood BM, Cleland PG, Haggie MHK, Lewis LS, Macfarlane JT, Taqi A, Whittle HC (1979) An epidemic of meningococcal infection at Zaria, Northern Nigeria. 2. The changing clinical pattern. Trans R Soc Trop Med Hyg 73: 563-566
16. Guirguis N, Hafez K, El Kholy MA, Robbins JB, Gotschlich EC (1983) Bacterial meningitis in Egypt: analysis of CSF isolates from hospital patients in Cairo, 1977-1978. Bull WHO 61: 517-524
17. Hailemeskel H, Tafari N (1978) Bacterial meningitis in childhood in an African city. Acta Paediatr Scand 67: 725-730
18. Haroche G, Lang T, Rea D, Lafaix C (1985) Hemophilus influenzae type B vaccine (letter). N Engl J Med 312: 53-54
19. Joubert MJ, Stephanov S (1977) Computerized tomography and surgical treatment in intracranical suppuration. J Neurosurg 47: 73-78
20. Lapeyssonie L (1963) La méningite cérébro-spinale en Afrique. Bull WHO [Suppl] 28: 3-114
21. Longe AC, Omene JA, Okolo AA (1984) Neonatal meningitis in Nigerian infants. Acta Paediatr Scand 73: 477-481
22. Macfarlane JT, Anjorin FI, Cleland PG, Hassan-King M, Tor-Agbidye S, Wali SS, Weir WRC, Whittle HC, Yahaya HN, Greenwood BM (1979) Single injection treatment of meningococcal meningitis. I. Long-acting penicillin. Trans R Soc Trop Med Hyg 73: 693-697
23. Mahapatra AK, Bhatia R, Banerij AK, Tandon PN (1984) Subdural empyema in children. Indian Pediatr 21: 561-567
24. Mohammed I, Obineche EN, Onyemelukwe GC, Zaruba K (1984) Control of epidemic meningococcal meningitis by mass vaccination I. Further epidemiological evaluation of groups A and C vaccines in northern Nigeria. J Infect 9: 190-196

25. Montefiore D, Alausa OK, Sobayo E (1978) Pyogenic meningitis in Ibadan, Nigeria. Scand J Infect Dis 10: 113-117
26. Naraqi S (1980) Aetiology of acute bacterial meningitis in the highlands and islands of Papua New Guinea. Papua New Guinea Med J 23: 108-110
27. Nazir BM, Wamola IA (1983) A bacteriological study of cerebrospinal fluids for pyogenic meningitis in Kenyatta National Hospital, Nairobi. East Afr Med J 60: 453-458
28. Olumide AA, Adeloye A (1981) Intracranial abscess in Ibadan, Nigeria. East Afr Med J 58: 231-239
29. Omanga U, Tady M, Ntihinyurwa M, Badibanga, Disengomoka I (1976) Méningites purulentes chez l'enfant drépanocytaire. A propos de 47 observations. Ann Pédiatr 23: 741-745
30. Omanga U, Ntihinyurwa M, Shako D, Muaku MM, Shango L, Lungambi M (1980) Aspects étiologiques et évolutifs des méningites purulentes de l'enfant à Kinshasa. Analyse de 471 cas. Méd Afr Noire 27: 25-34
31. O'Toole RD, Thornton GF, Mukherjee MK, Nath RL, De S, Das AK, Manji PM, Mondal A (1973) The clinical spectrum of central nervous system infections in Calcutta. J Indian Med Assoc 61: 82-86
32. Peltola H, Käyhty H, Virtanen M, Mäkelä PH (1984) Prevention of hemophilus influenzae type B bacteremic infections with the capsular polysaccharide vaccine. N Engl J Med 310: 1561-1566
33. Puddicombe JB, Wali SS, Greenwood BM (1984) A field trial of a single intramuscular injection of long-acting chloramphenicol in the treatment of meningococcal meningitis. Trans R Soc Trop Med Hyg 78: 399-403
34. Puthucheary SD, Rapport RL (1982) Intracranial abscess due to *Streptococcus milleri:* a report of seven cases. Singapore Med J 23: 33-36
35. Rey M, Ouedraogo L, Diop Mar I, Danjou G, Beutter J, Sow A, Acar JF (1975) Traitement de la méningite cérébrospinale épidémique à méningocoque par injection intramusculaire unique de chloramphenicol en suspension huileuse. Afr Méd 14: 615-618
36. Saliou P, Ouedraogo L, Muslin D, Rey M (1977) L'injection unique de chloramphenicol dans le traitement de la méningite cérébro-spinale en Afrique tropicale. Med Trop 37: 189-193
37. Sirinavin S, Chiemchanya S, Visudhipan P, Lolekha S (1983) Cefuroxime treatment of bacterial meningitis in infants and children. Antimicrob Agents Chemother 25: 273-275
38. Shann F, Barker J, Poore P (1985) Chloramphenicol alone versus chloramphenicol plus penicillin for bacterial meningitis in children. Lancet II: 681-684
39. Thong ML, Puthucheary SD, Omar A (1983) Haemophilus influenzae meningitis. Asean J Clin Sci 4: 47-56
40. Upunda G, Yudkin J, Brown GV (1983) Guidelines to drug usage. Macmillan, London, p25
41. Vimont-Vicary P, Rogerie F (1983) Epidemie de méningite cérébro-spinale à Neisseria meningitidis dans la region sanitaire de Ruhengeri (Rwanda). Méd Trop 43: 155-161
42. Wali SS, Macfarlane JT, Weir WRC, Cleland PG, Ball PAJ, Hassan-King M, Whittle HC, Greenwood BM (1979) Single injection treatment of meningococcal meningitis. II. Long-acting chloramphenicol. Trans R Soc Trop Med Hyg 73: 698-702
43. Weinman D (1967) Intracranial abscess. Ceylon Med J 12: 57-66
44. Weinman D, Samarasinghe HHR (1972) Subdural empyema. Aust NZ J Surg 41: 324-330
45. Whittle HC, Davidson NM, Greenwood BM, Warrell DA, Tomkins A, Tugwell P, Zalin A, Bryceson AD, Parry EH, Brueton M, Duggan M, Rajkovi'c AD (1973) Trial of chloramphenicol for meningitis in northern savanna of Africa. Br Med J III: 379-381
46. Whittle HC, Evans-Jones G, Onyewotu I, Adjukiewicz A, Turunen U, Crockford J, Greenwood BM (1975) Group-C meningococcal meningitis in the northern savanna of Africa (letter). Lancet I: 1377
47. Yang S-Y (1981) Brain abscess: a review of 400 cases. J Neurosurg 55: 794-799
48. Gaston MH, Verter JI, Woods G, Pegelow C, Kelleher J, Presbury G, Zarkowsky H, Vichinsky E, Iyer R, Lobel JS, Diamond S, Holbrook CT, Gill FM, Ritchey K, Falletta JM (1986) Prophylaxis with oral penicillin in children with sickle cell anemia. A randomized trial. New Engl J Med 314: 1593-1599
49. Eskola J, Peltola H, Takala AK, Käyhty H, Hakulinen M, Karanko V, Kela E, Rekola P, Rönnberg PR, Samuelson JS, Gordon LK, Mäkelä PH (1987) Efficacy of Hemophilus influenzae type B polysaccharide - diphtheria toxoid conjugate vaccine in infancy. New Engl J Med 317: 717-722

13 Infections of the Respiratory Tract

Acute Respiratory Infections in Childhood

Acute respiratory infections (ARI) are a leading cause of death in developing countries. Figures published by UNICEF indicate that 97% of deaths of children less than 1 year old, and 98% of deaths of children aged 1-5 years occur in third world countries, where 75% of the world's population live. More than 15 million children under the age of 5 years die each year, the vast majority of them in developing countries [29]. According to a recent estimate, ARI-related mortality accounts for 4 million deaths per year among children under the age of 5 years, i.e., 27% of all deaths in that age-group [80].

Table 13.1 shows that the mortality from ARI is much higher in developing than in developed countries [11]. The annual mortality (deaths/100000) from ARI in infants was 65 in the United States compared to 2136 in India; in preschool children the figures were 5 and 393, respectively [70]. These striking differences cannot be put down to a difference in overall incidence. The available data suggest that the overall incidence of ARI - five to eight episodes per child and year in an urban area according to a WHO estimate - are not much different in developing and developed countries [79].

Table 13.1. Mortality due to acute respiratory infections in the world, 1970-1973: infants and children (per 100000 population). From [11]

Continent	Number of countries	Infants	Children 1-4 years	Children 5-14 years
Africa	6	1454.1	467.0	21.6
America	23	886.5	82.8	7.9
North	2	146.3	8.0	1.5
Middle	13	1495.0	149.3	16.5
South	8	1110.5	112.6	10.8
Asia	111	822.1	132.3	15.5
1st group[a]	2	130.5	9.7	1.9
2nd group[b]	9	1242.4	204.1	23.4
Europe	28	390.3	15.3	2.1
Oceania	3	177.6	9.3	1.2
1st group[c]	2	159.8	8.2	1.1
2nd group[b]	1	566.1	33.6	2.7
Total	71	762.2	101.3	8.4

[a] Israel and Japan.
[b] Developing countries.
[c] Australia and New Zealand.

It appears that it is the incidence and severity of lower respiratory tract infection (LRTI), particularly that of pneumonia, which makes the great difference. The annual incidence of LRTI is apparently twice as high in developing countries than in an urban area of the USA - 70-100 episodes per 1000 vs. 30-40 per 1000 children under the age of 5 years [79]. In malnourished children or in infants, figures of up to 500 episodes per 1000 annually have been reported [79]. The severity of pneumonia can also be judged by the case-fatality rate. "While only a very small proportion of children suffering from pneumonia or bronchopneumonia die in developed countries, the case-fatality rate is 5 to 10% in the hospitals of large towns and over 10% in some rural areas of the developing countries" [79]. Malnutrition is associated with both increased incidence and increased mortality from pneumonia [8, 16, 25, 35, 60]. Other risk factors of mortality from ARI which have been delineated less clearly include malaria, measles, diarrheal disease and low birth weight. In a prospective study of the clinical pattern of lower ARI in hospitalized children in Port Moresby, Papua New Guinea, all 129 children had clinical and/or radiological evidence of pneumonia [75]. The majority of children had concurrent disease; 62% were malnourished, 16% had gastroenteritis, 16% had measles, and 15% had otitis media. The additional diagnosis of bronchitis was made in 12% of children [75]. The mortality among 406 children aged 1 month to 3 years who were admitted to a district hospital in Zimbabwe in the years 1982 and 1983 was studied retrospectively [16]. Whereas the overall rate was 10.3%, mortality was 14.4% in malnourished children in contrast to 6.7% in children with normal weight, and 6.1% in 148 infants with normal weight as opposed to 18.8% in 48 underweight infants [16].

Control of ARI, particularly of pneumonia, in childhood is therefore crucial. Chretien et al. [14] describe the situation as follows:

"The lack of recognition of the problem, the lack of funds, and the lack of programs have led to an international tragedy of almost unprecedented magnitude. Properly organized research programs into the etiologic agents involved in acute respiratory infections, together with data collection on other contributing factors, are needed so that effective programs of prevention and treatment can be mounted."

The complex etiology and the wide spectrum of clinical syndroms produced by different agents form one of the problems that complicate any strategy of control.

During recent years, studies on the bacterial etiology of pneumonia in hospitalized children in developing countries have accumulated evidence that about half of the more severe cases of pneumonia are caused by bacteria. *Streptococcus pneumoniae* and *Hemophilus influenzae* are isolated most commonly, agents that are uncommon causes of childhood pneumonia in developed countries (Table 13.5).

From this evidence the conclusion has been drawn that early use of antibiotics at the primary health care level is likely to prevent the death of many children from bacterial pneumonia due to *S. pneumoniae* or *H. influenzae* [62].

While prevention of ARI by vaccination is clearly desirable, and research is intensified in that field, the available polysaccharide vaccines against *S. pneumoniae* and *H. influenzae* are known for their poor immunogenicity in infants and young children. Vaccines against the major viruses causing ARI are still in the stage of research. Thus, prevention of ARI by vaccination is, at present, limited to measles, pertussis, and diphtheria, which are included in the WHO Expanded Programme on Immunization.

Improved case management of ARI, by means of early use of antibiotics in moderate or severe ARI and supportive treatment, has, therefore, been emphasized by the WHO control program of ARI [78, 79]. The ARI control program of improved case management is intended to be integrated into existing primary health care (PHC) and mother and child care services. The case management guidelines that have been developed address the community health worker with less than 6 months' formal training, the providers of PHC with 6 months' to 2 years' formal training, and the first referral center.

The management guidelines circumvent the problems of the conventional classification of ARI into upper, middle, and lower respiratory tract syndromes by adopting an action-oriented approach. Five clinical symptoms - the ability to drink fluids, retractions, respiratory rate, wheezing and stridor - serve to assess the degree of illness. Three grades are distinguished: (1) mild illness that can be treated at home with supportive measures alone, (2) moderate illness which is treated with antibiotics on an outpatient basis, and (3) severe illness that requires referral for hospitalization [9]. These guidelines and the associated teaching material are currently being field-tested. Figure 1 shows a flowchart that has been in use in the village health posts of Papua New Guinea since the mid 1970s. For comparison, more detailed guidelines in the form of a block system which were tested in PHC centres in Kenya are shown in Fig. 2. The antibiotics suggested in these guidelines are considered in the section on treatment of pneumonia (p. 122).

Data concerning the efficiency of this strategy are limited to four studies! In the Narangwal project in India, community health workers were trained to give penicillin to all children with cough and fever or labored respiration. This resulted in reduction in mortality from pneumonia by 50% compared to the level observed before [43]. A community survey conducted in rural villages in Papua New Guinea revealed mortality of 25% in children who did not come for treatment as compared to a rate of 5% in children treated with penicillin [67]. Lastly, in a controlled trial carried out a hospital in Cameroon, half of the children admitted to hospital with cough and fever but no clinical signs of pneumonia were treated with procaine penicillin, the other half received no antibiotic treatment [24]. Three of 50 children treated with penicillin developed clinically diagnosed pneumonia, in contrast to 26 of the 50 children given no antibiotic. An Indian study showed that effective case management of ARI among low-birth-weight infants can be introduced into existing PHC services [81]. The case fatality was 8.7 per 100 attacks of moderate or severe ARI in the intervention area in comparison to 24.6 in the control area [81].

The authors of a recent review of the control of ARI in childhood in developing countries [9] give the following warning: "Well-designed controlled studies of antibiotic intervention are urgently needed, since the value of any simple approach cannot be determined either on the basis of etiologic studies or from reports of uncontrolled investigations. Several controlled studies sponsored by WHO are currently underway."

Common Cold

The common cold is due to a variety of respiratory viruses and requires no treatment with antibiotics. Nasal obstruction may interfere with feeding in infants and the nose should be cleared from secretions with a soft tissue or a cloth formed into a wick.

In purulent rhinitis of more than 2 weeks' duration, the diagnosis of sinusitis (see p. 245) should be considered.

Purulent rhinitis in newborns is often staphylococcal in origin (treatment with cloxacillin), but syphilitic rhinitis, a common sign of congenital syphilis, should be considered in the differential diagnosis. The discharge, which is clear initially, becomes serosanguineous or even purulent with flecks of blood later. In infants, beta-hemolytic streptococci and chlamydiae may also be a cause of purulent rhinitis.

Nasal diphtheria commonly presents with a uni- or bilateral purulent, blood-stained discharge. For treatment see section on diphtheria (p. 136).

Tonsillitis, Pharyngitis

Respiratory viruses, coxsackie viruses (herpangina), Epstein-Barr virus (infectious mononucleosis), group A beta-hemolytic streptococci, and *Corynebacterium diphtheriae* are all among the causative agents. The clinical distinction of tonsillitis or pharyngitis due to group A beta-hemolytic streptococci from the viral forms may be impossible. The typical tonsillar exudate may be missing in streptococcal sore throat; on the other hand, a tonsillar exudate may be present in viral sore throat. Routine throat culture for beta-hemolytic streptococci is often recommended, but the culture may be falsely negative or, if positive, may not distinguish between asymptomatic carriage and infection. Nevertheless, the clinical picture is distinctive in the majority of cases. Clinically, treatment for presumed streptococcal sore throat should be instituted when the sore throat is associated with enlarged, tender lymph nodes, when there is evidence of spread to the middle ear, and in ill children. The risk of possible overtreatment of nonstreptococcal pharyngitis seems justified in areas where rheumatic fever and rheumatic heart disease are highly prevalent. Treatment of streptococcal sore throat aims at the prevention of complications like peritonsillar or retropharyngeal abscess, otitis media, sinusitis, cervical adenitis, streptococcal septicemia, rheumatic fever, and glomerulonephritis. Tonsillitis or pharyngitis associated with erosions or ulceration of the alveolar lining of the gums at some site should suggest necrotizing ulcerative gingivitis and be treated with penicillin (prevention of noma).

The membranes seen in faucial diphtheria (see section on diphtheria below for treatment) are usually firmly adherent and may spread rapidly to the soft palate and the uvula. In contrast, the pseudomembranes of infectious mononucleosis can be wiped off easily and are limited to the tonsils.

Treatment

Penicillin is the drug of choice for streptococcal tonsillitis and pharyngitis. Treatment with a single i.m. injection of 1.2 megaunits (MU) benzathine penicillin (0.6 MU in children with less than 27 kg body weight) is simple and effective. Treatment with a single injection of both benzathine penicillin and procaine penicillin, 0.6 MU each, is less painful and also effective. Alternatively, a 10 days' course of phenoxymethylpenicillin (1.5 g in adults, 750 mg in children aged 1-6 years, and 375 mg in infants daily in three divided doses) should be prescribed. With oral treatment, a course of 10 days, or 7 days at the very minimum, is required for prevention of rheumatic fever. Treatment with erythromycin is recommended for patients allergic to penicillin. Sulfonamides, co-trimoxazole, and tetracyclines are less effective, since a proportion of strains are resistant to these agents and the elimination of sensitive bacteria is inferior.

Rheumatic Fever

Rheumatic fever (RF) and rheumatic heart disease (RHD) are now as common in many developing countries as they were in the USA or Europe at the beginning of the century. RHD accounts for 25%-40% of all cardiovascular disease in most of the developing world [2]. Young age at the first attack and late presentation to hospital contribute to the severe carditis typical of RF in developing countries [26, 42]. Frequently, patients are first diagnosed at a recurrent attack of RF or with cardiac failure due to established rheumatic valvular disease. Advanced valvular disease develops rapidly, and "juvenile mitral stenosis" is common in many developing countries.

Treatment

Treatment of RF comprises bed rest, anti-inflammatory agents, and antibiotics. Aspirin, 100 mg/kg/day in children and 6-8 g/day in adults, is regarded as the mainstay of treatment. The tolerance of high doses of aspirin is very poor. The early use of corticosteroids is recommended to prevent carditis. Prednisone, 50-100 mg/day in adults and 2 mg/kg/day in children is given initially. The dose may then be tapered in the 3rd week of treatment, while aspirin is introduced (prevention of rebound). Eradication of group A streptococci from the throat is also required. Treatment is recommended even if the throat culture is negative at the time of admission. Treatment can be performed with a single i.m. injection of 1.2 megaunits (MU) benzathine penicillin (0.6 MU in children weighing less than 27 kg), or with a 2 weeks' course of phenoxymethylpenicillin, 0.4 MU (ca. 250 mg) twice daily.

Secondary prophylaxis against recurrent attacks of RF is strongly recommended. One of the following regimens should be chosen:

1. Intramuscular injection of 1.2 MU benzathine penicillin (0.6 MU in children weighing less than 27 kg) once every month. This regimen is associated with the

lowest failure rate, 0.4%, compared with failure rates of 3%–5% with oral regimens. Hypersensitivity reactions are uncommon, particularly in children, during treatment of patients with no history of allergy to penicillin. Intramuscular benzathine penicillin is, however, strictly contraindicated in patients with a known allergy to penicillin.
2. Phenoxymethylpenicillin, 0.2–0.25 MU twice daily.

Prophylaxis is generally continued into adult life, and lifelong prophylaxis is recommended for patients with RHD. Immediate treatment with penicillin in every suspected case of streptococcal infection is apparently an effective prophylaxis of relapse in RHD, but controlled studies are lacking.

Prophylaxis of bacterial endocarditis is required in patients with RHD at the time of invasive diagnostic or therapeutic procedures (tooth extractions, etc.) in addition to secondary prophylaxis of RF (see section on prophylaxis of bacterial endocarditis, p. 79).

Peritonsillar Abscess, Retropharyngeal Abscess, and Ludwig's Angina

Peritonsillar and retropharyngeal abscess and Ludwig's angina are considered together because the range of causative agents is similar. The main pathogens are mixed anaerobic bacteria of the normal mouth flora (peptostreptococci, *Bacteroides* spp., *Fusobacterium* spp.). *Staphylococcus aureus,* group A beta-hemolytic streptococci, and *Hemophilus influenzae* are further causes of these abscesses.

Treatment with clindamycin (or lincomycin) is suggested. Benzylpenicillin or chloramphenicol may be used when neither of these agents is available. Parenteral treatment may be indicated initially. The combination of penicillin and metronidazole may be useful if foul-smelling pus is present. Puncture or surgical incision with drainage of the abscess is also required. Aspirated pus should be examined microscopically and by aerobic and anaerobic culture. In case of a peritonsillar abscess, tonsillectomy should be performed later, since repeated attacks are common.

Croup

The croup syndromes are caused by measles virus, parainfluenza viruses, influenza viruses, and respiratory syncytial virus, by *Corynebacterium diphtheriae,* and by *Hemophilus influenzae* in the case of acute epiglottitis.

Severe croup due to measles or diphtheria is apparently common in many parts of the developing world [77]. Of 469 children admitted to a respiratory unit for upper airway obstruction in Durban, South Africa, over 6 years, 326 children had laryngotracheobronchitis (LTB), measles preceded the LTB in 75% of cases, and 115 children had diphtheria (77). Sixty-seven percent of the children with LTB required intubation, and in 85% there was radiological evidence of lower respiratory tract involvement (77). Twenty-three percent of all children with LTB died. No case of acute epiglottitis was seen in 6 years [77].

Maintenance of a patent airway and of adequate oxygenation is most important. Most children with croup seen in temperate climates, where nonmeasles viral croup predominates, respond to humidification of air, mild sedation, oxygen, and inhalation of racemic epinephrine. Intubation is rarely necessary. The use of glucocorticosteroids is controversial. Increasing restlessness, cyanosis, tachycardia (>160/min), and severe respiratory distress indicate the need for tracheostomy or intubation. Artificial ventilation becomes necessary with extensive involvement of the lower respiratory tract.

Antibiotics are usually not indicated in mild viral croup. Their use is, however, recommended for severe forms of LTB associated with measles and malnutrition. "The most popular antibiotic for uncomplicated LTB remains ampicillin, but in cases with lower respiratory tract disease and especially postmeasles, more potent drugs are required. An anti-staphylococcal agent combined with an aminoglycoside are usually effective against mixed superinfections commonly seen in malnourished children" [77].

Acute epiglottitis is feared for its rapid development of airway obstruction. An artificial airway is needed in the majority of cases. The underlying infection with *Hemophilus influenzae* is treated with ampicillin or chloramphenicol.

Acute Bronchitis

Acute bronchitis, which is most commonly associated with an upper respiratory tract infection, is usually viral in origin. Secondary bacterial infection with pneumococci, *Hemophilus influenzae*, or *Staphylococcus aureus* occurs in some cases. In contrast to pneumonia, the incidences of tracheobronchitis and of bronchiolitis were not higher in malnourished children than in children with normal weight in a 2-year ambulatory study of lower respiratory tract infections in Colombia [8]. Acute bronchitis is also seen in pertussis, in measles, and in typhoid fever. In young children, a sudden attack of what appears to be acute bronchitis without concurrent symptoms of an upper respiratory tract infection may in fact be due to an inhaled foreign body.

Antibiotic therapy is usually not indicated in the typical mild case. If in infants and young children there is any suspicion of bronchopneumonia, or if there are signs of respiratory distress like tachypnea or retractions, pneumonia should be assumed and be treated accordingly.

In children with frequent attacks of acute bronchitis, other diseases like allergy, foreign body aspiration, and bronchiectasis should be considered.

Acute Exacerbation of Chronic Bronchitis

Antibiotic therapy of chronic bronchitis is limited to treatment of acute exacerbations in order to prevent further impairment of lung function. Exacerbations are characterized by increased cough with purulent sputum, often accompanied by fever. They are usually caused by secondary bacterial infections; *Hemophilus influ-*

enzae and pneumococci are the commonest pathogens. *Staphylococcus aureus, Branhamella catharralis, Klebsiella pneumoniae* and *Pseudomonas aeruginosa* are infrequently isolated.

Specimens of sputum for microscopy and culture may be obtained early in the morning, prior to breakfast. Interpretation of findings is difficult, since bacteria of the normal mouth flora frequently contaminate the culture. Some authorities, therefore, recommend either washing of flecks of purulent sputum in physiologic saline or liquefaction with 1% pancreatine or *n*-acetyl cysteine before the specimen is processed. The presence of purulent sputum is diagnostically more important than bacteriological results.

Treatment

Ampicillin, its congeners, tetracyclines, and co-trimoxazole are the antibiotics used for treatment of acute exacerbations. The penicillins have to be given in high dosage in order to compensate for their poor penetration into sputum (ampicillin 3 g, amoxicillin 1.5 g, bacampillin 2.4 g, daily in adults).

Resistance of pneumococci to tetracycline and of *H. influenzae* to ampicillin and tetracyclines occurs in some cases. Defervescence, disappearance of purulent sputum, and decrease in the amount of sputum, as well as improvement of lung function, indicate response to treatment. The duration of treatment depends on the resolution of symptoms and usually comprises 1–2 weeks.

Bronchiolitis

Bronchiolitis is a disease of children under the age of 2 years; half of the cases are seen in infants below the age of 6 months. Respiratory syncytial virus is the most common causative agent worldwide.

Supportive treatment consists in administration of oxygen, humidification of air, maintenance of hydration, digitalization in case of cardiac failure, and, infrequently, artificial ventilation for respiratory failure. Glucocorticosteroids may be used in severe cases.

Pneumonia

Special Aspects in Developing Countries

In developing countries, in contrast to the situation in developed countries, primary community-acquired pneumonias are much more common than secondary pneumonias. Throughout the tropics, lobar or segmental pneumonias, usually caused by *Streptococcus pneumoniae,* are one of the leading causes of morbidity in previously healthy adults. In many parts of the tropics, lobar pneumonia is the commonest cause of medical admissions to the hospital, accounting for 9%–10% of adult medical admissions [74].

Owing to scarcity of hospital beds, primary pneumonias are often managed in outpatient clinics. This is contrary to the practice in the USA and Europe of treating almost every patient with pneumonia in hospital. In the outpatient clinic, diagnosis and treatment of pneumonia are based on the typical clinical features and on chest X-ray, if available. The management of uncomplicated cases is frequently delegated to medical assistants and nurses. The limited facilities and the high workload enforce standardized treatment regimens with benzylpenicillin. Single-dose and short-course regimens have to be administered when patients are unable to make daily journeys to the clinics. Such regimens are not considered in USA/European recommendations. In the absence of modern diagnostic procedures and without bacteriological laboratory, the response or nonresponse to benzylpenicillin has to be used as the main diagnostic criterion for further management of the case.

Nonresponse to benzylpenicillin which is found in about 10% of cases, is a strong argument in favor of admission to hospital (see Table 13.2). It renders further investigations necessary as well as a change to more potent antibiotics.

The classification of responsive and nonresponsive pneumonias is thus a useful concept in the tropics where medical facilities are limited and where lobar pneumonias due to *Streptococcus pneumoniae* prevail.

The situation is, however, more complex in infancy and childhood, where acute respiratory infections (ARI) are responsible for an excessive mortality in children under 5 years of age. The magnitude of the problem and the strategy of improved case management proposed in the WHO ARI control program have been outlined at the beginning of this chapter. Briefly, the goal of the proposed selective, action-oriented approach is the identification of those children with moderate to severe

Table 13.2 Response to treatment with penicillin (various regimens) in adult patients with acute pneumonia

Location of study [Reference]	No. of patients	Sex male/female (%)	Age (commonest age-group, years)	Lobar/segmental pneumonia on chest X-ray (*n*)	Response to penicillin (*n*)	Response to penicillin (%)
Salisbury, Zimbabwe [69]	109	89/11	20–29	109[a]	102	93.6
Lae, Papua New Guinea [21]	64	81/19	10–29	56	57	89.1
Nairobi, Kenia [66]	105	72/28	20–30	91 (of 92 X-rays taken)	98	93.3
Goroka, Papua New Guinea [51]	104	60/40	Not stated	104[a]	95	91.3
Kamasi, Zambia [4]	502	72/28	Mean age: 39	414 (of 459 X-rays taken)	440	87.7

[a] One of the criteria for admission to the study.

respiratory tract infection who are likely to benefit from antimicrobial therapy. Clinical signs of respiratory distress which are easily recognizable by community and primary health workers are used to this purpose. Examples of such management guidelines are shown in Figs. 13.1 and 13.2. It is stressed that guidelines of this kind have to be modified according to local conditions. Careful evaluation of their practicability and efficacy in pilot studies and field tests is necessary, with the goal of the reduction of childhood mortality from ARI, before they can be introduced into primary health care services on a wider scale.

Etiology

Data from developing countries on the etiology of pneumonias are limited. For comparison, Table 13.3 gives a rather long list of agents, yet still excluding some of the respiratory viruses, namely adenoviruses, parainfluenza viruses type 1, 2, 3, the respiratory syncytial virus, and parasitic agents like *Paragonismus spp.*

The lobar and segmental pneumonias are usually caused by *Streptococcus pneumoniae* (Table 13.4). In primary, community-acquired pneumonias, not responding to penicillin, *Staphylococcus aureus, Klebsiella pneumoniae, Hemophilus influenzae,* and, according to one study from Nigeria, *Mycoplasma pneumoniae* should be considered as the likely pathogens [22, 41, 66] (Table 13.4).

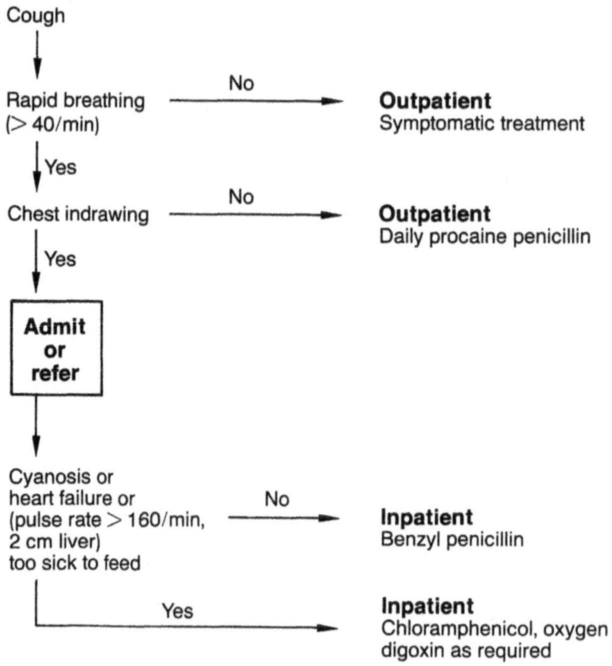

Fig. 13.1. Flowchart for clinical management of acute respiratory infections in children under 5 years of age in Papua New Guinea. All children with fever are given antimalarials. From [78]

A. A child with a:

		Signs and symptoms		Treatment
Mild Acute Respiratory Infection →	has one or more of the following →	1. Clear nasal discharge 2. Blocked nose 3. Red throat without membrane or white patches 4. Hoarseness, stridor or wheezing — but no nasal flaring or chest retraction 5. Cough — not seriously disturbing	→ and is sent home on the following treatment	1. No antibiotic or antimicrobial 2. Paracetamol for pain or fever 3. Moist air for hoarseness or stridor 4. Syrup franol for recurrent wheeze

B. A child with a:

Moderate Acute Respiratory Infection →	has one or more of the following →	1. Purulent exudate, nose or throat 2. Ear ache or discharge 3. Fever of 40°C 4. Respiratory rate of 40 - 70 5. Cough with purulent sputum 6. Seriously disturbing cough 7. Whooping cough	→ and is sent home to be reviewed after 24 h on the following treatment	1. Oral penicillin for purulent exudate, nose or throat 2. Procaine penicillin for a) Respiratory rate of 40 - 70 b) Cough with purulent sputum or disturbing cough c) Temperature of 40°C 3. Procaine penicillin + sulphadimidine for a) Ear ache or discharge

C. A child with a:

Severe Acute Respiratory Infection →	has one or more of the following →	1. Nasal flaring 2. Chest retractions 3. Wheezing with chest retraction or nasal flaring 4. Stridor or hoarseness with chest retraction 6. Blue lips, hands, or tongue 7. Whooping cough + flaring or retraction 8. Too sick to feed	→ and is referred for hospital admission and treatment, where the following is expected	1. Benzylpenicillin for chest retraction 2. Moist oxygen for stridor 3. Lumbar puncture for first convulsions 4. Subcutaneous adrenalin for recurrent wheeze

Definitions

Nasal flaring: active movement of lateral walls of lower end of the nose during breathing.

Chest retraction: indrawing of muscles between the ribs, below the ribs, above the ribs, above the sternum, and above the clavicles during breathing.

Seriously disturbing cough: a cough occurring every few minutes, interfering with the child's sleep, and often associated with vomiting.

Advice to parents with a child with ARI

1. Continue breast-feeding if the child still breast-feeds. Otherwise continue with normal feeds.
2. Give plenty of oral fluids.
3. In case of high body temperature, do the following:
   a) dress the child with light clothing — to allow heat to escape.
   b) smear lukewarm film of water over the body — it reduces fever. This can be done many times.
4. Moist air - keep a bowl of hot water (safe distance from child) and hold the child near it for short periods.

Fig. 13.2. Management of acute respiratory infection. From [71]

Table 13.3. Etiologic agents responsible for community- and hospital-acquired pneumonias and laboratory tests used in diagnosis. From (15)

Agent	Culture	Serology	Pulmonary biopsy, lavage	Other laboratory tests[a]
Streptococcus pneumoniae (1)	+	+[b]		
Staphylococcus aureus (1, 2)	+			
Hemophilus influenzae (1)	+			
Klebsiella pneumoniae (1, 2)	+			
Escherichia coli, Proteus spp (2)	+			
Pseudomonas aeruginosa (2)	+			
Serratia spp. (2)	+			
Anaerobic bacteria (1, 2)	+			
Legionella pneumophilia (1, 2)	+	+		1,5
Streptococcus pyogenes (1)	+			
Francisella tularensis (1)	+	+		
Yersinia pestis (1)	+	+		2
Nocardia asteroides (2)	+			6
Mycobacterium tuberculosis (1,2)	+			3
Mycoplasma pneumoniae (1)	(+)	+		
Chlamydia psittaci (1)	+	+		
Coxiella burnetii (1)		+		
Influenza (1, 2)	+	+		
Cytomegalovirus (2)	+	+	+	
Varicella zoster (2)	+	+		
Coccidioides immitis (1)	+	+	+	
Histoplasma capsulatum (1)	+	+	+	
Aspergillus spp. (2)	+		+	
Candida spp. (2)	+		+	
Cryptococcus neoformans (2)	+		+	
Zygomycetes (2).	+		+	
Pneumocystis carinii (2)			+	4

(1), community-acquired; (2), hospital-acquired. [a] On sputum or tracheobronchial secretion: [1] fluorescent antibody, [2] Gram stain, [3] acid-fast or fluorescence, [4] methenamine-silver, [5] Dieterle silver, [6] modified acid-fast stain; [b] CIE (counterimmune electrophoresis).

Interstitial, "atypical" pneumonias are caused mainly by *Mycoplasma pneumoniae*, whereas *Chlamydia psittacosis* and *Coxiella burnetii* are rare causes in developed countries. The etiology of this type of pneumonia has, however, not been studied in developing countries, presumably because lobar pneumonia is a much more common clinical problem. *Mycoplasma pneumoniae* infection was diagnosed by culture or serology in 12 of 74 adult patients with lobar pneumonia [41]. Mixed infection with *Streptococcus pneumoniae* was present in six of the 12 patients.

Legionella pneumophila may play a role as pathogen, but no case could be identified in two African studies which, together, involved 179 adult patients with pneumonia [41, 54].

Several studies of a viral etiology of pneumonia in infancy and childhood were

Table 13.4. Etiology of pneumonia in adults in Africa and Papua New Guinea

Country [Reference]	Nairobi, Kenya [66]	Zaria, Nigeria [41]	Port Moresby, Papua New Guinea [22]		
No. of patients	105	74	632		
Sex male (%)	72	80	76		
female (%)	28	20	24		
Age	Commonest age group: 20–30 years	Mean age: 28.5 years	83% in range 15–45 years		
Chest X-ray					
Lobar pneumonia	91 } 92 X-rays taken	74	379		
Bronchopneumonia	1	0	130		
Normal	0	0	123		
Mortality (%)	1.9	6.8	1.9		
Pathogens					
Type of specimen	Sputum and blood cultures combined	Sputum, pleural fluid, lung aspirate, blood culture, and serology combined	Sputum	Blood culture	Lung aspirate
No. of samples	n. a.	n. a.	447 (154)[b]	553	24
Streptococcus pneumoniae	39[a]	37 (7)[b]	257 (129)[b]	51	15
Hemophilus influenzae	0	0	196 (115)[b]	2	3
Staphyloccus aureus	1	1	53 (40)[b]	2	0
Beta-hemolytic streptococci	0	0	38 (30)[b]	0	0
Klebsiella spp.	2	3	0	0	0
Salmonella paratyphi	0	1	0	0	0
Escherichia coli	1	0	0	0	0
Mycobacterium tuberculosis	1	2			
Mycoplasma pneumoniae	n. i.	12 (6)[b]	n. i.	n. i.	n. i.
Other bacterial species	0	0	27 (21)[b]	0	0
Influenza A	n. i.	3 (2)[b]	n. i.	n. i.	n. i.
No pathogens	60	23	49	498	6

n. a., not appropriate; n. i., not investigated.

[a] Plus 12 cases where Gram's stain of sputum showed pneumococci but which were negative on culture.

[b] No. of mixed infections in brackets.

undertaken in developing countries with a subtropical or tropical climate [3, 8, 12, 49, 68]. The results indicate that the spectrum of viruses is similar to that known from studies in temperate climates. Viral infection, mainly due to respiratory syncytial virus, parainfluenza viruses, adenoviruses, and influenza virus, was diagnosed by culture and/or serology in between 17% and 50% of cases. The role of *Mycoplasma pneumoniae* as a pathogen of childhood pneumonia has not been delineated clearly. *M. pneumoniae* infection was implicated in 5% of cases in a serological survey conducted in several subtropical and tropical countries in the 1960s [12]. A recent study in Colombia found serological evidence of *M. pneumoniae* infection in less than 1% of LRTI seen in 401 ambulatory children [8]. A smaller Indian study detected *M. pneumoniae* infection serologically in seven of 47 children with LRTI [31].

On the other hand, studies using percutaneous needle biopsy of the lung have accumulated evidence of bacterial infection in up to 40%–50% of hospitalized infants and children with moderate to severe pneumonia [19, 20, 25, 27, 34, 36, 44, 45, 53, 55, 57, 59, 62, 64, 72]. *Streptococcus pneumoniae, H. influenzae,* and *Staphylococcus aureus* are the pathogens most commonly isolated (Table 13.5).

Staphylococcus aureus was the commonest pathogen in several of the studies

Table 13.5. Etiology of pneumonia in infancy and childhood: bacteriological findings in lung aspirates

Country, year of publication, [Reference]	No. of Children	No. of cultures yielding						Case-fatality rate (%)
		Streptococcus pneumoniae	*Hemophilus influenzae*	*Staphylococcus aureus*	Beta-hemolytic streptococci	Mixed culture (*Strep. pneumoniae + H. influenzae*)	No growth	
No history of recent antibiotic therapy								
Brazil, 1974 [55]	37	15	3	1	–	–	17	ND
Colombia, 1976 [25]	71[b]	4	4	5	–	–	56	ND
Nigeria, 1977 [64]	88	31	9	8	3	–	34	5
Brazil, 1980 [27]	58	21	8	–	–	4	25	ND
Papua New Guinea, 1983 [53]	18	7	1	–	–	–	10	ND
Gambia, 1986[a] [72]	51	22	8	1	1	4	22	3
Total (%)	323 (100)	100 (31.0)	33 (10.2)	14 (4.6)	4 (1.2)	8 (2.5)	164 (50.8)	
No statement as to history of recent antibiotic therapy								
India, 1969 [34]	18[c]	–	4	–	–	2	12	11
Chile, 1971 [44]	25	6	–	1	–	–	18	ND
Nigeria, 1979 [20]	56	10	–	8	2	–	18	ND
Nigeria, 1986[a] [45]	56[d]	15	1	3	1	2	36	20
Total (%)	155 (100)	31 (20.0)	5 (3.2)	12 (7.8)	3 (1.9)	4 (2.6)	84 (54.2)	
Positive history of recent antibiotic therapy (% of children)								
Chile, 1966 [57]	125 (64)	26	18	15	–	–	58	ND
Chile, 1971 [44]	505 (43)	5	12	144	3	–	277	ND
Colombia, 1976 [25]	31 (100)[b]	1	–	3	–	–	27	ND
India, 1978 [59]	50 (82)[e]	14	–	14	2	–	19	48
India, 1981 [36]	70 (29)	18	2	5	4	–	34	3
Nigeria, 1981 [19]	73 (37)	6	2	22	2	–	33	8
Papua New Guinea, 1984 [62]	83 (12)	6	14	1	–	18	35	11
Total (%)	937	76 (8.1)	48 (5.1)	204 (21.8)	11 (1.2)	18 (1.9)	483 (51.5)	

[a] Figures of bacteriological yield include yieldings of blood culture and/or counter-immunoelectrophoresis. Figures of no growth indicate no. of sterile lung aspirates.
[b] Includes a proportion of cases of ARI other than pneumonia.
[c] 18 episodes in 16 children.
[d] Cases of measles pneumonia only.
[e] Newborn only.

which included children with a history of recent antibiotic therapy [19, 44]. *Staphylococcus aureus* was grown from 144 of 505 lung aspirates in a study from Chile where only cases of bronchopneumonia were investigated [44]. Cases of lobar pneumonia had been excluded. In contrast, *Streptococcus pneumoniae* was the commonest pathogen in a study of lobar pneumonia in The Gambia (no history of recent antibiotic therapy) [72]. *Staphylococcus aureus* was as common as *Streptococcus pneumoniae* in the 50 Indian newborns studied [59].

Mixed infection with *H. influenzae* and *Streptococcus pneumoniae* was noted in several studies [62, 72]. It was the commonest finding in a study from Papua New Guinea, where it was predominantly seen in infants [62]. Single infection with either of the two agents occurred more commonly in children beyond the age of 12 months. The age difference was statistically significant in case of *Streptococcus pneumoniae* [62].

Possible viral etiology was investigated in parallel in three of these studies [25, 34, 62]. In the Indian study, viruses were isolated in nine of 18 instances [34]. In addition, evidence of current *Mycoplasma pneumoniae* infection was found in further 2 cases. Viruses were isolated in 18 of 62 cases where viral cultures from nasopharyngeal aspirates were performed in the study from Papua New Guinea [62]. Bacterial-viral interaction was suggested by isolation of bacterial pathogens from lung aspirates or blood culture in four of the nine and 12 of the 18 cases respectively [34, 62]. There were 131 cases of pneumonia in the third, Colombian, study [25]. Thirty-eight of them were associated with measles, 22 were attributed to a viral etiology, 15 to a bacterial etiology, and five to a mixed viral-bacterial etiology. No cause could be identified in the remaining 51 cases [25].

The case-fatality rates in these studies of moderate to severe pneumonia range from 3% to 48% (Table 5). The highest mortality rates were observed in measles pneumonia, 20%, and in newborns, 48% [45, 59].

In pneumonias of severe protein-energy malnutrition, both gram-negative bacteria - *E. coli, Klebsiella pneumoniae, Pseudomonas aeruginosa,* and others - and opportunistic pathogens such as *Pneumocystis carinii* are frequently isolated [7, 10, 20, 33]. *E. coli, Proteus* spp., *Enterobacter* spp., *Acinetobacter* spp., and *Streptococcus faecalis,* which are often recovered from sputum, are only very rarely the true pathogens, except in leukemia and severe protein-energy malnutrition.

Treatment of Community-Acquired (Primary) Pneumonia

International Standard Therapy

Adults: *Phenoxymethylpenicillin* 500 mg every 6 h is sufficient only for mild cases.
Benzylpenicillin:
- 1 megaunit (MU) every 6 h (USA),
- 5-10 MU 2-3 times daily (Europe) *or*

Procaine penicillin 1.2 MU every 12 h for 10 days. A switch to oral penicillin is possible 48 h after defervescence.
Alternatives: erythromycin, cefazolin.

Children: An oral cephalosporin with activity against *Hemophilus influenzae* (cefaclor) is recommended for mild cases. Amoxicillin or erythromycin are alternatives for oral treatment. Parenteral cephalosporins should be used to treat severe cases in hospital.

Treatment in Developing Countries with Limited Resources

Parenteral penicillin is the drug of choice in all age groups, with the tendency to reduce the dosage and the duration of therapy compared to the regimens used in the USA and Europe.

Adults

Outpatient Treatment. A dose of 0.6 to 1.2 MU procaine penicillin i.m. daily for 5 days is a widely used regimen. The response within 24–48 h is then taken as criterion for further management.

In the late 1960s, certain combinations of short- and long-acting penicillins (e.g., Triplopen) were recommended and are still used. These combinations, either given as a single dose [69] or repeated on day 3 and, if necessary, on day 5 [63], were found to be effective in 90% and 85% of uncomplicated cases, respectively. Dose of Triplopen: one vial i.m.

A single injection of 1.2 MU benzathine penicillin may also be used, but is inferior because of low serum levels and risk of protracted allergy.

Single-dose treatment with 500 mg doxycycline was shown to be as effective as the daily administration of 0.6 MU procaine penicillin [47], but 500 mg doxycycline constitutes an overdosage with a considerable risk of side effects.

Criteria for Admission to Hospital. The main indications for hospitalization are respiratory failure, cardiac failure, shock, jaundice, septicemia, endocarditis, pericarditis, meningitis, complications such as empyema and lung abscess, and severe underlying disease apart from nonresponse to ambulatory treatment.

Duration of Therapy of Uncomplicated Lobar Pneumonia. Treatment for 10 (–14) days, often performed in hospital for the whole time, is recommended in the USA and Europe. Experience from developing countries tells us that such an extended duration of treatment is usually not necessary.

The following treatment policy of lobar pneumonia in hospitalized patients is reported from Zambia:

"All patients were initially treated with intramuscular benzylpenicillin 0.5–1.0 million units six hourly or with procaine penicillin 0.6 million units daily. ... Patients who did not show clinical evidence of improvement within about 48 hours of starting penicillin treatment were given an additional antibiotic, either gentamicin or kanamycin (according to availability) ... Those who had sputum, blood culture, or strong radiological evidence suggestive of staphylococcal pneumonia were given intramuscular cloxacillin. ... In patients who responded promptly to penicillin the treatment was changed to oral phenoxymethylpenicillin 250 mg six hourly when body temperature returned to normal. Most patients were discharged from hospital then and given oral phenoxymethyl penicillin for at least five days. The mean duration of inpatient treatment was 3.4 days in the penicillin responsive patients and 8.7 days in those requiring additional antibiotics" [4].

A study from Papua New Guinea demonstrated that short-course chemotherapy (SCC) of lobar pneumonia was as effective as the more traditional regimens for hospitalized patients with moderately severe pneumonia [51]. The treatment regimens that were compared are shown in Table 13.6. The average length of treatment required was 2.4 days (mean for benzylpenicillin and chloramphenicol combined). Patients with severe pneumonia, as evidenced by cyanosis, shock, bacteremia, extrapulmonary infections, or involvement of more than one lobe, also responded to SCC. Recurrence of persistent fever necessitating further treatment was more frequent, but the difference was not statistically significant.

In a similar study for Nigeria, patients with concomitant disease (hepatosplenomegalic schistosomiasis, tropical splenomegaly syndrome) required antibiotics for a significantly longer period than patients without concomitant disease, 3.75 days as opposed to 2.54 days [5].

Five-day courses of procaine penicillin aluminium monostearate, 1.2 MU once a day, and of benzylpenicillin, 0.5 MU every 6 h, were equally effective in the management of uncomplicated cases of lobar pneumonia [21].

Table 13.6. Duration of treatment of lobar pneumonia: comparison of four treatment schedules. From [51]

Type of regimen	SCC	Standard	SCC	Standard
Treatment schedule	Benzylpenicillin		Chloramphenicol	
	4 MU/day until 24 h after defervescence of fever followed by:		1.5 g day orally until 24 h after defervescence of fever followed by:	
	No further treatment	Procaine penicillin 1 MU i.m. daily until discharge	No further treatment	Chloramphenicol 1.5 g daily until discharge
No. of patients	52	52	51	48
Days in hospital (mean ± 1 SD)	6.8 ± 4.1	6.7 ± 3.4	6.5 ± 2.7	5.6 ± 3.6
Duration of treatment (days)	$2.1^a \pm 1.2$	$6.2^a \pm 2.69$	$2.8^b \pm 1.4$	$5.3^b \pm 2.95$

SCC, short-course chemotherapy; MU, megaunits.
[a, b] Difference is statistically significant.

Children

Outpatient Treatment. Consistent with the classification of ARI into mild, moderate, and severe illness, children with pneumonia of moderate severity are managed as outpatients. Three antibiotics have been suggested for outpatient treatment of childhood pneumonia in a recent review (Table 13.7). Of these three agents, daily procaine penicillin is the treatment of choice covering *Streptococcus pneumoniae* and *Hemophilus influenzae* with the exception of beta-lactamase producing strains. This recommendation is in contrast to the widely accepted opinion that benzylpenicillin is not sufficiently active against *H. influenzae.* The minimum inhibitory concentration (MIC) of benzylpenicillin is comparatively low, 1 μg/ml, and experience from Papua New Guinea tells us that daily procaine penicillin is effective in most instances [39]. Clinical studies addressing this point have, however, not been conducted. The daily dosage of 0.2–0.4 MU that is often used is somewhat lower than the 50000 units/kg that are conventionally recommended.

When an oral antibiotic has to be given, either because the primary health workers are not trained to give i.m. injections, or in order to prevent transmission of AIDS in endemic areas, the choice lies between ampicillin and co-trimoxazole. Ampicillin is not only more expensive, but has also a more limited spectrum of ac-

Table 13.7. First-line antibiotics for outpatient treatment of acute respiratory infections in children (WHO recommendation). From [9]

Characteristic	Summary of data on indicated antibiotic		
	Procaine penicillin	Ampicillin	Co-trimoxazole
Pathogen sensitivity			
Hemophilus influenzae	Good (but up to 25% of strains resistant)	Good (but up to 25% of strains resistant)	Very good
Streptococcus pneumoniae	Very good (resistant strains in some countries)	Very good (resistant strains in some countries)	Very good
Staphylococcus aureus	Poor	Poor	Good
Group A streptococci	Very good,	Very good	Good
Chlamydiae, *Pneumocystis*	Resistant	Resistant	Good
Toxicity			
Mild but common	Rash	Diarrhea, rash	Rash
Grave but rare	Fatal anaphylaxis (1/250000)	Fatal anaphylaxis (1/250000)	Megaloblastic anemia related to dose; aplastic anemia; Stevens-Johnson syndrome
Route of administration	i.m.	p.o.	p.o.
Dose	50000 units/kg daily	25 mg/kg every 6 h	4 mg/kg every 12 h
Cost[a]	$ 0.20	$ 0.40	$ 0.08

[a] Cost is for 5 days of therapy in a child weighing 10 kg.

tivity against the more common pathogens of childhood pneumonia and may select resistant strains of *H. influenzae*. The major argument against the use of co-trimoxazole is its potential to induce resistance. The frequency of resistance to co-trimoxazole in gram-negative enteric bacilli like salmonellae or the pathogens of urinary tract infections is considerably higher in developing countries, where co-trimoxazole is used on an much larger scale and often indiscriminately, than in developed countries. This risk is probably less in more remote, rural areas.

Other agents that should not be used for oral treatment of childhood pneumonia are either less effective due to resistance (tetracyclines, sulfonamides, phenoxymethylpenicillin) or more expensive (erythromycin) or carry the risk of side effects (tetracyclines in children less than 7 years old, chloramphenicol). Chloramphenicol is generally not recommended for outpatient treatment of pneumonia. It should be kept in reserve for treatment of severe infections.

Treatment in Hospital. Several antibiotics have been suggested for treatment of severe ARI in children (Table 13.8). Intramuscular benzylpenicillin, given every 6 h, should be the standard treatment for severe pneumonia (cough, intercostal recessions, but no cyanosis). Chloramphenicol is only given to children who are very severely ill.

Table 13.8. Antibiotics recommended for the treatment of acute respiratory infections in hospitalized children (modified WHO recommendation). From [9]

Characteristic	Summary of data on indicated antibiotic				
	Chlor-amphenicol	Oxacillin	Gentamicin	Ampicillin	Benzyl-penicillin
Pathogen sensitivity					
Hemophilus influenzae	Very good	Poor	Fair	Good[a]	Fair
Streptococcus pneumoniae	Good	Fair	Poor	Very good	Very good
Staphylococcus aureus	Fair	Good	Good	Poor	Poor
Group A streptococci	Good	Good	Fair	Very good	Very good
Chlamydiae	Poor	Resistant	Resistant	Resistant	Resistant
Gram-negative enteric bacteria	Good	Resistant	Very good	Fair	Resistant
Route of administration	i.m./i.v./p.o.	i.m./i.v.	i.m./i.v.	i.m./i.v./p.o.	i.m./i.v.
Dose mg/kg	25 mg/kg every 6 h	50 mg/kg every 6 h	2.5 mg/kg every 8–12 h	50 mg/kg every 6 h	50000 units/kg every 6 h
Cost[b]	$ 1.00	$ 6.00	$ 5.00	$ 20.00	$ 0.50

[a] However, up to 30% of strains produce beta-lactamase and are resistant.

[b] Cost is for 5 days of therapy in a child weighing 10 kg.

This subgroup of children with severe pneumonia was identified in a randomized clinical trial as follows: "Children were admitted to the trial if they had cough and intercostal recession, pulse rate over 160/min. with hepatomegaly; inability to feed, bronchial breathing, grunting, cyanosis, severe chest X-ray changes, or a total white cell count over 30000 cells/µl" [61]. The efficiency of treatment with chloramphenicol alone or with benzylpenicillin and chloramphenicol was compared in 748 children admitted to the trial.

"Chloramphenicol succinate was given i.m. in a dose of 25 mg/kg every 6 hours until the child began to improve and then 25 mg/kg of chloramphenicol palmitate was given orally every 6 hours. Benzylpenicillin was given i.m. every 6 hours in a dose of 250000 units to children weighing 3-9 kg, and 500000 units to children weighing 10 to 19 kg. All children were given quinine or amodiaquine" [61].

Monotherapy with chloramphenicol was as effective as the combination of chloramphenicol and benzylpenicillin. Mortality was 13% for chloramphenicol alone and 17% for the combination. The need to change treatment arose in 1% and 2% respectively [61].

Chloramphenicol should only be used in infants more than 2 months old. Neonatal pneumonia which is more likely to be due to gram-negative enteric bacilli or to *Staphylococcus aureus* should be treated with cloxacillin and gentamicin.

Likewise, staphylococcal pneumonia should be treated with cloxacillin or another penicillinase-resistant penicillin. The addition of an aminoglycoside is often recommended.

Nonresponse to Initial Therapy

In uncomplicated cases of pneumococcal pneumonia defervescence takes place within 48 h after the first dose of benzylpenicillin. Nonresponse to treatment with penicillin is observed in about 10% of adult patients with lobar pneumonia (Table 13.2). A supervening, pyogenic complication, such as empyema, lung abscess, pericarditis, may be the reason for nonresponse. Pulmonary tuberculosis with a superimposed pneumonia should always be considered if there is only partial improvement. Bronchial obstruction by aspiration of a foreign body, tuberculous lymph nodes, or bronchial carcinoma may be the underlying cause of a nonresolving pneumonia. Amebic liver abscess may mimic a right-sided basal pneumonia. In endemic areas, pulmonary plague, anthrax, and melioidosis should be considered as differential diagnoses from the onset in extremely ill patients. Tropical pulmonary eosinophilia may rarely present as an acute pneumonia not responding to penicillin.

Concomitant malaria should be excluded or treated in endemic areas.

Finally, a pathogen other than *Streptococcus pneumoniae* may be the cause of a nonresponding lobar pneumonia. *Klebsiella pneumoniae, Staphylococcus aureus, Hemophilus influenzae,* and *Mycoplasma pneumoniae* are the most likely agents. A pneumonia caused by *Streptococcus pneumoniae* with intermediate penicillin resistance (MIC between 0.1 to 1.0 µg/ml), however. can be expected to respond to the usual regimens [30].

An antibiotic other than benzylpenicillin should be used if a patient does not improve within 24–48 h of treatment with penicillin, and when a patient is extremely ill on admission. Chloramphenicol which covers all the main aerobic and anaerobic pathogens is still widely used in that situation when resources are limited. Cloxacillin, when available, should be used instead of chloramphenicol when there is blood culture or radiological evidence of staphylococcal pneumonia.

The combination of tetracycline or doxycycline with clindamycin, which is active against anaerobic bacteria, *Staphylococcus aureus*, *Mycoplasma*, *Chlamydia*, and *Coxiella burnetii*, but not against *Klebsiella pneumoniae*, is a less toxic alternative to chloramphenicol. In Europe, parenteral cephalosporins (cefazolin, cefoxitin, cefotaxime) are widely used in nonresponding pneumonias. Erythromycin or, alternatively, rifampicin is added if infection by *Legionella pneumophila* is suspected.

Alternatives to Benzylpenicillin

Alternatives to benzylpenicillin are listed in Table 13.9.

Supportive Therapy

Adequate fluid intake must be maintained in order to improve expectoration. Intravenous fluids and electrolyte replacement may be necessary in severe dehydration with shock. Otherwise, fluids should be given orally or by nasogastric tube. Care must be taken not to overhydrate small infants and children with excessive amounts of intragastric or intravenous fluids (risk of pulmonary edema and respiratory failure).

Fever is an useful indicator of the clinical course and, in adults, antipyretics should be avoided if possible. Children with high fever should be given antipyretics like paracetamol. Cooling measures like sponging or application of cool compresses may also be used.

Further supportive measures are directed toward pain relief with codeine (in the absence of respiratory distress), correction of severe anemia by blood transfusions, and adequate nutrition. In infants, breast-feeding should be continued.

Expectorants, cough suppressants, or mucolytic agents are not needed in the treatment of pneumonia. Bronchodilators may be required for treatment of concomitant wheeze in young children.

Prevention

Malnutrition, overcrowding and poor hygiene could all be put forward as possible targets for primary preventive measures. Important means of prevention of mortality from acute respiratory infections in children are the measles and pertussis vaccines included in the WHO Expanded Programme on Immunization. The use of the expensive pneumococcal vaccine which now contains 23 of 84 serotypes is

Table 13.9. Alternatives to benzylpenicillin

Drug(s)	Evaluation
Cheap	
Phenoxymethylpenicillin	Mild adult lobar pneumonia only, not sufficiently active against *Hemophilus influenzae* (MIC 4 μg/ml)
Benzathine penicillin	Possible in adult lobar pneumonia, serum levels presumably only sufficient for fully sensitive strains of *Streptococcus pneumoniae* and inadequate for *H. influenzae*
Tetracycline (doxycycline)	Dental discoloration is possible in children less than 7 years old, resistant pneumococci may be frequent
Co-trimoxazole	Effective in mild to moderate pneumonia, but inferior to penicillin, potential of induction of resistance, drug of choice for *Pneumocystis carinii*
Sulfonamides	Less effective, widespread resistance of *H. influenzae,* poor activity against *Staphylococcus aureus* and many other pathogens, poor tolerance
Chloramphenicol	Risk of aplastic anemia, risk of grey syndrome in newborns, should be kept in reserve for failure of initial therapy with penicillin and for severely ill patients
Metronidazole	Only active against anaerobic bacteria including the rare penicillin-resistant *Bacteroides fragilis,* therefore indicated only together with penicillin in aspiration pneumonia and chronic lung abscess
More expensive	
Erythromycin	Poor absorption possible, has to be given 4 times daily
Ampicillin	Broad-spectrum antibiotic which is not needed in adult lobar pneumonia, selection of resistant strains of gram-negative enteric bacilli and of *H. influenzae*
Cloxacillin	Drug of choice in staphylococcal pneumonia, only active against gram-positive cocci, resistance of *Staphylococcus aureus* possible
Aminoglycosides	Not indicated in lobar pneumonia, poor clinical results with monotherapy have been reported even in infections by sensitive strains of *Klebsiella pneumoniae, Pseudomonas aeruginosa,* and *Staphylococcus aureus*
Clindamycin/lincomycin	Side effects (diarrhea, pseudomembraneous enterocolitis), may be drug of choice in aspiration pneumonia or chronic lung abscess
Oral cephalosporins	Expensive, poor activity against *H. influenzae* (except cefaclor) and against gram-negative enteric bacilli
Parenteral cephalosporins	Broad-spectrum antibiotics which are not needed in common lobar pneumonia, expensive, two to four daily injections, often regarded as drugs of choice in Europe and United States because they cover the main pathogens of primary and secondary pneumonias
Rifampicin	High activity against *Staphylococcus aureus,* anaerobic bacteria, and *Legionella pneumophila,* to be used only in selected cases (risk of side effects, resistance, and monotherapy of tuberculosis)

recommended only for groups at special risk so far (splenectomy, sickle cell disease). Field trials on the efficacy of the vaccine in reducing childhood mortality from acute lower respiratory tract infections (ALRI) were carried out in Papua New Guinea [52]. The overall mortality was reduced by 19% in the vaccinated group of children compared to the control group. The reduction in mortality in instances where ALRI was the sole cause of death was estimated at 59% and 50% for children vaccinated before the age of 5 and 2 years, respectively [52]. Influenza vaccines are of little protective value, both in the tropics and in Europe and the USA. They can be recommended only for patients at high risk.

The most practical approach to reduction of mortality from pneumonia in the tropics to date, therefore, seems to lie in early and effective treatment within a system of primary health care.

Treatment of Special Types of Pneumonia

Staphylococcal Pneumonia

Pneumonia due to *Staphylococcus aureus* is far less common in adults than in infants and young children. In a study from India, staphylococcal pneumonia was diagnosed in two of three infants less than 1 year old, regardless of the radiological findings; lobar consolidation, bronchopneumonia, and pneumatoceles, in that order of frequency, were all observed in bacteremic cases [50]. Measles, malnutrition, and neglected staphylococcal infections at other sites, especially the skin, are all important predisposing factors.

Healthy adults may contract staphylococcal pneumonia postoperatively, as a nosocomial infection, or following an episode of influenza. Staphylococcal pneumonia may also arise from hematogenous spread in septicemia, and in association with pyomyositis [58].

Constitutional symptoms are usually severe; children look toxic and ill within hours after the onset of symptoms. If antibiotic treatment is not instituted in time, pneumatoceles are likely to develop in infants with the subsequent risk of empyema, pneumothorax, and pyopneumothorax. In adults an abscess-forming pneumonia is the common presentation of staphylococcal pneumonia. The abscesses are rapidly destructive, but do not progress to pneumatoceles.

Penicillin-sensitive strains of *Staphylococcus aureus* respond best to benzylpenicillin. Cloxacillin is regarded as the drug of choice for penicillin-resistant strains. Chloramphenicol, erythromycin, or a combination of both can be substituted if cloxacillin is not available. The addition of gentamicin may prove useful.

First-generation cephalosporins (cefazolin and others) are as effective as cloxacillin. Modern cephalosporins are usually less active against *Staphylococcus aureus*. Clindamycin, fusidic acid, vancomycin, or rifampicin may be administered if there is no improvement following the initial therapy or if methicillin (=cloxacillin)-resistant strains are suspected. Combination therapy, either with cloxacillin or with two of the latter drugs, is then advisable.

Measles Pneumonia

The etiology of the pneumonia which is a frequent cause of death in severe measles is not uniform. The lungs of 21 children who died within 28 days of the onset of the measles rash were studied [37] and four diagnostic categories were distinguished. Active measles pneumonia was seen in five cases, adenovirus infection and herpes simplex virus infection was present in five and six cases respectively, and suppurative pneumonia in the remaining five cases. Death occurred early in four of the five cases of measles pneumonia, 2 to 4 days after the onset of the rash. The fifth death occurred later, after 15 days. Measles virus could not be recovered in any of the five cases, but giant cells were present in all of them. Staphylococcal abscesses were found in two of the five cases of suppurative pneumonia. Similarly, only nine of 26 cases showed the features of bacterial bronchopneumonia in an earlier autopsy study from Uganda [18]. Interstitial pneumonia was found in the other 17 cases, in 10 of which giant cells were present.

Thus, it appears that only a proportion of deaths due to pneumonia in measles are preventable by antimicrobial therapy and that active measles infection or intercurrent infection with adenovirus or herpes simplex virus is frequently found in measles pneumonia.

On the other hand, a bacterial etiology could be established in about 40% of 56 Nigerian children with measles pneumonia by means of a combination of percutaneous needle biopsy, blood culture and countercurrent immunoelectrophoresis findings [45]. The commonest pathogen was *Streptococcus pneumoniae* (17 of 56 cases). *Staphylococcus aureus* accounted for three cases and *Hemophilus influenzae* and *Streptococcus pyogenes* for one each. The authors summarize: "More severe changes on chest X-ray on presentation were associated with higher rates of bacterial identification and worse outcome at 2 weeks. Chest X-ray changes were worse and bacterial identification was more common in poorly nourished children" [45].

Bacterial findings may be different in children who have been treated with penicillin initially. "Gram negatives and staphylococci" were recovered from lung aspirates or transtracheal aspirates of several children in a center in Nairobi where the majority of children with measles came from peripheral clinics where they had been treated unsuccessfully with penicillin [28].

Evidence from developed countries suggests that antimicrobial prophylaxis does not decrease the rate of bacterial complications in measles [76], but data from developing countries concerning the value of antimicrobial prophylaxis in severe measles are not available.

Antimicrobial treatment was studied in 301 children with measles and pneumonia in Nigeria [48]. Ninety-seven children with clinically mild pneumonia were divided into a group treated with penicillin and a control group. Mortality was similar for both groups. Seventeen of the 47 children in the control group were later given penicillin for persistence of fever or symptoms of increasing pneumonia; 16 responded promptly. The remaining 204 children with more severe pneumonia were treated either with penicillin or with chloramphenicol. The mortality was 10% for children treated with penicillin and 20% of the group treated with chloramphenicol.

In conclusion, first-line treatment of pneumonia in measles should be carried out with parenteral penicillin. A clinical trial of broad-spectrum therapy with cloxacillin and an aminoglycoside to cover for staphylococci and gram-negative bacteria appears indicated in those children who do not respond to penicillin.

Klebsiella Pneumonia

Pneumonia due to *Klebsiella pneumoniae* is uncommon, accounting for 1%–3% of lobar pneumonias in adults. It may occur in debilitated patients or in alcoholics following aspiration, but, in developing countries, is also observed in previously healthy adults.

Patients are usually severely ill, shocked and centrally cyanosed when compared to pneumococcal lobar pneumonia. The sputum is mucoid and red rather than rusty. On the chest X-ray, the affected lobe is densely opacified, often with bulging of the fissure. The diagnosis should be suspected in particular in cases of nonresponse to treatment with benzylpenicillin.

Chloramphenicol is used for treatment either alone or in combination with an aminoglycoside, which should not be given as monotherapy. Tetracycline, ampicillin, and their analogues are ineffective, and the efficacy of co-trimoxazole is doubtful. Third-generation cephalosporins (cefotaxime) may be life-saving. Other parenteral cephalosporins are less effective but are equivalent alternatives to chloramphenicol. Quinolones (ciprofloxacin) are promising agents.

Even with optimal treatment, a slow response to therapy is characteristic.

Aspiration Pneumonia

Aspiration of saliva, blood, or vomit may occur in coma or alcoholic intoxication, during epileptic convulsions or in severe tetanus, and during general anesthesia. The main pathogens of aspiration pneumonia are penicillin-sensitive gram-negative rods *(Bacteroides melaninogenicus),* often in mixed infection with anaerobic streptococci. Foul-smelling sputum, caused by the anaerobic infection, is typical. Most cases respond well to benzylpenicillin. Clindamycin, metronidazole, and chloramphenicol should be reserved for resistant cases.

Aspiration pneumonia, which may rapidly destroy the lung parenchyma, will present on chest X-ray as an acute abscess-forming pneumonia, sometimes similar to cavitating pulmonary tuberculosis.

Acute Abscess-Forming Pneumonia

Acute abscess-forming pneumonia is an acute pneumonia in which the chest X-ray shows either single or multiple abscesses in the lung.

Staphylococcus aureus and *Bacteroides* spp., mainly the penicillin-sensitive *Bacteroides melaninogenicus,* are the usual pathogens. *Klebsiella pneumoniae, Pseudomonas pseudomallei* and the penicillin-resistant *Bacteroides fragilis* are only rarely encountered.

Benzylpenicillin is the drug of choice. If one suspects staphylococcal pneumonia, cloxacillin should be used from the onset. Clindamycin and chloramphenicol cover both *Staphylococcus aureus* and the anaerobic pathogens, including *Bacteroides fragilis.*

Interstitial Pneumonias

Interstitial, "atypical" pneumonias are characterized by the insidious onset of malaise, pyrexia, headache, and cough and the paucity or absence of physical signs over the lungs, in the presence of central, dense opacifications on X-ray. The incidence of the main pathogen, *Mycoplasma pneumoniae,* falls within the range of 5%-15% of primary pneumonias [31, 41] whereas Q fever and psittacosis are only rarely found. Interstitial pneumonias, regardless of their etiology, do not respond to penicillins or cephalosporins. Tetracycline or erythromycin are the drugs of choice.

Pneumocystis carinii is the cause of interstitial plasma cell pneumonitis, which develops in premature and malnourished infants and presents with increasing tachypnea and fever. Co-trimoxazole is the only effective antibiotic.

Legionnaires' Disease

Legionnaires' disease, caused by *Legionella pneumophila,* presents as a pneumonia which does not respond to penicillins, cephalosporins, or aminoglycosides. The association with diarrhea, prostration, and confusion and the predominant incidence in patients with underlying disease are typical features. The diagnosis is usually established by serology. Only erythromycin or rifampicin are effective.

Secondary, Hospital-Acquired Pneumonias

Infections by resistant hospital strains of *Klebsiella* spp., *Pseudomonas aeruginosa,* and *Staphylococcus aureus,* often mixed with *Streptococcus pneumoniae* and anaerobic bacteria, are frequent in secondary, hospital-acquired pneumonias. These pneumonias can be classified according to whether or not prior antibiotics had been given.

No Prior Antibiotics

International standard therapy consists of treatment with first- to third-generation cephalosporins.

When resources are limited, chloramphenicol, possibly combined with an aminoglycoside, should be used. If there is a history of aspiration, benzylpenicillin may be used instead. Ampicillin and tetracyclines do not cover the main pathogens and may select resistant strains.

Prior Antibiotic Therapy

Prior antibiotics are discontinued and the therapy is changed to other agents.

When the causative pathogen is unknown, very broad combinations (cefotaxime + azlocillin or cefotaxime + amikacin) should be used, if available. Otherwise, the combination of chloramphenicol with an aminoglycoside (preferably amikacin) must be tried. The combination of doxycycline with clindamycin or erythromycin may provide an alternative in less severe cases.

When the causative agent is known, directed therapy according to the sensitivity results should be performed as in septicemia (see Table 9.8 of chapter on "Septicemia").

Postoperative Pneumonia

Postoperative pneumonia is commonly due to *Staphylococcus aureus,* whereas pneumococci, streptococci, and *Klebsiella* spp. are only infrequently encountered.

Treatment with a penicillinase-resistant penicillin is appropriate, but is ineffective in the occasional case of *Klebsiella* infection. Treatment with first- and second-generation cephalosporins covers the spectrum of possible agents more completely.

Aspiration pneumonia with mixed aerobic-anaerobic bacteria is also common after operations.

Postinfluenza Pneumonia

Bacterial superinfection with *Staphylococcus aureus* or *Streptococcus pneumoniae,* resulting in tracheobronchitis or pneumonia, can complicate influenza in debilitated patients. Treatment with cloxacillin or another penicillinase-resistant penicillin is adequate.

Efficacy of Antimicrobial Agents in Different Types of Pneumonia

The efficacy of antimicrobial agents contained and not contained in the WHO list in different types of pneumonia is described in Tables 13.10 and 13.11. For comparison, the European standard therapy is summarized in Table 13.12.

Empyema

The major predisposing factors of empyema in adults are pulmonary tuberculosis, chest trauma, and poorly treated pneumonias. In infants and children, empyema is commonly a complication of acute bronchopneumonia and measles and is associated with malnutrition and severe anemia [1, 38, 56]. The majority of cases of empyema in childhood, according to the culture results of aspirated pus, are caused by *Staphylococcus aureus,* whereas *Streptococcus pneumoniae* accounts for only 5%–8% of cases [1, 38, 56]. On investigation by counter-immunoelectrophoresis, a considerable proportion of culture-negative cases may turn out to be due to *Streptococcus pneumoniae* [40].

Table 13.10. Efficacy of antibiotics of the WHO list in different types pneumonia

	Lobar/segmental pneumonia	Interstitial, "atypical" pneumonia	Postinfluenza pneumonia	Measles pneumonia	Staphylococcal pneumonia	Aspiration pneumonia	Postoperative pneumonia	Acute abscess-forming pneumonia	Secondary pneumonia hospital-acquired
Benzylpenicillin/phenoxymethylpenicillin	!	Ø	!	!	2b, f	!	2b, f	!	Ø
Ampicillin	1c	Ø	2d	2d	2d, f	1c	2b, f	1c	Ø
Cloxacillin	2a	Ø	1a	1a	!	2a	!	2a	2d
Chloramphenicol	1b, c	1b	1b	1b	1b, c	1b, c	1b	1b, c	1b
Erythromycin	1a	!	1a	1a	2a	1a	1c, d	1a	2b
Gentamicin	Ø	Ø	2d	2d	2b, e	Ø	Ø	Ø	2e
Metronidazole	Ø	Ø	Ø	Ø	Ø	2e	Ø	2b	Ø
Sulfadimidine	2a	Ø	2a	2b	Ø	Ø	Ø	Ø	Ø
Co-trimoxazole	2b	Ø	2b	2b	2b	2b	2b	2b	Ø
Tetracycline/doxycycline	1c	!	1c	2a	2b, f	2b, f	2b, f	2a, f	Ø
Amikacin	Ø	Ø	2d	2d	2d	Ø	2d	2d	2e

!, Drug of choice. 1, Effective drugs: 1a, expencive compared to other effective agents; 1b, use not advisable if less toxic agents available; 1c, range of activity is too broad; 1d, resistance is possible; 1e, use in combination therapy. 2, Limited value: 2a, less effective compared to group 1; 2b, efficacy is doubtful; 2c, efficacy is unknown; no studies available despite in vitro sensitivity; 2d, use only after isolation of a sensitive organism; 2e, use in combination therapy; 2f, resistance is possible. Ø Ineffective, ? Effect unknown

The diagnosis is established radiologically and by aspiration of pus from the pleural cavity. If thick, odorless pus is withdrawn, a Gram's stain can differentiate between *Streptococcus pneumoniae, Staphylococcus aureus,* and other rare pathogens. Benzylpenicillin is the drug of choice for *Streptococcus pneumoniae* and penicillin-sensitive staphylococci. Cloxacillin, chloramphenicol, or erythromycin, in that order of preference, should be used for penicillin-resistant staphylococci. Aspiration of foul-smelling pus is a sign of a mixed anaerobic infection, and treatment with penicillin should be instituted. The rare penicillin-resistant *Bacteroides fragilis* responds to clindamycin or metronidazole. If serous to semiserous, clear fluid is aspirated, tuberculous pleuritis is the most likely cause and an antituberculous triple regimen should be started. A bloody, serous to semiserous, odorless aspirate may be the sign of an amebic abscess. Metronidazole is then the drug of choice, but a bacterial superinfection may be present additionally (Gram's stain!).

Antibiotic treatment should be continued at least until 1 week after the patient has become afebrile. The institution of antituberculous treatment may result in a major improvement in patients who failed to respond, or responded only partially, to 2 weeks' initial therapy.

Supportive treatment of empyema consists in repeated aspirations of pus or, better, in closed chest tube drainage with underwater seal, physiotherapy, correction of anemia, and nutrition.

Table 13.11. Efficacy of further antibiotics (not on WHO list) in different types of pneumonia

	Lobar/segmental pneumonia	Interstitial, "atypical" pneumonia	Postinfluenza pneumonia	Measles pneumonia	Staphylococcal pneumonia	Aspiration pneumonia	Postoperative pneumonia	Acute abscess-forming pneumonia	Secondary pneumonia, hospital-acquired
Clindamycin/lincomycin	1a	2a	1a	1a	1a	!	2d	!	2d
Rifampicin	1b	1a	1b	1b	1b	1b	2d	1b	2d
Cephalosporins									
Oral	2a	Ø	1a	1a	1a	2a	2d	2a	2d
First-generation	1a	Ø	1a	1a	1	1a	1	1a	2a
Second-generation	1a	Ø	1a	1a	1a	1a	1a	1a	2a
Third-generation	1a, c	Ø	1a, c	1a	1a, c	1a	1a	1a	!e
Amoxicillin/clavulanic acid	1a, c	Ø	1a	1a	1c	1a	1a	1a	2a
Acylaminopenicillins	1c	Ø	2b	2b	2b	2b	2b	2b	2b, e
Fusidic acid	2b	Ø	2b	2b	1a, d	Ø	Ø	2d	Ø
Vancomycin	1a, c	Ø	2b	2b	1a	Ø	Ø	2d	Ø
Quinolones (ofloxacin, ciprofloxacin)	2b	2b	?	?	?	?	?	?	?
Imipenem	1c	Ø	1c	1c	1c	1a	1a	1a	1a

Key: see Table 13.10.

Table 13.12. European standard therapy of different types of pneumonia

Type of pneumonia	Drug of choice	Alternatives
Lobar/segmental pneumonia	Benzylpenicillin	Cefazolin, erythromycin
Postinfluenza pneumonia	Cefazolin	Cefuroxime, cefamandole, doxycycline + clindamycin
Interstitial, "Atypical" pneumonia	Doxycycline	Erythromycin, rifampicin
Acute abscessforming pneumonia	Cefoxitin	Clindamycin
Aspiration pneumonia	Cefoxitin	Clindamycin + cefotaxime
Postoperative pneumonia	Cefoxitin, cefuroxime	Cefotaxime + azlocillin
Secondary pneumonia (no prior antibiotics)	Cefoxitin, cefuroxime	Cefotaxime (+ gentamicin)
Secondary pneumonia (prior antibiotics)	Cefotaxime + azlocillin + gentamicin, cefotaxime + aminoglycoside	Imipenem (+ rifampicin)

Chronic Lung Abscess

The symptoms of chronic lung abscess - hemoptysis, foul-smelling sputum, weight loss, pyrexia, and anemia - are of long duration, weeks to months, in contrast to the acute presentation of an abscess-forming pneumonia. Chronic lung abscess is commonly seen in adults in the third to fifth decades of life. Predisposing factors are poor oral hygiene, betel chewing, aspiration of a foreign body, and alcoholism.

The diagnosis is made by exclusion of cavitating pulmonary tuberculosis, amebic lung abscess, and cavitated peripheral bronchial carcinoma, all of which can appear very similar radiologically.

Various aerobic bacteria, *Klebsiella pneumoniae, E. coli, Pseudomonas aeruginosa,* and *Proteus* spp., isolated from sputum or lung aspirate, represent oral contaminants or secondary invaders of doubtful significance. The majority of cases are due to mixed infection by *Bacteroides melaninogenicus* and other *Bacteroides* spp. together with peptostreptococci and certain other streptococci. A minority of cases are caused by *Staphylococcus aureus* [6].

Treatment is primarily medical, with postural drainage, physiotherapy, and antibiotics. Benzylpenicillin, covering the common anaerobic pathogens, is the drug of choice. Chloramphenicol and the widely used tetracyclines are also effective. If there is no response to benzylpenicillin, clindamycin should be used which covers anaerobic bacteria including *Bacteroides fragilis* and *Staphylococcus aureus.* The combination of penicillin with metronidazole constitutes an alternative to clindamycin (inactive against *Staphylococcus aureus* but active against *Entamoeba histolytica* and anaerobic bacteria).

In addition, rifampicin will produce good results in anaerobic and staphylococcal lung abscess if the condition is mistaken for tuberculosis.

A certain number of patients have to undergo lobectomy or pneumonectomy, often as emergency operation for massive hemoptysis. Lung abscess was the commonest cause of massive hemoptysis in Nigeria [1].

Bronchiectasis

Bronchiectasis is a relatively common complication of pneumonia in measles and whooping cough in developing countries [37]. Tuberculosis is another important cause of bronchiectasis (1).

The principles of treatment comprise postural drainage, antibiotic treatment of acute exacerbations, and, in selected patients, partial or total lung resection.

The efficacy of antibiotic treatment is, however, generally limited. Acute exacerbations are often caused by *Streptococcus pneumoniae* or *Hemophilus influenzae,* but mixed infections and infections with *Staphylococcus aureus,* gram-negative enteric bacilli, *Pseudomonas aeruginosa,* and anaerobic bacteria are more frequent than in acute exacerbations of chronic bronchitis. Guidance of antibiotic therapy by sputum culture is, therefore, preferable.

Initial treatment with ampicillin, co-trimoxazole, or tetracyclines is suggested, to be modified later according to sensitivity results. In the absence of facilities for anaerobic culture, the addition of metronidazole may be worthwhile in nonresponding cases. The use of metronidazole is advocated for patients with foul-smelling sputum. Since many antibiotics diffuse only poorly into bronchial secretions and pus, administration by inhalation presents an therapeutic alternative. Aminoglycosides like neomycin (200 mg/ml) or gentamicin (10 mg/ml), thiamphenicol (125 mg/ml), polymyxin B (10 mg/ml), or bacitracin (250 units/ml) are used for this purpose.

Long-term treatment with co-trimoxazole (regular blood counts!) or tetracyclines may be tried in severe cases.

Whooping Cough

Whooping cough continues to be a major cause of morbidity and mortality in developing countries [73]. It occurs at an earlier age than in industrialized countries. As in industrialized countries the highest case-fatality ratio, up to 10%, is observed in children less than 1 year old, and especially in infants under the age of 6 months. Like measles, whooping cough has a detrimental effect on the nutritional state of the child, and may precipitate protein-energy malnutrition in children with an already borderline nutritional state [46]. *Bordetella pertussis* accounts for the vast majority of cases in developing countries, *B. parapertussis* being of little importance in nonimmunized populations. (The vaccine confers immunity to *B. pertussis* only.)

The diagnosis of whooping cough is usually made clinically (paroxysms of cough terminating in the typical "whoop"). Small infants, however, present with attacks of apnea and cyanosis rather than with a typical whoop. The lymphocytosis which is characteristic of the late catarrhal and early paroxysmal state is also often not seen in infants less than 6 months old. The definitive diagnosis of whooping cough requires the isolation of *B. pertussis*, which is usually only carried out for research purposes.

Treatment

Antibiotics are of only limited value. They do not alter the clinical course of the disease, once the paroxysmal stage has been entered. However, erythromycin, ampicillin and tetracyclines (which should not be used in children less than 7 year old), can eliminate the organism from the respiratory secretions even in the paroxysmal stage, thereby rendering the patient noninfectious after 4–6 days. A 2-week course of erythromycin, as the least toxic agent, is often administered to patients in the USA and Western Europa.

Antibiotic treatment is needed when suppurative complications like otitis media and pneumonia supervene (onset of fever) which are frequently due to *Hemophilus influenzae* or *Streptococcus pneumoniae,* and less often due to *Staphylococcus aureus* and group A beta-hemolytic streptococci. Treatment with parenteral benzylpenicillin or chloramphenicol (in case of severely ill patients) is reasonable when resources are limited. Use of ampicillin, of cloxacillin in case of suspected staphylococcal bronchopneumonia, or of a second-generation cephalosporin which is active against all four aforementioned agents may be preferred depending upon availability.

Supportive measures include maintenance of adequate hydration and nutrition, prevention of aspiration in young infants, provision of supplemental oxygen where needed, and treatment of convulsions. Sedatives may be used. Cough suppressants like codeine may even aggravate the patchy atelectasis associated with whooping cough.

Prevention

Chemoprophylaxis with erythromycin (40 mg/kg/day) is usually offered to all exposed susceptible individuals in affluent communities. In developing countries, attention should be directed at all exposed susceptible infants and treatment should be started at the first sign of illness (coryza, mild conjunctivitis, fits of cough, or edema of the eyelids), or earlier, when possible. Treatment with erythromycin, ampicillin, or tetracyclines (if neither of the first two antibiotics is available) may still abort the illness or reduce its severity when started at the early catarrhal stage.

Prevention of whooping cough by *immunization* is clearly desirable and feasible. Since there is no protection of the young infant by maternal antibodies, vaccination should be started at an early age, as proposed in the WHO Expanded Programme on Immunization.

Diphtheria

Diphtheria continues to be an important cause of morbidity and mortality in developing countries. This is in contrast to the low incidence of diphtheria in industrialized countries, where immunization with diphtheria toxoid is widespread.

The causative organism, *Corynebacterium diphtheriae,* is an aerobic, pleomorphic, often club-shaped, gram-positive rod. Loeffler's medium (containing coagulated serum) and potassium tellurite containing media are used for isolation. The three biotypes of *C. diphtheriae* are distinguished by colony morphology on tellurite media and by biochemical characteristics. Strains are examined for toxigenicity by immunoprecipitation (Elek's plate) or by inoculation of guinea pigs. Toxigenic strains elaborate the exotoxin that is absorbed into the bloodstream and is responsible for the toxemia of diphtheria and the complications of myocarditis and paralysis.

Diphtheria presents as a life-threatening infection of the upper respiratory tract in many developing countries [13, 17, 23, 65]. Cutaneous diphtheria, particularly occurring as a secondary infection of septic skin lesions and of wounds, is common in several parts of the tropics (see Table 16.7 in the chapter "Skin and wound Infections"). Although cutaneous diphtheria is usually mild and not associated with major systemic toxicity, its epidemiological importance is twofold, firstly cutaneous carriage may serve as a major reservoir for transmission, and secondly, in the absence of immunization programmes, cutaneous infection results in acquisition of immunity by absorption of small amounts of toxin [32].

The diagnosis of tonsillar-pharyngeal diphtheria is made clinically, based upon the presence of a thick grayish-white membrane which is firmly adherent to the underlying mucosa. Swabs are taken from underneath the membrane, i.e., from the ulcer base, but treatment should never await bacteriological confirmation.

Treatment

Diphtheria antitoxin (DAT) is the mainstay of treatment. As with any equine serum, an intradermal and/or conjunctival test prior to the i.m. or i.v. administration is recommended. Desensitization may be required if either test proves to be positive. The dose of DAT depends upon the extent of the diphtheric membrane and the severity of the illness. A dose of 20000 units may be given if the membrane is still limited to the tonsil, while up to 80000 units are given in case of more extensive lesions.

Continuing production of toxin is prevented by antibiotics. Benzylpenicillin, 1.2–2.4 mega units (MU) daily in divided doses or procaine penicillin, 1.2 MU in adults and 0.6 MU in children, i.m. once daily for 7 days, is used. Patients sensitive to penicillin are given erythromycin (40 mg/kg/day in children, 1–2 g daily in adults). Further recommendations include strict bed rest for 2 weeks (risk of myocarditis), regular ECG (arrhythmias like atrioventricular block, left bundle-branch block), treatment with digitalis in case of congestive heart failure, use of steroid in cases with severe toxemia and shock or laryngeal obstruction, and gavage feeding in case of soft palate paralysis. Tracheostomy or intubation may be required in case of severe laryngeal obstruction (croup). Isolation procedures are followed until two consecutive cultures of throat and nose are negative for *C. diphtheriae.*

Immunization is necessary when the patient has recovered from the acute illness. Susceptible contacts of patients with diphtheria are treated with penicillin or erythromycin. Administration of a low dose of DAT has been suggested if regular follow-up is not possible.

Prevention

Immunization is the best way to reduce morbidity and mortality. Diphtheria toxoid is included in the Expanded Programme on Immunization that is promoted by WHO and UNICEF in developing countries.

References

1. Adebonojo SA, Grillo IA, Osinowo O, Adebo OA (1982) Suppurative diseases of the lung and pleura: a continuing challenge in developing countries. Ann Thorac Surg 33: 40-47
2. Agarwal BL, Agrawal R (1986) Rheumatic fever: clinical profile of the initial attack in India. Bull WHO 64: 573-578
3. Agarwal SC, Bardoloi JNS, Walia BNS, Jamdar S, Mahajan RC, Sehgal S (1975) Serological survey for antibodies to respiratory viruses in acute respiratory illnesses in infants and children in Chandigarh, Northern India. Indian J Pathol Microbiol 18: 132-137
4. Allen SC (1984) Lobar pneumonia in Northern Zambia: clinical study of 502 adult patients. Thorax 39: 612-616
5. Awunor-Renner C (1979) Length of antibiotic therapy in inpatients with primary pneumonias. Ann Trop Med Parasitol 73: 235-240
6. Ayyagari A, Pancholi VK, Goswami A, Malik SK, Agarwal KC (1981) Anaerobic bacteria in pleuropulmonary infections. Indian J Med Res 74: 809-814
7. Berkowitz FE (1983) Infections in children with severe protein-energy malnutrition. Ann Trop Pediatr 3: 79-83
8. Berman S, Duenas A, Bedoya A, Constain V, Leon S, Borrero I, Murphy J (1983) Acute lower respiratory tract illnesses in Cali, Colombia: a two-year ambulatory study. Pediatrics 71: 210-218
9. Berman S, McIntosh K (1985) Selective primary health care: strategies for control of disease in the developing world. XXI. Acute Respiratory Infections. Rev Infect Dis 7: 674-691
10. Brown KH, Gilman RH, Gaffar A, Alamgir SM, Strife JL, Kapikian AZ, Sack RB (1981) Infections associated with severe protein-energy malnutrition in hospitalized infants and children. Nutr Res 1: 33-46
11. Bulla A, Hitze KL (1978) Acute respiratory infections: a review. Bull WHO 56: 481-498
12. Chanock R, Chambon L, Chang W, Ferreira F, Gharpure P, Grant L, Hatem J, Imam I, Kalra S, Lim K, Madalengoitia J, Spence L, Teng P, Ferreira W (1967) WHO respiratory disease survey in children, a serological study. Bull WHO 37: 363-369
13. Chintu C, Bathirunathan N, Patel JM (1978) Diphtheria at the University Teaching hospital, Lusaka. East Afr Med J 55: 36-38
14. Chretien J, Holland W, Macklem P, Murray J, Woolcock A (1984) Acute respiratory infections in children. A global public health problem. N Engl J Med 310: 982-984
15. Cohen S, Goldstein E (1983) Pneumonia in the hospital. In: Kass EH, Platt R (eds) Current therapy in infectious disease 1983-1984. Decker, Philadelphia p 75
16. Collings DA, Martin KS (1985) A retrospective analysis of childhood pneumonia in a district hospital. Centr Afr J Med 31 152-155
17. Denis F, Prince-David M, M'Boup S, Diop Mar I (1983) Antibiotic resistance of Corynebacterium diphtheriae in West Africa. J Infect 7: 279-281
18. DeBuse PJ, Lewis MG (1970) Pulmonary complications of measles in Uganda. J Trop Pediatr 16: 197-203
19. Diakparomre MA, Obi JO (1981) Aetiological diagnosis of pneumonia in childhood by lung puncture. Niger J Paediatr 8: 61-64
20. Diallo AA, Silverman M, Egler LJ (1979) Bacteriology of lung puncture aspirates in malnourished children in Zaria. Niger Med J 9: 421-423
21. Douglas RM, Riley ID (1971) Treatment of pneumonia in New Guinea. A controlled trial of crystalline penicillin and procaine penicillin aluminium monostearate. Med J Aust 1: 1230-1233
22. Douglas RM, Devitt L (1973) Bacteriological findings in 632 adults with particular reference to *Haemophilus influenzae*. Med J Aust 1: 42-49
23. Dutta JK, Ayyagari A, Gautam AP, Chadha SK, Ray SN (1976) A comparative study of bacteriologically proved and clinically diagnosed (culture negative) cases of diphtheria. J Indian Med Assoc 67: 241-245
24. Eigner FD (1981) Cough and fever in children. Trop Doct 11: 31
25. Escobar JA, Dover AS, Duenas A, Leal E, Medina P, Arguello A, de Gaiter M, Greer DL, Spillman R, Reyes MA (1976) Etiology of respiratory tract infections in children in Cali, Colombia. Pediatrics 57: 123-130

26. Fadahunsi HO (1981) Rheumatic fever in Nigeria children: a prospective study. Ann Trop Pediatr 1: 115–118
27. Ferreira OS, Carvalho MRdC (1980) Estudo comparativo entre os resultados das culturas do material colhido pelo "swab" faringeo e pela punção pulmonar, em 58 crianças portadoras de pneumonia aguda. J Pediatr 48: 211–215
28. Forbes CE, Scheifele DW (1973) Management of measles at Kenyatta hospital, Nairobi. East Afr Med J 50: 159–161
29. Grant JP (1984) The state of the world's children 1984. Oxford University Press, New York
30. Gratten M, Naraqi S, Hansman D (1980) High prevalence of penicillin-insensitive pneumococci in Port Moresby, Papua New Guinea. Lancet II: 192–195
31. Gupta U, Vaishanava S, Gulati AK (1975) *Mycoplasma pneumoniae* infections in infants and children. Indian J Med Res 63: 385–389
32. Hewlett EL (1985) Selective primary health care: strategies for control of disease in the developing world. XVIII: Pertussis and diphtheria. Rev Infect Dis 7: 426–433
33. Hughes WT, Price RA, Sisko F, Havron WS, Kafatos AG, Schonland M, Smythe PM (1974) Protein-calorie malnutrition. A host determinant for *Pneumocystis carinii* infection. Am J Dis Child 128: 44–52
34. Hughes JR, Sinha DP, Cooper MR, Shah KV, Bose SK (1969) Lung tap in childhood. Bacteria, viruses, and mycoplasmas in acute lower respiratory tract infections. Pediatrics 44: 477–485
35. James JW (1972) Longitudinal study of the morbidity of diarrhoeal and respiratory studies in malnourished children. Am J Clin Nutr 25: 690–694
36. Kalra SK, Sasidharan T, Vatwani V, Sarkar P (1981) Lung puncture: a diagnostic aid in childhood pneumonia. Indian Pediatr 18: 727–730
37. Kaschula RO, Druker J, Kipps A (1983) Late morphologic consequences of measles: a lethal and debilitating lung disease among the poor. Rev Infect Dis 5: 395–404
38. Khan AA, Gupta BM, Olonga J, Maguire MJ (1978) Empyema in children in tropics. Acta Paediatr Scand 67: 731–734
39. Kucers A, Bennett NMcK (1979) The use of antibiotics, 3rd edn. Heinemann, London, p 12 und p 57
40. Kumar L, Ayyagari A (1984) The etiology of lobar pneumonia and empyema thoracis in children. Indian Pediatr 21: 133–138
41. Macfarlane JT, Adegboya DS, Warrell MJ (1979) *Mycoplasma pneumoniae* and the aetiology of lobar pneumonia in Northern Nigeria. Thorax 34: 713–719
42. Markowitz M (1981) Observations on the epidemiology and preventability of rheumatic fever in developing countries. Clin Ther 4: 240–251
43. McCord C, Kielmann AA (1978) A successful programme for medical auxiliaries treating childhood diarrhoea and pneumonia. Trop Doct 8: 220–225
44. Mimica I, Donoso E, Howard JE, Ledermann GW (1971) Lung puncture in the etiological diagnosis of pneumonia. A study of 543 infants and children. Am J Dis Child 122: 278–282
45. Morton R, Mee J (1986) Measles pneumonia: lung puncture findings in 56 cases related to chest X-ray changes and clinical features. Ann Trop Pediatr 6: 41–46
46. Morley D (1973) Paediatric priorities in the developing world. Butterworth, London, p 231–247
47. Nagpal VP, Harries JR (1974) Single-dose doxycycline in the treatment of lobar pneumonia: comparative trial against procaine penicillin. East Afr Med J 51: 778–781
48. O'Donovan C, Barua KN (1973) Measles pneumonia. Am J Trop Med Hyg 22: 73–77
49. Ong SB, Lam KL, Lam SK (1982) Viral agents of acute respiratory infections in young children in Kuala Lumpur. Bull WHO 60: 137–140
50. Puri RK, Khanna KK (1977) Bacterial pneumonias in infancy and childhood. Indian J Pediatr 44: 150–156
51. Rée GH, Davis M (1983) Treatment of lobar pneumonia in Papua New Guinea: short course chemotherapy with penicillin or chloramphenicol. J Infect 6: 29–32
52. Riley ID, Alpers MP, Gratten H, Lehmann D, De C Marshall TF, Smith D (1986) Pneumococcal vaccine prevents death from acute lower respiratory tract infections in Papua New Guinea. Lancet II: 877–881

53. Riley I, Carrad E, Gratten H, Gratten M, Lovuru K, Philipps P, Pratt D, Rose A, Siwi H, Smith D (1983) The status of research on acute respiratory infections in children in Papua New Guinea. Pediatr Res 17: 1041-1043
54. Rolfe M (1986) A study of Legionnaire's disease in Zambia. Ann Trop Med Parasitol 80: 325-328
55. Rozov T, Sakane PT, Carneiro Novaes JR, Gelli DS, Takimoto S, Senaga S (1974) Contribuição ao estudo da etiologia das pneumopatias da infancia, por meio sa aspiração pulmonar transcutanea. Pediatr Pract 45: 221-232
56. Santhanakrishnan BR, Thirumoothy MC, Raju VB (1972) Empyema in children. Indian Pediatr 9: 805-811
57. Schuster AC, Pino MC, Neira MS, Vildosola CSM, Faíní LM (1966) La punción biopsia pulmonar como metodo diagnostico de las neumopatias de la infancia. Pediatria 9: 9-12
58. Scrimgeour EM, Kaven J (1982) Severe staphylococcal pneumonia complicating pyomyositis. Am J Trop Med Hyg 31: 822-826
59. Shakunthala KSV, Mallikarjuna RG, Urmila S (1978) Diagnostic lung puncture aspiration in acute pneumonia of newborn. Indian Pediatr 15: 39-44
60. Shann F, Germer S (1979) Childhood pneumonia at Goroka Hospital. Papua New Guinea Med J 22: 72-75
61. Shann F, Barker J, Poore P (1985) Chloramphenicol alone versus chloramphenicol plus penicillin for severe pneumonia in children. Lancet II: 684-686
62. Shann F, Germer S, Hazlett D, Gratten M, Linnemann V, Payne R (1984) Aetiology of pneumonia in children in Goroka hospital, Papua New Guinea. Lancet II: 537-541
63. Sibellas M (1966) Treatment of lobar pneumonia in out-patients. J Trop Med Hyg 69: 94-96
64. Silvermann M, Stratton D, Diallo A, Egler LJ (1977) Diagnosis of acute bacterial pneumonia in Nigerian children. Value of needle aspiration of lung and of countercurrent immunoelectrophoresis. Arch Dis Child 52: 925-931
65. Singh M, Saidali A, Bakhtiar A, Arya LS (1985) Diphtheria in Afghanistan - a review of 155 cases. J Trop Med Hyg 88: 373-376
66. Slack RCB, Stewart JD, Lewis CLH, Cameron DI, Carvalho GR, Mangalji Z, Lowry PR, Wanjala S (1976) Acute pneumonias in adults in Nairobi. East Afr Med J 53: 480-483
67. Smith D, Riley I, Etao E, et al. (1982) Patterns of morbidity, mortality and health services utilization associated with acute respiratory infection, Asaro Valley, 1980-1981. Proceedings 18th annual symposium Medical Society of Papua New Guinea, Port Moresby. In: Shann F, Germer S, Hazlett D, Grattey M, Linnemann V, Payne P (1984) Aetiology of pneumonia in children in Garoka Hospital, Papua New Guinea. Lancet II: 537-541
68. Sobelavsky O, Sebikari SRK, Harland PSEG, Skrtic N, Fayinka OA, Soneji AD (1977) The viral etiology of acute respiratory infections in children in Uganda. Bull WHO 55: 625-631
69. Sutton DR, Wicks ACB, Davidson L (1970) One-day treatment for lobar pneumonia. Thorax 25: 241-244
70. Steinhoff MC, John TJ (1983) Acute respiratory infections in children in India. Pediatr Res 17: 1032-1035
71. Wafula EM, Tukei PM, Bell TM, Nsanze H, Ndinya-Achola JO, Haziett TG, Ademba GR, Pamba A (1984) How should primary health workers diagnose and treat acute respiratory infections (ARI)? East Afr Med J 61: 736-744
72. Wall RA, Corrah PT, Mabey DCW, Greenwood BM (1986) The etiology of lobar pneumonia in The Gambia. Bull WHO 64: 553-558
73. Walsh JA, Warren KS (1979) Selective primary health care. An interim strategy for disease control in developing countries. N Engl J Med 301: 967-974
74. Warrell DA (1976) Respiratory tract infections in the tropics. Practitioner 215: 740-744
75. Watt P (1984) Acute lower respiratory tract infections in children admitted to Port Moresby General Hospital. Papua New Guinea Med J 27: 95-102
76. Weinstein L (1955) Failure of chemotherapy to prevent the bacterial complications of measles. N Engl J Med 253: 679-683
77. Wesley AG, Buchanan N, Cane RD (1982) Laryngotracheobronchitis: the profile and management in developing countries (a review). J Trop Pediatr 28: 81-84
78. WHO Memorandum (1981) Clinical management of acute respiratory infections in children. Bull WHO 59: 707-716

14 Infections of the Gastrointestinal Tract

Diarrheal Diseases

Special Aspects in Developing Countries

In developing countries, diarrheal diseases constitute one of the leading causes of morbidity and mortality in children under 5 years of age.

Snyder and Merson [58] who analyzed active surveillance data from longitudinal, prospective, community-based studies, estimated the morbidity and mortality in 1980 from diarrheal disease in children under 5 years of age in Latin America, Asia (excluding China), and Africa, to be as high as 794–1000 million episodes, with 4.6 million deaths. Within this age group, younger children are more vulnerable (Figs. 14.1 and 14.2).

Malnutrition, lack of safe water supply, poor sanitary and domestic hygiene, and crowded housing conditions are all among the factors contributing to the morbidity and mortality from diarrheal diseases in developing countries, whereas climate is not. There were no differences in morbidity rates between the areas studied in the temperate zone (all in Latin America) and those in the tropical zone (all in Africa and Asia) [58]. Not surprisingly, salmonella carrier rates reflecting the general standard of environmental and domestic sanitation are high in developing countries (Table 14.1).

Early age of weaning, bottle- instead of breast-feeding, and measles are special risk factors for diarrheal diseases in young children.

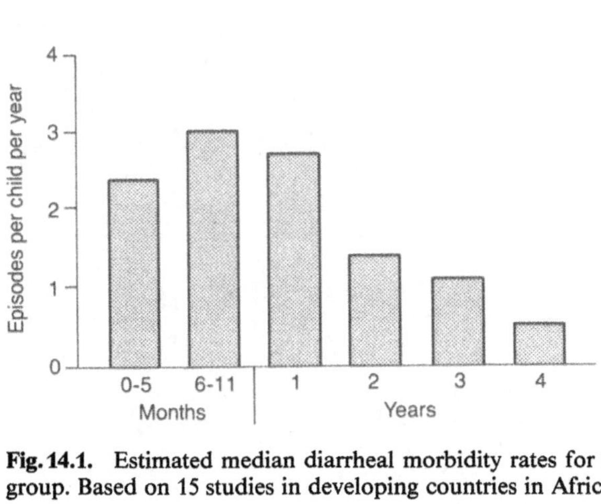

Fig. 14.1. Estimated median diarrheal morbidity rates for children under 5 years of age, by age group. Based on 15 studies in developing countries in Africa, Asia, and Latin America. From [58]

79. WHO Technical Advisory Group (1984) A programme for controlling acute respiratory infections in children: memorandum from a WHO meeting. Bull WHO 62: 47-58
80. Anonymous (1987) Global mortality from acute respiratory infections among children aged below 5 years. Bull WHO 65: 114-116
81. Datta N, Kumar V, Kumar L, Singhi S (1987) Application of case management to the control of acute respiratory infections in low birth weight infants: a feasibility study. Bull WHO 65: 77-82

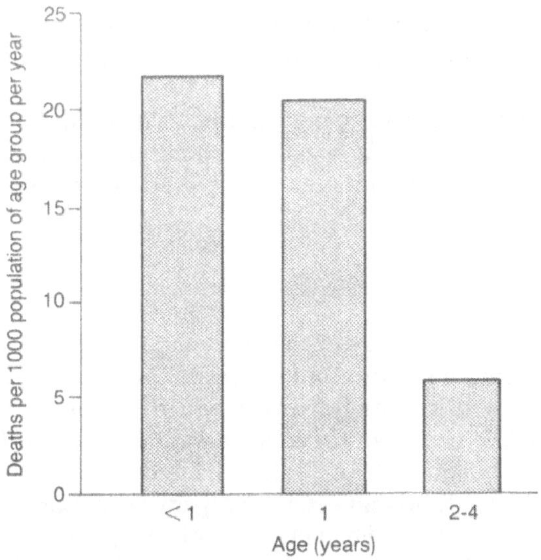

Fig. 14.2. Estimated median diarrheal mortality rates for children under 5 years of age. Based on 12 studies with active surveillance in developing countries in Asia and Latin America. From [58]

Table 14.1. Salmonella and shigella carrier rates

Country [Reference]	Togo [7]	Ivory Coast [14]	Indonesia [22]	Thailand [51]
No. of persons investigated	5075[a]	1732	464	931
Type of area	Rural (13 villages in different climatic zones)	Suburban (four villages at the outskirts of Abidjan)	Urban (central Jakarta)	Rural (five villages in central Thailand)
Carrier rates				
All *Salmonella* spp. (%)	6.6 (2–19.8)	5.9	8.4	3.3
Salmonella typhi (%)	0.1	0.3	0	0
Shigella spp. (%)	0.6	0.2	ND	0.8

ND, no data.
[a] Each village was visited twice, once during the rainy and once during the dry season; 6051 stool specimens were examined. Carrier rates were calculated from the number of positive stool specimens.

The negative nutritional impact of diarrheal diseases, acute weight loss, growth retardation, and malnutrition have long been recognized; diarrheal episodes are frequently associated with anorexia and nutrient malabsorption.

In 1978, WHO mounted a major program for the control of diarrheal diseases. Particular emphasis was put on the early use of oral rehydration therapy and appropriate feeding during the diarrheal episode and convalescence in an effort to reduce the diarrheal disease mortality in young children. Whereas the efficacy of oral rehydration therapy is recognized, delivery of it remains a problem, for a variety of reasons. For one thing, prepacked glucose or sucrose electrolyte solutions will not reach a considerable proportion of rural people through existing health systems in developing countries. Alternative strategies, (e.g. oral rehydration therapy with simple sugar-salt solutions prepared by village health workers, home-based oral rehydration therapy) are, therefore, being investigated [11, 33].

Etiology

The largest study of the etiology of endemic diarrhea in developing countries was conducted in rural Bangladesh. The results are shown in Table 14.2. The data of a study of pediatric diarrhea, which was undertaken at Bangui, Central African Republic, are given for comparison (Table 14.3). The findings of several other studies are summarized in Tables 14.4 and 14.5.

Length of study (seasonal pattern of diarrhea caused by rotavirus or shigellae), type of study (hospitalized patients, patients coming to a diarrhea treatment center, or long-term surveillance for episodes of diarrhea), age (infants, older children, or adults), epidemiological differences, and the diagnostic methods employed all influence the results. Investigation of asymptomatic controls often reveals asymptomatic infection or a carrier state. Multiple infections occur in a proportion of cases.

Table 14.2. Percentage of patients with diarrhea associated with enteric pathogens by age in the periods February 1977–January 1978 (year 1) and February 1978–January 1979 (year 2)

Pathogen identified	Age group, study year (no. of patients)							
	<2 years		2-9 years		⩾10 years		All ages	
	1 (3500)	2 (2614)	1 (1815)	2 (1391)	1 (2824)	2 (2347)	1 (8139)	2 (6352)
Enterotoxigenic *Escherichia coli*								
ST	12	17	12	15	16	22	14	18
ST/LT	6	7	6	7	14	19	9	11
LT	5	4	3	1	3	2	4	2
Total	23	28	22	24	33	42	26	32
Rotavirus	0	46	0	12	0	9	0	24
Vibrio cholerae	2	2	29	34	14	14	12	14
Non-group 0:1 vibrios	8	5	6	4	8	11	7	7
Shigella spp.	4	5	8	10	4	5	5	6
Entamoeba histolytica	1	0	13	0	8	0	4	0
Giardia lamblia	<1	0	4	0	4	0	2	0
Group F vibrio-like organisms	4	<1	2	<1	2	<1	3	<1
Salmonella spp.	<1	<1	<1	<1	<1	<1	<1	<1
Vibrio parahaemolyticus	0	<1	0	0	<1	<1	<1	<1

ST, strain producing heat-stable enterotoxin; LT, strain producing heat-labile enterotoxin.
This study was conducted at a diarrhea treatment center (Matlab) in rural Bangladesh. From [5]

Table 14.3. Enteropathogens associated with diarrhea in children according to age group (March 1981-February 1982)

Pathogen identified	% of diarrheic children in following age group from whom enteropathogen was isolated:						
	0-5 months (*n* = 274)	6-11 months (*n* = 298)	12-17 months (*n* = 231)	18-23 months (*n* = 124)	2-5 years (*n* = 203)	6-15 years (*n* = 67)	Total (*n* = 1197)
Shigella spp.	1	3	3.4	4	3.4	1.4	2.7
Salmonella spp.	2.5	2.3	4.7	7.2	1.9	4.4	3.4
Salmonella typhi	0	0.3	0	0.8	3.9	10.4	1.4
Enteropathogenic *E. coli*	14.5	12.7	8.2	12.9	NI	NI	12.1
Campylobacter jejuni	9.1	14	13.4	8	9.3	2.9	10.7
Enterotoxigenic *E. coli*							
ST	0.8	1.6	1.1	0	1.2	0	1
LT	1.9	1.6	1.5	2.1	0	0	1.5
ST and LT	0	3.2	0	0	0	4	0.9
Entamoeba histolytica	1	2	2.5	8	12.8	22.3	5.5
Giardia lamblia	1	0.6	5.1	4	6.4	2.9	3.9
Schistosoma mansoni	0	0	0	0	1.4	17.9	1.2
Trichuris trichiura	0	0	0.4	3.2	0.9	4.4	0.8
Strongyloides stercoralis	0	0	1.2	2.4	3.4	7.4	1.5
Rotavirus	27.7	28.1	15.1	6.4	3.9	0	17.6
Unknown etiology	40.5	30.6	43.4	41	51.5	22	35.8

NI, Not investigated; ST, strain producing heat-stable enterotoxin; LT, strain producing heat-labile enterotoxin.
This study was conducted at the National University Hospital of Bangui and an urban dispensary at Bangui, Central African Republic; 746 of the children investigated were hospitalized at the National University Hospital. From [19]

Rotavirus and enterotoxin-producing strains of *E. coli* have been recognized as major causes of infantile diarrhea in developing countries during the past decade. They are more common than shigellae, salmonellae, or *V. cholerae* 01 in infants.

The role that enteropathogenic strains of *E. coli* play in the etiology of pediatriac diarrhea is still being discussed controversially. These strains were, however, isolated with considerable frequency in some studies, e.g., those from the Central African Republic and Kenya (Tables 14.3 and 14.5).

The importance of another group of viruses, Norwalk and Norwalk-like agents, as a cause of diarrhea in developing countries has not yet been determined [6].

Campylobacter jejuni has been isolated quite frequently both from patients with diarrhea and from healthy persons (Tables 14.3-14.6) [21, 47]. Other rarer bacterial pathogens are *Yersinia enterocolitica, Vibrio parahaemolyticus, Vibrio cholerae* non-O1, *Aeromonas hydrophila, Plesiomonas shigelloides,* and clostridia [24, 28, 45].

Entamoeba histolytica and *Giardia lamblia* usually cause a more chronic diarrhea than the viruses and bacteria mentioned already. *Cryptosporidium* infection has recently been recognized as a further cause of acute, watery diarrhea in infants and children in developing countries [62, 67].

Table 14.4. Enteric pathogens isolated in children and adults (%)

Study location [Reference]	Calcutta, India [55]	Dacca, Bangladesh [60]	Pacatuba, Northeastern Brazil [23]
Design of study	Hospital inpatients with acute diarrhea	Patients with diarrhea (65% outpatients, 4% inpatient ward, 31% i.v. treatment center)	Surveillance for episodes of diarrhea of approx. 300 people living in a rural village district
Duration of study	27 months	12 months	30 months
Patients (*n*)	356	3550	150 (of 1097 episodes of diarrhea)
Age (%)			
<1 year	4	25	18% of people surveyed were under 5 years of age
1-2 years	7	27	
>14 years	47	26	
Pathogen identified	% (*n* positive/ *n* studied)	% (*n* studied)	% (*n* positive/ *n* studied)
Enterotoxigenic *E. coli*	6	20 (3123)	21 (31/149)
Rotavirus	8	19 (3280)	19 (24/124)
Shigella spp.	2	12 (3550)	8 (12/149)
Campylobacter jejuni	5 (2/40)	14 (3038)	8 (3/40)
Vibrio cholerae 01	30	6 (3550)	0
Salmonella spp.	0	1 (3550)	0
Yersinia enterocolitica	0	NLF	NLF
Enteroinvasive *E. coli*	0	NLF	2 (3/149)
Enteropathogenic *E. coli*	6	NLF	5 (6/130)
Vibrio parahaemolyticus	6	0	0
Entamoeba histolytica	NLF	6 (2246)	2 (3/150)
Giardia lamblia	NLF	6 (2246)	7 (10/150)
Strongyloides stercoralis	NLF	ND	5 (8/150)
Multiple infections	9	ND	17 (26/150)

NLF, not looked for; ND, no data.

Diagnosis

A pathogen can be identified in up to 70%-80% of diarrhea episodes when all existing sophisticated laboratory methods are used (Table 14.6). However, even if these methods are available, the results are not present at the time when the patient is first seen.

Regardless of the etiology, assessment of the degree of dehydration (Tables 14.7 and 14.8) and initiation of rehydration therapy are the most important means for prevention of death from acute dehydrating diarrheas, especially in young children. Severe dehydration (>10% of body weight) developing within hours after the onset of symptoms is characteristic for cholera, but is also seen in some cases caused by enterotoxigenic strains of *E. coli*.

Cultures are not necessary in mild, self-limiting diarrhea. Inspection of a stool

Table 14.5. Enteric pathogens isolated in children with diarrhea (%)

Study location [Reference]	Vellore, India [35]		Bangkok, Thailand [29]		Yogyakarta, Indonesia [59]		Kivu, Zaire [10]				Kenya [40]			Mexico City, Mexico [43]	
Duration of study	12 months		2.5 months		12 months		1 month				12 months			22 months	
Children (*n*)	Hospitalized		Hospitalized		Hospitalized		Outpatients		Hospitalized[c]		Hospi-talized[d]	Hospitalized and outpatients		Outpatients	
	D[a]	C	D	C	D	C	D	C	D	C	D	D[e]	D[f]	D	C
	50	30	105	90	338	69	271	203	84	17	160	104	116	340	98
Age (%)															
<2 years	100%		All aged 0–6 years		87%	ND	54%	26%	ND	ND	All aged 0–5 years			94%	72%
2–4 years							18%	22%							
>4 years							28%	44%							
Pathogen identified															
Rotavirus	26	0	23	1	38	7[b]	4	0	0	0	41	24	16	17	3
Enterotoxigenic *E. coli*	ND	ND	18	9	12	8[b]	10	2	5	7	ND	ND	ND	7	4
Shigella spp.	6	0	9	1	4	2	1	1	4	0	11	7	17	14	0
Salmonella spp.	8	0	6	0	6	0	3	1	2	2	3	2	1	12	0
Enteropathogenic *E. coli*	50	30	ND	ND	ND	ND	4	4	ND	ND	18	8	15	ND	ND
Campylobacter jejuni	ND	ND	4	2	ND	ND	14	3	24	9	ND	ND	ND	ND	ND
Yersinia enterocolitica	ND	ND	0	1	ND	ND	1	0	0	0	ND	ND	ND	ND	ND
Vibrio cholerae 01	4	0	2	0	2	0	5	<1	12	0	ND	ND	ND	0	0
Giardia lamblia	0	0	ND	ND	<1	0	4	ND	10	1	ND	ND	ND	1	0
Entamoeba histolytica	0	0	ND	ND	<1	4	2	ND	1	0	ND	ND	ND	2	0
Multiple infections	14	0	11	1	8	0	ND	ND	ND	ND	ND	ND	ND	5	0

D, Children with diarrhea; C, controls (asymptomatic); ND, not looked for, no data.
[a] Children with dysentery were excluded from the study.
[b] These children all subsequently developed diarrhea.
[c] With severe malnutrition.
[d] Nairobi.
[e] Kisumu; surveys done in Jan., May, and Nov. for 1 year.
[f] Mombasa; surveys done in Jan., May, and Nov. for 1 year.

Table 14.6. Common microbial pathogens of diarrheal disease. Modified from [44, 50]

Pathogen	Usual type of diarrhea	Site of colonization	Pathogenesis	Methods of routine diagnosis
Rotavirus	Watery diarrhea	Small intestine	Invasion	Electron microscopy; ELISA
Norwalk and Norwalk-like agents	Diarrhea	Small intestine	Invasion	Electron microscopy
E. coli				
Enterotoxigenic	Watery diarrhea	Small intestine	Enterotoxin (LT and/or ST)	Detection of enterotoxin by tissue culture (LT), rabbit ileal loop assay (LT), and suckling mouse assay (ST)
Enteroinvasive	Dysentery	Large intestine	Invasion	Sereny test[b]
Enteropathogenic	Diarrhea	Small intestine	?	Serotyping
Salmonella spp. (other than *S. typhi* and *S. paratyphi*)	Diarrhea[a], less commonly, dysentery	Small and large intestine	Invasion and ? enterotoxin	Culture (stool-selective media)[b]
Shigella spp.	Diarrhea; more commonly, dysentery	Large intestine	Invasion and ? enterotoxin	Culture (stool-selective media)[b]
Vibrio cholerae O1	Watery diarrhea	Small intestine	Enterotoxin	Culture (TCBS agar)
V. cholerae non-O1	Diarrhea	Small intestine	Enterotoxin	Culture (TCBS agar)
V. parahaemolyticus	Diarrhea[a], dysentery	Small intestine	Invasion and ? enterotoxin	Culture (TCBS agar)
Aeromonas hydrophila	Diarrhea	Unknown	? enterotoxin	Culture (stool-selective media)
Campylobacter jejuni	Diarrhea[a], dysentery	Large intestine	Invasion	Culture (special culture conditions)[b]
Yersinia enterocolitica	Diarrhea	Large intestine	Invasion and ? enterotoxin	Culture (cold enrichment)
Staphylococcus aureus	Diarrhea[a]	Do not colonize	Enterotoxin	Isolation from food necessary
Bacillus cereus	Diarrhea[a]	Small intestine	Enterotoxin	Isolation from food necessary
Clostridium difficile	Diarrhea, dysentery	Large intestine	Enterotoxin	Detection of enterotoxin by tissue culture[b]
Clostridium perfringens				
Type A (food poisoning strains)	Diarrhea[a]	Small intestine	Enterotoxin	Isolation from food necessary
Type C (pigbel strains)	Enteritis necroticans	Small intestine	B toxin	Culture (stool-selective media)

ST, strains producing heat-stable enterotoxin; LT strains producing heat labile enterotoxin; ELISA, enzyme-linked immunosorbant assay.

[a] Agents of food poisoning.

[b] Fecal leukocytosis is commonly present.

Table 14.7. Assessment of dehydration and of fluid deficit in **infants.** From [66]

Signs and symptoms	Degree of dehydration (% loss of body weight)		
	Mild[a] (4%-5%)	Moderate (6%-9%)	Severe, with or without shock (⩾10%)
Appearance	Thirsty, alert, restless	Thirsty; restless, or lethargic but irritable when touched, or drowsy	Drowsy; limp, cold, sweaty, and often cyanotic extremities; may be comatose
Skin elasticity	Normal	Decreased - mild to moderate (+ to ++)	Marked decrease (+++)
Eyes	Normal	Sunken (detectable)	Grossly sunken
Tears	Present	Absent	Absent
Anterior fontanelle (when open)	Normal	Depressed	Depressed
Mucous membranes	Moist (±)	Dry	Very dry
Radial pulse	Normal (good volume)	Rapid/weak	Rapid, feeble, on rare occasions impalpable
Fever (rectal temperature)	Nil (±)	Present	Present
Urine flow	Normal	No urine for several hours, or small quantity of dark urine passed	No urine for several hours (⩾8 h usually)
Respiration	Normal	Deep (±), rapid	Deep/rapid
Heart (auscultation)	Normal	Tachycardia	Tachycardia, feeble
Estimated fluid deficit	40-50 ml/kg (e.g., 240-300 ml in a 6-kg child)	60-90 ml/kg (e.g., 360-540 ml in a 6-kg child)	100-120 ml/kg (e.g., 600-720 ml in a 6-kg child)

[a] A baby with mild dehydration may not have any obvious physical signs and may present only with a history of watery diarrhea and intense thirst.

specimen and microscopy for detection of fecal leukocytosis and parasites are useful screening methods.

Acute watery diarrhea without blood is commonly caused by enterotoxins produced by *E. coli* or *V. cholerae* and by viral agents (rotavirus in young children!), less commonly so by salmonellae or *G. lamblia*. Early in the course of the disease, shigellae infection may also produce watery diarrhea.

Acute dysentery, with blood and mucus in the stool, fever, abdominal cramps, and tenesmus, is commonly due to shigellae, but salmonellae, *Campylobacter jejuni*, and rarer pathogens, such as enteroinvasive *E. coli* and *V. parahaemolyticus*, may produce a similar clinical syndrome.

Chronic or recurrent bloody diarrhea without marked constitutional symptoms is typical of amebic dysentery, but ulcerative colitis or subacute shigellosis may present in the same way.

Table 14.8. Assessment of the degree of dehydration and amount of fluid deficit in **older children and adults.** From [66]

Signs and symptoms	Degree of dehydration (% loss of body weight)		
	Mild (4%-5%)	Moderate[a] (6%-9%)	Severe[a] (≥10%)
Appearance	Alert, restless, thirsty	Alert, postural hypotension, thirsty	Usually conscious, apprehensive, with cold, sweaty, often cyanotic extremities, wrinkled skin of fingers and toes, muscle cramps
Skin elasticity	Normal	Decreased (+ to ++)	Marked loss (+++)
Eyes	Normal	Sunken (detectable)	Grossly sunken
Mucous membranes	Moist ±	Dry	Very dry
Radial pulse	Normal (good volume)	Rapid/weak	Feeble or even impalpable
Systolic blood pressure	Normal	Normal→low	<90 mm systolic, often unrecordable
Urine flow	Normal	May or may not be diminished	No urine passed for several hours, and bladder empty
Respiration	Normal	Deep, ± rapid	Deep/rapid
Estimated fluid deficit	40-50 ml per kg	60-90 ml per kg	100-110 ml per kg

[a] Cholera and cholera-like diarrhea may lead to moderate or severe dehydration within several hours of onset of watery diarrhea. Signs of shock, e.g., cold sweaty extremities, weak and rapid or impalpable radial pulse, very low or unrecordable blood pressure, are common in severe cases.

In contrast, chronic diarrhea with frothy, foul-smelling stools containing no blood, should lead one to consider *Giardia lamblia* infection apart from other causes of a malabsorption syndrome.

Fecal leukocytosis (methylene blue stain) is indicative of infection with invasive bacteria like shigellae, salmonellae, and *Campylobacter jejuni,* which produce inflammatory changes of the bowel wall. Culture is useful in these cases, since some of them may require chemotherapy. Antibiotic-associated colitis (due to *C. difficile*), ulcerative colitis, and Crohn's disease (which are rare in the tropics) are other causes of fecal leukocytosis.

Microscopy may also detect cysts, oocysts, and trophozoites of *G. lamblia, Cryptosporidium,* and *E. histolytica.*

Dark-field microscopy has been used for the rapid, presumptive diagnosis of cholera and of *Campylobacter jejuni* infection.

Management

Rehydration

Rehydration is the essential component in the treatment of acute diarrheal diseases.

During the past decade, oral rehydration therapy has been established as the treatment for mild and moderate dehydration [25, 53]. Use of i.v. fluids is indicated for treatment of severe dehydration with shock, where immediate resuscitation is necessary, and for treatment failures of oral rehydration therapy (fluid loss continues at a rate faster than the oral intake).

WHO promotes a standard formulation of oral rehydration solution (ORS) for all age groups including infants (Table 14.9).

Guidelines on the use of ORS for rehydration and maintenance therapy as recommended by WHO are given in Tables 14.10 and 14.11.

Standard ORS and plain water in a ratio of 2:1 has been shown to be an effective regimen for treatment of dehydrating diarrhea in neonates [46].

Substitution of glucose by other carbohydrates may be more practical for reasons of cost, stability and availability. Treatment with ORS containing sucrose (40 g/l) or rice powder (30 g/l) was just as successful as standard ORS [39, 49].

Some pediatricians in developed countries are concerned about the comparatively high sodium content of WHO ORS (risk of hypernatremia in infants). Early feeding with tolerated foods and fluids, however, recommended to counteract the deleterious nutritional impact of diarrheal episodes (Fig. 14.3), provides a source of free water. If feeding has been discontinued, it should be reintroduced as soon as the child wishes to eat, usually when rehydration is complete. During maintenance therapy, ORS is given to replace stool losses volume by volume, whereas the usual daily fluid requirement may be met by continued breast-feeding, or half-strength milk formula and plain water in non-breast-fed infants. Furthermore, hypernatremia in children with acute diarrhea is less common in developing than in developed countries [25] and, in a recent study from Bangladesh [52] hyponatremic diarrhea was associated with a higher mortality than hypernatremic diarrhea.

Table 14.9. WHO oral rehydration solution (which is made up to 1 liter)

Standard formula		Improved formula[a]	
Glucose	20.0 g	Glucose	20.0 g
Sodium chloride	3.5 g	Sodium chloride	3.5 g
Sodium bicarbonate	2.5 g	Trisodium citrate dihydrate	2.9 g
Potassium chloride	1.5 g	Potassium chloride	1.5 g

[a] Trisodium citrate dihydrate is as effective as sodium bicarbonate for correction of acidosis [12]. For this reason, and because of the additional advantage of improved stability with longer shelf life, WHO and UNICEF now recommend the production and use of oral rehydration salts containing 2.9 g trisodium citrate dihydrate instead of 2.5 g sodium bicarbonate where feasible [12].

Table 14.10. Guide for rehydration therapy in **infants.** From [66]

Admission weight (approx. age)	Degree of dehydration and type and amount of fluid		
	Mild ORS (ml in 4 h)	Moderate[a] ORS (ml in 6 h)	Severe with shock i.v. Ringer's lactate[b] plus ORS
3 kg (1-2 months)	150	300	Give i.v. Ringer's lactate at a rate of 30 ml/kg in 1 h, then 20 ml/kg/h 2 h more. Stop i.v. and start ORS to make up the remaining deficit: 50 ml/kg over the next 3 h[c]
5 kg (3-4 months)	250	500	
7 kg (6-9 months)	350	700	
10 kg (1-2 years)	500	1000	
12 kg (2-3 years)	600	1200	

ORS, oral rehydration solution containing (mmol/l): sodium 90, potassium 20, chloride 80, bicarbonate 30, glucose 111.

[a] Although the estimated deficit in moderate dehydration is up to 90 ml/kg, the fluid amounts have been calculated on the basis of 100 ml/kg to allow a small extra amount for anticipated purging during rehydration.

[b] If Ringer's lactate is not available, use an alternative.

[c] Example: a baby weighing 5 kg will receive 150 ml i.v. Ringer's lactate during the first hour, then 100 ml/h for 2 h; this is followed by 250 ml ORS in the next 3 h.

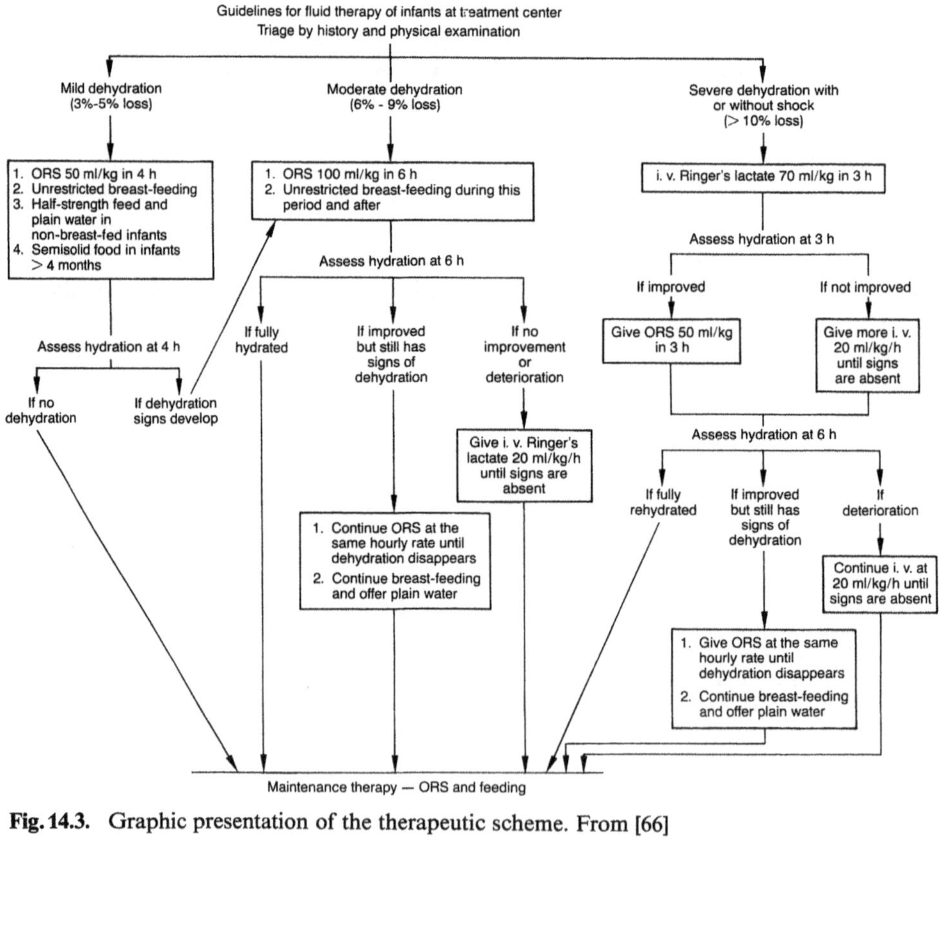

Fig. 14.3. Graphic presentation of the therapeutic scheme. From [66]

Table 14.11. Guide for rehydration and maintenance therapy in **adults and older children.** From [66]

Therapy	Dehydration status	Type of fluid	Estimated fluid deficit	Rate of administration
Rehydration	Severe	Ringer's lactate i.v. (or alternatives)	100 ml/kg body weight	(a) Give half the estimate (e.g., 2.5 l in a 50-kg adult) as fast as possible (simultaneous i.v. drips at two different sites may be needed in a "pulseless" patient) until a good volume radial pulse is obtained; give the total initial deficit in 2–3 h. (b) If signs of dehydration are absent then go on to maintenance; if not, continue i.v. administration until the signs disappear.
	Moderate or mild	ORS	Moderate: 80 ml/kg body weight Mild: 50 ml/kg body weight	Adults, 750 ml/h; children, 200–300 ml/h. Replace fluid as fast as the patient can take it (average time of rehydration, 3–5 h). Higher rates may be required for continued purging. If signs worsen or intractable vomiting prevents oral intake, then give i.v. as under severe dehydration.
Maintenance		ORS		Replace stool losses volume for volume. If they cannot be measured, estimate the losses using the clinical status of the patient as a guide. Adults, up to 750 ml/h; children, up to 200–300 ml/h. If stool output is higher, vomiting is intractable, or signs of dehydration appear, then i.v. Ringer's lactate (20–40 ml/kg depending on severity) until signs disappear; then give ORS.

ORS, oral rehydration solution containing (mmol/l): sodium 90, potassium 20, chloride 80, bicarbonate 30, glucose 111.

Simple sugar-salt solutions can be used where standard ORS is not available [33]: Four heaped teaspoons sugar and one level teaspoon salt in a liter-bottle of water or three heaped teaspoons sugar and half a level teaspoon salt in a 750-ml bottle of water are proportions with which reliable results have been obtained [32, 33].

Drugs Not Recommended for the Treatment of Diarrhea

The following drugs are not recommended for the treatment of diarrhea:

1. Orally administered, nonabsorbable antibiotics, such as neomycin and colistin, are of no practical clinical value in the treatment of bacterial diarrheas.
2. Neither the poorly absorbed sulfonamides (sulfaguanidine) nor the well-absorbed short-acting sulfonamides (sulfadimidine) are useful for the treatment of salmonella enteritis. Most strains of *Shigella* spp. and *E. coli* are now sulfonamide resistant.
3. Adsorbents, such as charcoal, pectine, and kaolin, are of only minimal efficacy, if any.
4. Antiperistaltic agents such as tincture of opium, atropine, metoclopramide, loperamide (Imodium), or diphenoxylate (Lomotil) may potentiate clinical disease in shigellosis and salmonellosis by increasing the intestinal transit time. Overdosage may lead to paralytic ileus in infants.
5. Use of hydroxyquinolines, such as clioquinol or diiodoquine is *not* indicated in "nonspecific" diarrhea. Subacute myelo-optic neuropathy (the SMON syndrome) has been ascribed particularly to clioquinol, but serious acute and chronic neurotoxicity has also been observed with other 8-hydroxy-quinolines. These agents are luminal amebicides and thus of no value in amebic dysentery or amebic liver abscess. A short course at the correct dosage (!) may be advised for eradication of asymptomatic amebiasis, but treatment with diloxanide furoate (no serious side effects are known) is preferable, if at all indicated, in areas where *E. histolytica* infection is endemic.

Use of Antibiotics

Antibiotics are *not* indicated in the majority of cases of diarrhea. They are, however, very often used indiscriminately for diarrheal illnesses. A study from Thailand [29] showed that 29% of children with diarrhea had received antibiotics prior to admission to hospital. According to the recommendations of WHO (see below) only 11%, at the most, should have been treated, i.e., those with shigellosis (9%) and *V. cholerae* (2%). Similarly, 96% of patients with diarrhea presenting at the hospital run by the International Centre for Diarrhoeal Disease Research in Dacca, Bangladesh, were given antibiotics. After introduction of guidelines for the use of antibiotics, the rate fell to 43% [60].

Indiscriminate use of antibiotics has undoubtedly contributed to the emergence of multiresistant *Salmonella*, *Shigella*, and *V. cholerae* in many parts of the world. This renders effective, empiric treatment difficult.

In 1981, WHO [66] recommended antibiotics only for severe shigellosis and severe cholera:

Antibiotic therapy
Antibiotics are not needed for the treatment of acute diarrhoea, except in:

- severe bacillary dysentery persisting for 3 or more days, particularly that due to *Shigella dysenteriae,* and
- severe cholera.

For severe bacillary dysentery, the drugs of choice are ampicillin or trimethoprim/sulfamethoxazole (TMP/SMX):

Ampicillin - 100 mg/kg/day (oral or parenteral) in 4 divided doses for 5 days (note: ampicillin resistance may be common in some areas).

TMP/SMX - Children: TMP 10 mg/kg/24 h and SMX 50 mg/kg/24 h in 2 divided doses for 5 days. Adults: TMP 160 mg and SMX 800 mg twice daily.

Second-choice drugs are:

Nalidixic acid - 55 mg/kg/day in 4 divided oral doses for 5 days.

Tetracycline - 50 mg/kg/day in 4 divided doses for 5 days (Note: tetracycline resistance may be common in some areas).

Chloramphenicol - may also be used, but because of adverse reactions should be avoided if possible (dosage: 50 mg/kg/day for 5 days).

For severe cholera, the drug of choice is:

Tetracycline - Children: 50 mg/kg/day in 4 divided doses for 3 days. Adults: 500 mg every 6 hours for 48 hours.

USA/European recommendations are usually less restrictive, in that all cases of shigellosis and cholera should be treated with antibiotics. The use of antibiotics in nontyphoidal salmonella infections is controversial. There is, however, general agreement, that carriers and mild cases of gastroenteritis should not be treated. Use of antibiotics may lead to prolonged excretion of salmonella, may convert carriage into symptomatic disease, may be followed by clinical and bacteriological relapses, and, last but not least, favors selection of resistant strains. Eradication of the postenteritic carrier state is, however, possible with modern quinolones (ciprofloxacin, ofloxacin). This may change medical opinions in future.

Dysentery

In the situation of large numbers of cases of endemic diarrhea, the WHO recommends restricting the use of antibiotics to cases of dysentery.

The following practice is reported from the rural diarrhea treatment center in Matlab, Bangladesh: "Because laboratory findings are not available at the time of admission, the assessment of dehydration and the choice of treatment were not influenced by the identification of an enteropathogen. Only persons whose histories indicated a dysenteric illness were given ampicillin or trimethoprim-sulfamethoxazole on admission" [5].

Data from Dacca [20] on the etiology of dysentery (defined as passing blood and mucus in the stool) and fecal leukocytosis (on microscopical examination) support this concept; patients with hematophagous *Entamoeba histolytica* had been excluded (Table 14.12).

Dysentery caused by enteroinvasive *E. coli* responds to treatment for shigellosis.

In areas where salmonella infections are more prevalent than in Bangladesh, some cases of dysentery will be due to salmonellae and other pathogens (Table 14.13) [64]. Persistent fever after rehydration requires special attention.

Table 14.12. Bacterial isolates of acute non-amoebic dysentery in Dacca, Bangladesh. From [20]

	Adults ($n=76$)		Children ($n=56$)		All subjects ($n=132$)	
Shigella species	56	(74%)	52	(93%)	108	(81%)
V. parahaemolyticus	2	(3%)	0	-	2	(1%)
Enteroinvasive *E. coli*	1	(1%)	1	(2%)	2	(1%)
No recoverable bacterial pathogen	16[a,b]	(21%)	2[b]	(4%)	18	(13%)
Other	1[c]	(1%)	1[c]	(2%)	2[c]	(1%)

[a] The difference between the adult and children group is $p<0.01$ by Chi Square.
[b] Includes 4 adults with Group F vibrio-like isolates (not a known cause of dysentery).
[c] Includes one child with Salmonella E and one adult with ulcerative colitis.

Table 14.13. Enteric pathogens isolated from 200 children with mucoid or bloody diarrhea (Thailand, January 1984–May 1984) [64]

Pathogens	No. of pathogens isolated per episode of diarrhea		
	One ($n=86$)	Two or more ($n=82$)	Total ($n=168$)
Shigella spp.	47	41	88
P. shigelloides	7	36	43
Aeromonas spp.	6	25	31
Enterotoxigenic *E. coli*	6	25	31
Campylobacter spp.	8	16	24
Salmonella spp.	2	17	19
G. lamblia	3	16	19
Enteroinvasive *E. coli*	3	7	10
Vibrio spp.[a]	1	7	8
E. histolytica	2	5	7
Enteropathogenic *E. coli*	1	1	2
Total	86	196	282

[a] *V. parahaemolyticus* (4), non-O1 *V. cholerae* (3), *V. cholerae* El Tor Inaba (1).

Malaria, shigellosis (dysentery with fecal leukocytosis) and parenteral diarrhea (bacterial infections like otitis media, pneumonia, etc. are often accompanied by diarrhea in children) must be considered. Other differential diagnoses are typhoid fever and, in some areas, septicemia due to nontyphoidal salmonellae. Febrile diarrhea due to salmonellae is normally self-limited, but carries a risk of septicemia by invasion of the bloodstream, particularly in infants. For this reason, prompt treatment with ampicillin or co-trimoxazole is often recommended for infants with high fever and "toxemia", whereas in adults one can usually wait for 48 h to see whether the fever comes down. Spiking fever for more than 48 h, splenomegaly, appearance of septic metastases and leukocytosis are typical of septicemia complicating salmonella enteritis.

The majority of diarrhea episodes in developing countries are cared for by nonmedical personnel, if at all. Simple flowcharts for training purposes and repeated reference are, therefore, needed. Such a flowchart for the management of acute diarrhea by village health care workers is presented in Fig. 14.4. It provides a useful

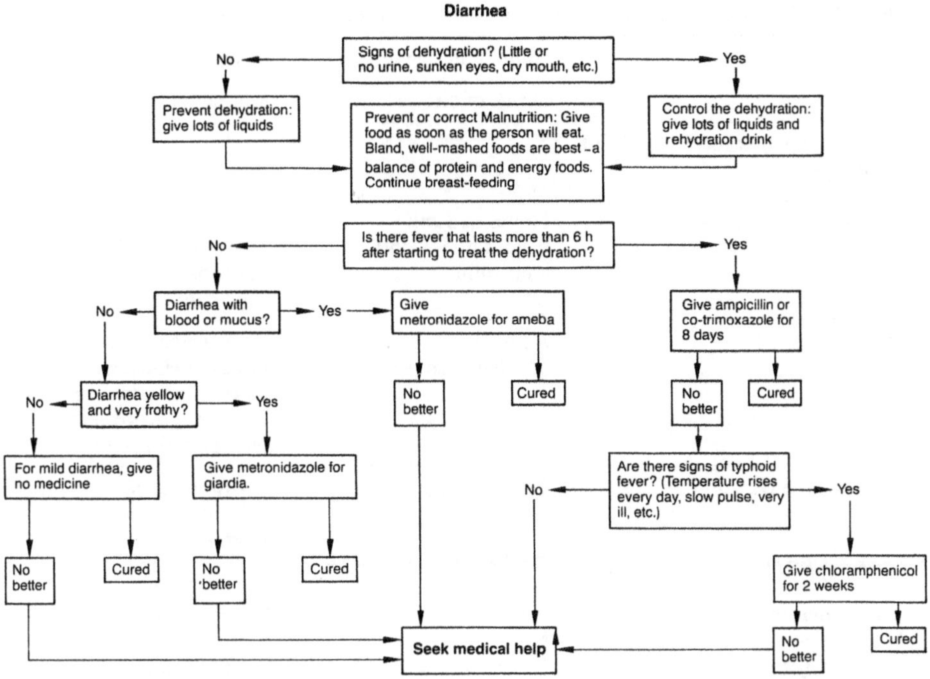

Fig. 14.4. Care of an adult person with acute diarrhea by village health care workers (no doctor available). Modified from Werner [65]

supplementation to the graphic presentation of the treatment of dehydration, already shown in Fig. 14.3.

Chemotherapy for Pathogens Known to Cause Diarrheal Disease

Chemotherapy for diarrhea according to the known pathogens is shown in Table 14.14, and the dosage in Table 14.15.

Salmonella enteritis should be treated in children with sickle cell disease (risk of osteomyelitis) and in patients with AIDS or malignancy, especially lymphomas (risk of septicemia). Treatment of salmonella enteritis is often considered in infants because of the increased risk of septicemia in that age group. High, initial fever with "toxemia" or persistence of fever should guide the use of antimicrobial agents. Forty-two of 45 strains of several nontyphoidal salmonella serotypes which had been recovered from 2990 samples of blood, CSF, and stool at the University Teaching Hospital in Lusaka, Zambia, were isolated from infants less than 2 years old. Of the 42 isolates, 12 were from blood, the other 30 from stool [15]. Chronic salmonella septicemia associated with *Schistosoma mansoni* infection often responds to antimicrobial therapy only after treatment for schistosomiasis.

Emergence of multiresistant salmonellae has been reported from many parts of the world. 201 clinical isolates of *S. typhimurium* from different parts of India

Table 14.14. Chemotherapy for pathogens known to cause diarrheal disease

Pathogen	Antibiotics usually recommended	
	Drugs of choice	Alternatives
Salmonella spp. (except *S. typhi* and *S. paratyphi*)		
Carrier state	None	
Enteritis, food poisoning	None, as a rule	
Septicemia, abscesses	Chloramphenicol[a], ampicillin[a]	Co-trimoxazole[a], ciprofloxacin, ofloxacin[a]
Shigella spp.	Ampicillin, co-trimoxazole	Tetracycline, single-dose tetracycline, 2.5 g, should be restricted to adults and older children; norfloxacin, oflox-acin, ciprofloxacin
Vibrio cholerae O1	Tetracycline	Co-trimoxazole, chloramphenicol, quinolones
Campylobacter jejuni	Erythromycin in severe cases	
Yersinia enterocolitica	Tetracycline	Co-trimoxazole
Vibrio parahaemolyticus	None	
Aeromonas hydrophila	None, as a rule	Co-trimoxazole
Clostridium difficile		
Antibiotic-associated colitis	Oral vancomycin	Metronidazole
Clostridium perfringens		
Food poisoning	None	
Enteritis necroticans (pigbel)	Benzylpenicillin	Metronidazole
Enterotoxigenic *E. coli*	None	
Enteroinvasive *E. coli*	As for *Shigella* spp.	
Enteropathogenic *E. coli*	None, as a rule	
Bacillus cereus		
Food poisoning	None	
Staphylococcus aureus		
Food poisoning	None	

[a] Dosage as for septicemia.

Table 14.15. Daily dosage of oral antibiotics in diarrheal diseases

	Adults	Children	Dose interval
Ampicillin	3-4 g	100 mg/kg	8 h
Co-trimoxazole	1.92 g	20-30 mg/kg	12 h
Tetracyclines	1-2 g	50 mg/kg[a]	6-8 h
Chloramphenicol	2-3 g	50 mg/kg	8 h
Erythromycin	1-2 g	30-40 mg/kg	6-8 h
Metronidazole	1.2-2.25 g	20-40 mg/kg	8 h
Vancomycin	0.5-1.5 g	20-40 mg/kg	4-6 h
Norfloxacin, pefloxacin	0.8 g	---[b]	12 h
Ofloxacin	0.2-0.4 g	---[b]	12 h
Ciprofloxacin	0.5-1 g	---[b]	12 h
Nalidixic acid	3-4 g	55 mg/kg[c]	6 h

[a] Only children above 7 years of age.
[b] Contraindicated in children.
[c] In Europe contraindicated in children.

(1979-1982) were tested. 167 of them were resistant to ampicillin, chloramphenical, kanamycin, streptomycin, sulfonamides, tetracycline, and trimethoprim [57]. Systemic infections with multiresistant strains require treatment with second- or third-generation cephalosporins, when sensitivity testing demonstrates resistance to all commonly used antibiotics [30]. Monotherapy with gentamicin must be relied upon when cephalosporins are not available. Quinolones may prove useful alternatives in future.

Shigellae are now resistant to tetracycline in most parts of the world. Single-dose tetracycline (2.5 g) has been shown to be effective in the treatment of shigella infections regardless of the in vitro sensitivity [42] and may be preferred in areas where resistance to ampicillin has become frequent [3]. Amoxicillin (which is better absorbed than ampicillin) is not reliable, presumably because the luminal levels achieved in the large intestine are too low. Epidemics due to multiresistant strains of *Shigella dysenteriae* serotype 1 have caused thousands of deaths in Central America, Bangladesh, and Somalia. An epidemic of bacillary dysentery in Zaire was caused by a strain resistant to ampicillin, chloramphenicol, streptomycin, sulfonamide, and tetracycline. During the epidemic the strain acquired resistance to co-trimoxazole, and nalidixic acid had to be used for treatment [36]. Nalidixic acid is, however, inferior to ampicillin and co-trimoxazole. Modern quinolones like ciprofloxacin will replace the weakly active and poorly tolerated nalidixic acid in the future.

Treatment with tetracycline shortens the duration of diarrhea in cholera. Excretion of *Vibria cholerae* O1 ceases within 24-48 h, as compared to 7 days in untreated cases. Derivatives of tetracycline with comparatively poor absorption are more effective than doxycycline, 25% of patients treated with single-dose doxycycline and 14% of patients treated with multiple doses of doxycycline continued to excrete *V. cholerae* for more than 3 days - both groups were then treated with tetracycline [48]. Furazolidone (400 mg daily for 3 days) is an effective alternative, but is no longer acceptable because of the high risk of toxic side effects.

Prophylactic use of antibiotics for contacts of cholera patients is controversial because of limited efficacy and the possibility of selecting resistant strains. During the cholera epidemic in Tanzania, tetracycline was widely used for both treatment and prophylaxis; strains resistant to tetracycline emerged within a few months [38]. During the cholera epidemic in eastern Zaire (which originated in Tanzania), tetracycline was reserved for treatment and sulfadoxine, a long-acting sulfonamide, was used for prophylaxis; resistance was not encountered [54]. The prophylactic effect of a single dose of sulfadoxine was limited to from day 3 to day 6 when given to contacts of cholera patients. Six doses of tetracycline given over 3 days were effective from day 2 to day 6 [9]. Occurrence of multiresistant strains of *V. cholerae* O1 has recently been reported from an endemic area in India [63].

Oral erythromycin achieves symptomatic response and eradication of *Campylobacter jejuni* from stool when given early. Start of treatment 4 days or more after onset of symptoms seems to have little influence on the natural course of enteritis.

Erythema nodosum, arthralgia, or polyarthritis (commonly seen in adult Europeans) argue in favor of treatment. Infections with *Y. enterocolitica* may also present as mesenteric adenitis. Septicemia should be treated with chloramphenicol or gentamicin.

Vibrio parahaemolyticus is commonly the cause of food poisoning associated with shellfish, mussels, and other seafood. Chemotherapy has so far not been shown to be of value.

Vibrio cholerae non-O1 may cause diarrheal diseases as severe as cholera and has occurred in areas where cholera is not epidemic. Infection is usually acquired by ingestion of contaminated seafood, but water itself has been implicated in some cases. Tetracycline, quinolones or co-trimoxazole may be used in severe cases.

Mild, self-limited disease is the rule with *Aeromonas hydrophila*. Treatment with co-trimoxazole or tetracycline is suggested only for patients with chronic, serious illness (chronic dysentery-like symptoms in infants) and for patients at high risk of septicemia (malignancy, especially leukemia, and hepatobiliary disease).

Discontinuation of all antibiotics alone suffices in mild cases of *Clostridium difficile*. Oral vancomycin (125–250 mg every 4–6 h) is the most effective available agent. Metronidazole provides a much cheaper, but less effective, alternative.

Enteritis necroticans (pigbel) is caused by the B toxin of *Clostridium perfringens*, elaborated by type C strains. Vaccination with *C. perfringens* type C B toxoid has led to a decrease in the incidence of the disease in the Highlands of Papua New Guinea [8]. Clinical cases should be treated with benzylpenicillin or metronidazole or both.

Chemotherapy of diarrhea caused by enterotoxin-producing *E. coli* is not recommended; the clinical presentation is often indistinguishable from other acute dehydrating diarrheas.

The clinical picture for enteroinvasive *E. coli* is indistinguishable from dysentery caused by shigella. Outbreaks due to strains of enteropathogenic *E. coli* (defined by serotyping) occurred in hospital nurseries during the 1940s and 1950s. Antimicrobial chemotherapy with oral colistin, neomycin, or gentamicin was often employed, apart from other measures, in order to control those outbreaks.

Prevention and Control

Oral rehydration therapy is widely accepted and promoted as a cost-effective means for the reduction of diarrheal disease mortality in young children. The limitations of oral rehydration therapy concern its lesser value in chronic or dysenteric diarrhea compared to acute dehydrating diarrhea, problems of delivery within existing health systems, and the assumption that morbidity rates of diarrheal diseases are not at all or only marginally affected [17].

Prevention of diarrheal diseases is clearly desirable. At present, there are no effective vaccines available on the market. Chemoprophylaxis is not considered a cost-effective means for control of diarrheal diseases, since according to the available evidence chemoprophylaxis of household contacts of known shigellosis or cholera cases would reduce the overall diarrhea incidence rates in children under 5 years of age only by less than 1% and the diarrheal death rates by 0.7%-2.0% at the most [13]. By using data on the frequency and severity of measles-associated diarrhea, it has been estimated that measles immunization at the age of 9-11 months could prevent 0.6%-3.8% of diarrhea episodes and between 6% and 26% of diarrhea deaths [18].

Potential interventions for the control of diarrheal diseases (Table 14.16) are reviewed in the WHO Diarrheal Diseases Control Programme. Reviews on the effectiveness, feasibility, and costs of some major interventions (e.g., measles immunization, chemoprophylaxis, breast-feeding, supplementary feeding programs, promotion of personal and domestic hygiene, water supply and sanitation) are published in the *Bulletin of WHO* [17].

Appendicitis

Appendicitis is less common in developing countries than in Europe or the USA. Acute appendicitis accounted for 18% of all acute abdominal emergencies in an African series, while 55% of cases were due to intestinal obstruction [1]. Likewise, only 103 proved cases of appendicitis in childhood were seen at a university hospital in Nigeria over a 10-year period [41]. Perforation was common, being noted in 25.5% of children at the time of operation [41].

The treatment of appendicitis is primarily surgical. Antibiotics are indicated when complications like perforation, peritonitis, intra-abdominal abscess, or pylephlebitis have occurred. When perforation resulting in peritonitis is suspected prior to the operation, antibiotic treatment should be started preoperatively. The antibiotic treatment of peritonitis resulting from perforated appendix is that of secondary peritonitis (see section on peritonitis, this chapter). Appendicitis may be mimicked by localized amebiasis. Treatment with metronidazole covers both *Bacteroides* species and *Entamoeba histolytica*. In cases of suspected ameboma a therapeutic trial with metronidazole is recommended.

Table 14.16. Potential interventions for reducing diarrheal morbidity or mortality among children under five years of age. From [17]

I. By case management

A. *Oral rehydration therapy*
1. Administration of oral rehydration in the home
2. Administration of oral rehydration at a medical facility.

B. *Non-oral rehydration therapy*
1. Administration of rehydration by intravenous or other routes at a medical facility.

C. *Appropriate feeding*
1. Promoting the appropriate feeding of children during diarrheal illness and convalescence.

D. *Chemotherapy*
1. Administration of therapeutic agents in the home
2. Administration of therapeutic agents at a medical facility.

II. By increasing host resistance to infection and/or illness and/or death

A. *Maternal nutrition*
1. Improving prenatal nutrition to reduce the incidence of low birth-weight.
2. Improving prenatal and postnatal nutrition to improve the quality of breast milk.

B. *Child nutrition*
1. Promoting exclusive breast feeding up to age 4–6 months and partial breast-feeding thereafter.
2. Improving weaning practices for children aged 4–18 months (introducing non-milk foods not later than the sixth month, continuing breast-feeding for as long as possible, and using nutritions and locally available weaning foods).
3. Supplementary feeding to improve the nutritional status of children aged 6–59 months.
4. Promoting the use of growth charts by mothers as an aid to proper child nutrition and child care.

C. *Immunization*
1. Rotavirus and/or cholera immunization (when effective and tested vaccines are available) of the child and/or mother.
2. Measles immunization to reduce measles-associated diarrhea.

D. *Chemoprophylaxis*
1. Chemoprophylaxis of children at special risk, such as contacts of known cases, to reduce the incidence and/or severity of diarrhea.

III. By reducing transmission of the pathogenic agents of diarrheal diseases

A. *Water supply and excreta disposal*
1. Constructing water supplies that improve the quality and availability of water for domestic purposes, and improved excreta disposal facilities; and providing the necessary educational support to ensure use and maintenance of these new facilities.

B. *Personal and domestic hygiene*
1. Promoting specific features of personal and domestic hygiene, such as hand-washing, by appropriate educational campaigns.

C. *Food hygiene*
1. Promoting improved practices for the preparation and storage of foods, both commercially and in the home, and especially emphasizing the hygienic preparation of weaning foods.

D. *Control of zoonotic reservoirs*
1. Control of infection of domestic and farm animals by pathogens causing diarrhea in man.

E. *Fly control*
1. Control of these especially flies breeding in association with human or animal faeces.

IV. By controlling and/or preventing diarrhea epidemics

A. *Epidemic surveillance investigation and control*
1. Improving the ability to identify and investigate an epidemic early in its course and the capacity to implement effective control activities.

Peritonitis

Primary peritonitis was common in the preantibiotic era, but is now very rare in North America and Europe. In developing countries, the incidence varies among geographical regions. For example, although accounting for only a few cases of childhood peritonitis in Calcutta, primary peritonitis accounted for 48 of 150 cases of childhood peritonitis seen in Madras over a 5 year period [7 a, 27]. Primary peritonitis was seen predominantly in girls, and was most often pneumococcal in origin, in another series from India [56]. Pneumococci could be identified in the vaginal discharge early in the course of the disease, 1-3 days after onset of symptoms [56]. Other causes of primary peritonitis are beta-hemolytic streptococci and *Staphylococcus aureus*. Primary peritonitis is a characteristic complication of the nephrotic syndrome and of cirrhotic liver disease with ascites. Gram-negative enteric bacilli, particularly *E. coli*, are frequently the cause of the latter. In neonates, primary peritonitis is frequently associated with umbilical sepsis [16].

Secondary perforating peritonitis can result from a wide variety of causes: transabdominal trauma, perforated peptic ulcer or appendix, and perforations of typhoid or amebic ulcers are some of the commoner ones. Intestinal obstruction due to invagination, volvulus, adhesions, strangulated hernia, ascaris infection, or another cause, gives rise to intestinal necrosis when not relieved and also results in peritonitis. Rupture of an amebic liver abscess and leakage from a hydatid cyst are further causes of peritonitis.

Secondary peritonitis is usually polymicrobial in origin. The spectrum of organisms comprises *Bacteroides fragilis* and other *Bacteroides* species, anaerobic streptococci, gram-negative enteric bacilli, particularly *E. coli*, *Fusobacterium* spp., and *Clostridia* spp.

Other forms of bacterial peritonitis are associated with peritoneal dialysis (commonly *Staphylococcus epidermidis* or *Candida* spp.) or pelvic inflammatory disease. Meconium peritonitis occurs in newborns with mucoviscidosis but is not bacterial. Tuberculous peritonitis, in contrast to the other types, presents as chronic rather than acute disease.

In the case of clinical suspicion of primary peritonitis, only explorative laparatomy can definitively exclude secondary peritonitis due to underlying pathology. However, when paracentesis reveals pneumococci or gonococci in the peritoneal exudate, the disease should be managed conservatively with antibiotics alone. Drainage is felt necessary in cases where a significant amount of exudate has already accumulated [56]. When paracentesis reveals no organisms, gram-negative rods, or mixed infection, explorative laparatomy is indicated for diagnosis and treatment of acute secondary peritonitis. In contrast, the management of secondary peritonitis is primarily surgical to identify and treat any underlying pathology. Definitive diagnosis of tuberculous peritonitis is usually achieved by peritoneal biopsy [31].

Commonly, isolation of the different organisms involved in secondary peritonitis is incomplete. In particular, the yield of anaerobic bacteria is low with routine processing of peritoneal exudate and of swabs taken at operation. Chemotherapy should not be directed at *E. coli* only, if this happens to be the only organism grown. It should cover the far more important anaerobic bacteria as well. Treatment should be modified when necessary according to the results of sensitivity testing.

Treatment

Primary peritonitis due to pneumococci or beta-hemolytic streptococci is treated with parenteral benzylpenicillin. In staphylococcal primary peritonitis, cloxacillin is the drug of choice. In secondary peritonitis, two or three antibiotics are usually combined to cover the mixed infection. Antibiotic treatment should be instituted prior to the explorative laparatomy in order to prevent hematogenous spread at the time of operation. Generally, secondary peritonitis is treated with broad-spectrum combination therapy. In Europe, treatment with cefotaxime + metronidazole, mezlocillin + metronidazole, or cefotaxime + azlocillin + metronidazole is the most modern recommendation. In Great Britain and the USA, aminoglycosides, which are active against enteric bacilli but not against anaerobic bacteria, form one part of combination treatment. Combinations widely used there are gentamicin (or tobramycin) + metronidazole, gentamicin + clindamycin, or gentamicin + chloramphenicol. Metronidazole or ampicillin is often added as third agent. Amikacin is recommended as a substitute for gentamicin in severely ill patients when infection with gentamicin-resistant enterobacteria is likely. In developing countries, monotherapy with ampicillin or tetracycline and combination treatment with benzylpenicillin + streptomycin is still practiced for reasons of cost, tradition and unavailability of other drugs. Data from developing countries concerning the efficacy of this treatment are not available. None of these three regimens, however, is reliably active against *Bacteroides fragilis,* which has been recognized as the most important pathogen in the polymicrobial infections of secondary peritonitis. From this viewpoint, and on the basis of the experience in the USA and Europe, combination treatment with chloramphenicol, an aminoglycoside, and metronidazole seems more advantageous. Treatment with metronidazole also covers the possible involvement of *Entamoeba histolytica* in the pathogenesis of peritonitis. If chloramphenicol is to be reserved for typhoid fever, combinations of ampicillin with metronidazole or, more expensive, a cephalosporin with metronidazole, may be used. Quinolones alone have no advantage in the treatment of secondary peritonitis (poor activity against *Bacteroides* spp. and certain *Streptococcus* spp.). Chemotherapy of peritonitis is generally systemic. The only exception is peritonitis associated with peritoneal dialysis, where antibiotics may, instead, be added to the dialysis fluid. Caution must be exercised when aminoglycosides or polymyxins are instilled into the peritoneal cavity, as neuromuscular blockade with respiratory arrest may ensue. Prostigmin and calcium gluconate are the antidotes to use. Chloramphenicol succinate, the parenteral formulation, is inactive when instilled locally.

Tuberculous peritonitis responds well to standard antituberculous treatment.

Treatment of peritonitis associated with pelvic inflammatory disease is that of pelvic inflammatory disease (see chapter "Obstetric and Gynecological Infections").

Pyogenic Liver Abscess

In developing countries, pyogenic liver abscesses are far less frequent than amebic liver abscesses. Common predisposing conditions of pyogenic liver abscesses (appendicitis, cholelithiasis, colon carcinoma) are rare in most tropical countries. Portal and biliary sepsis, staphylococcal septicemia, infected hydatid cysts, trauma, and chronic granulomatous disease are among the tropical predisposing factors of these abscesses, though a proportion of cases are cryptogenic [34, 37].

Anaerobic bacteria, *Bacteroides* spp., gram-positive cocci and others, and gram-negative enteric bacilli, frequently *E. coli* or *Klebsiella* spp., are the agents most frequently cultured. Mixed infection is common in these instances. *Staphylococcus aureus* is frequently the causative agent in children, presumably as a result of bacteremia [4, 26]. Examination of the aspirate from the abscess by Gram's stain and aerobic and anaerobic culture is recommended. Blood cultures are positive in 30%-60% of cases. In endemic areas, examination of the terminal milliliters of the aspirate for amebic trophozoites should also be performed, since clinical differentiation between amebic and pyogenic liver abscesses is difficult. Amebic serology, where available, can be of help. High titers are indicative of amebic liver abscess, but the diagnostic value of high titers is greater in nonendemic areas. Amebic abscesses may be superinfected by enteric bacilli.

The clinical presentation of fever and chills, abdominal pain, and tender hepatomegaly, sometimes accompanied by plain chest X-ray findings of an elevated right diaphragm with basilar atelectasis or pleural effusion, is often the only means available for diagnosis of a hepatic abscess under conditions of limited resources. Leukocytosis, a raised ESR, and raised alkaline phosphatase are characteristic laboratory findings. Visualization of the abscess, which may be multiple, by ultrasonography, scintigraphy, or computed tomography is preferable. Ultrasonography is particularly useful when aspiration has to be performed.

With the advent of metronidazole for treatment of amebic liver abscesses, diagnostic aspiration of an abscess is no longer recommended. Aspiration should, however, not be delayed if there is no response to initial treatment with metronidazole or in patients who present late with large abscesses at risk of imminent rupture. The aspirate is then subjected to the investigations mentioned above. "Pus" from an amebic abscess does not usually have a foul odor like pus from a pyogenic abscess. Initial treatment of a pyogenic liver abscess should be performed with metronidazole or another nitroimidazole to cover anaerobes but also ameba. It must be supplemented with broad-spectrum antibiotics to cover gram-negative enteric bacilli. An aminoglycoside, a parenteral cephalosporin, ampicillin, or chloramphenicol, in that order of priority, may be chosen according to availability. Initial chemotherapy of pyogenic liver abscesses in children should include cloxacillin to cover staphylococci. In Europe, mezlocillin or piperacillin or cefotaxime combined with metronidazole are preferred for treatment of hepatic abscesses. Debate continues as to whether medical management alone or the addition of needle aspiration, placement of a percutaneous catheter, or surgical drainage carries the greatest chance of cure [37]. Choice of procedure will in part depend on the facilities available.

Biliary Tract Infections

Cholelithiasis, the major cause of biliary tract infections in industrialized countries, is distinctly uncommon throughout most of the developing world. In Southeast Asia, recurrent pyogenic cholangitis, biliary abscess, and pylephlebitis are well-known complications of heavy infection with the liver flukes *Clonorchis sinensis* and *Opisthorchis viverini*. Biliary stasis due to bile duct obstruction and formation of stones in the intrahepatic bile ducts leads to these secondary pyogenic infections. Primary recurrent pyogenic cholangitis occurring in the absence of parasitic infection is common in the Chinese population in Hong Kong and Singapore.

The spectrum of organisms comprises mainly gram-negative enteric bacilli, different streptococci and enterococci, and anaerobic bacteria like *Bacteroides fragilis* and *Clostridium* spp. Mixed infection is common [61]. The isolation of the causative organisms is often impossible. Chemotherapy is directed at this spectrum of pathogens. For treatment, effective blood and tissue levels are as important as the drug concentration in the bile. Tetracyclines which are excreted into the bile are rendered inactive in bile, and are, therefore, no longer recommended for treatment of biliary tract infections. The biliary levels of co-trimoxazole are low and unreliable. Aminoglycosides the bile concentration of which are less than their respective plasma concentration are, on the other hand, more active in bile. Some of the modern beta-lactam antibiotics, such as cefoperazone, cefotaxime, ceftriaxone, and the acylaminopenicillins (e.g. mezlocillin), achieve sufficient concentrations in bile. Other beta-lactam antibiotics, like ampicillin, cefazolin, cefamandole, cefotiam, or cefoxitin, are less satisfactory in this regard.

Modern European recommendations advise to treat severe infections like cholangitis or cholangiohepatitis with cefotaxime, cefoperazone, or mezlocillin, possibly in combination with gentamicin. British and American recommendations still suggest to use ampicillin + gentamicin. Initial therapy may be modified later according to the bacteriological results.

References

1. Adesola AO (1979) Tropical surgery. McGraw Hill, New York, p 274
2. Ajao OG, Ajao AO (1982) "Idiopathic" intra-abdominal abscess. Trans R Soc Trop Med Hyg 76: 75-76
3. Arora DR, Midha NK, Ichhpujani RI, Chugh TD (1982) Drug-resistant shigellosis in north India. Indian J Med Res 76: 74-79
4. Arya LS, Ghani R, Abdali S, Singh M (1982) Pyogenic liver abscess in children. Clin Pediatr 21: 89-93
5. Black RE, Merson MH, Rahman ASMM, Yunus M, Alim ARMA, Huq I, Yolken RH, Curlin GT (1980) A two-year study of bacterial, viral, and parasitic agents associated with diarrhea in rural Bangladesh. J Infect Dis 142: 660-664
6. Black RE, Greenberg HB, Kapikian AZ, Brown KH, Becker S (1982) Acquisition of serum antibody to Norwalk virus and Rotavirus and relation to diarrhea in a longitudinal study of young children in rural Bangladesh. J Infect Dis 145: 483-489
7. Bockemühl J (1976) Salmonellosis and shigellosis in Togo (West Africa), 1971-1973. I. Carrier rates in the rural population. Tropenmed Parasitol 27: 112-120

7a. Chatterjee SK (1976) Peritonitis in children. J Indian Med Assoc 67: 104

8. Davis MW (1984) More on necrotizing enterocolitis: pigbel in Papua New Guinea (letter) N Engl J Med 311: 1126
9. Deb BC, Sen Gupta PG, De SP, Sil J, Sikdar SN, Pal SC (1976) Effect of sulfadoxine on transmission of Vibrio cholerae infection among family contacts of cholera patients in Calcutta. Bull WHO 54: 171-175
10. De Mol P, Hemelhof W, Butzler JP, Brasseur D, Tshimpaka Kalala, Vis HL (1983) Enteropathogenic agents in children with diarrhoea in rural Zaire. Lancet I: 516-518
11. De Zoysa I, Carson D, Feachem R, Kirkwood B, Lindsay-Smith E, Loewenson R (1984) Home-based oral rehydration therapy in rural Zimbabwe. Trans R Soc Trop Med Hyg 78: 102-105
12. De Zoysa I (1985) Abstract of: Islam MR, Samadi AR, Ahmed SM, Bhardan PK, Ali A (1984) Oral rehydration therapy: efficacy of sodium citrate equals to sodium bicarbonate for correction of acidosis in diarrhoea. Gut 25: 900-904. Trop Dis Bull 82: 711-712 (abstract no. 2500)
13. De Zoysa I, Feachem RG (1985) Interventions for the control of diarrhoeal diseases among young children: chemoprophylaxis. Bull WHO 63: 295-315
14. Dosso M, Rive J, Iwuji Nnaegbulem S, Douzoua C, Duchassin M (1981) Porteurs de salmonelles dans la population de 4 villages de la région d'Abidjan (République de Côte-d'Ivoire). Bull Soc Pathol Exot Filial 74: 477-488
15. Dube SD, Bhagwat AG (1983) Non-typhoidal salmonella infection in Zambian infants. Trans R Soc Trop Med Hyg 77: 336-337
16. Duggan MB, Khwaja MS (1975) Neonatal primary peritonitis in Nigeria. Arch Dis Child 50 (2): 130-132
17. Feachem RG, Hogan RC, Merson MH (1983 a) Diarrheal disease control: reviews of potential interventions. Bull WHO 61: 637-640
18. Feachem RG, Koblinsky MA (1983 b) Interventions for the control of diarrheal diseases among young children: measles immunization. Bull WHO 61: 641-652
19. Georges MC, Wachsmuth IK, Meunier DMV, Nebout N, Didier F, Siopathis MR, Georges AJ (1984) Parasitic, bacterial, and viral enteric pathogens associated with diarrhea in the Central African Republic. J Clin Microbiol 19: 571-575
20. Gilman RH, Spira WM, Rabbani GH, Al-Mahomod A (1980) Invasive E. coli and V. parahaemolyticus - a rare cause of dysentery in Dacca (correspondence). Trans R Soc Trop Med Hyg 74: 688-689
21. Glass RI, Stoll BJ, Huq MI, Struelens MJ, Blaser M, Kibriya AKMG (1983) Epidemiologic and clinical features of endemic Campylobacter jejuni infection in Bangladesh. J Infect Dis 148: 292-296
22. Gracey M, Iveson JB, Sunoto, Suharyono (1980) Human salmonella carriers in a tropical environment. Trans R Soc Trop Med Hyg 74: 479-482
23. Guerrant RL, Kirchhoff LV, Shields DS, Nations MK, Leslie J, de Sousa MA, Araujo JG, Correia LL, Sauer KT, McClelland KE, Trowbridge FL, Hughes JM (1983) Prospective study of diarrhoeal illnesses in northeastern Brazil: patterns of disease, nutritional impact, etiologies, and risk factors. J Infect Dis 148: 986-997
24. Guha Mazumder DN, Ghosh AK, De SP, Sikkar BK (1977) Vibrio parahaemolyticus infection in man. Indian J Med Res 66: 180-188
25. Hirschhorn N (1980) The treatment of acute diarrhea in children, an historical and physiological perspective. Am J Clin Nutr 33: 637-663
26. Karrar ZA, Abdullah MA (1985) Pyogenic liver abscess in children: a report of three patients and review of the literature. Ann Trop Pediatr 5: 97-102
27. Kumaravel MS, Ramakrishnan (1976) Peritonitis in children. J Indian Med Assoc 67: 93-95
28. Lalitha MK, Walter NM, Jesudason M, Mathan VI (1983) An outbreak of gastroenteritis due to Vibrio parahaemolyticus in Vellore. Indian J Med Res 78: 611-615
29. Leksomboon U, Echeverria P, Suvongse C, Duangmani C (1981) Viruses and bacteria in pediatric diarrhea in Thailand: a study of multiple antibiotic-resistant enteric pathogens. Am J Trop Med Hyg 30: 1281-1290
30. Lepage PH, Bogaerts J, Nsenguremyi F, Hitimana DG, Van Goethem C, Vandepitte J, Butzler JP (1984) Severe multiresistant Salmonella typhimurium systemic infections in Central Africa. Clinical features and treatment in a paediatric department. J Antimicrob Chemother 14: (Suppl B) 153-159

31. Lester TF, Tsega E (1976) Tuberculous peritonitis in Ethiopian patients. Trop Geogr Med 28: 169-174
32. Levine MM, Clements ML, Black RE, Hughes TP, Blum D, Adino R, Corrales G (1981) A practical, reliable method for preparing simple sugar/salt oral rehydration solution. J Trop Med Hyg 84: 73-76
33. Levine MM, Clements ML, Black RE, Hughes TE, Tome FC (1981) Oral rehydration with simple sugar/salt solutions as an alternative to rural areas when glucose/electrolyte solutions are unavailable. In: Holme T, Homgren J, Merson MH, Mollby R (eds) Enteric infections in children, new prospects for treatment and prevention. Elsevier, Amsterdam, pp 325-331
34. Mabogunje OA, Lawrie JH (1984) Bacterial liver abscesses in Nigeria. Trop Geogr Med 36: 159-163
35. Maiya PP, Pereira SM, Mathan M, Bhat P, Albert MJ, Baker SJ (1977) Aetiology of acute gastroenteritis in infancy and early childhood in southern India. Arch Dis Child 52: 482-485
36. Malengreau M, Molima-Kaba, Gillieaux M, De Feyter M, Kyele-Duibone, Mukolo-Ndjolo (1983) Outbreak of Shigella dysentery in eastern Zaire, 1980-1982. Ann Soc Belg Méd Trop 63: 59-67
37. McCorkell SJ, Niles NL (1985) Pyogenic liver abscess: another look at medical management. Lancet I: 803-806
38. Mhalu FS, Mmari PW, Ijumba J (1979) Rapid emergence of El Tor vibrio cholerae resistant to antimicrobial agents during first six months of fourth cholera epidemic in Tanzania. Lancet I: 345-347
39. Molla AM, Sarker SA, Hossain M (1982) Rice-powder electrolyte solution as oral therapy in diarrhoea due to Vibrio cholerae and Escherichia coli. Lancet I: 1317-1319
40. Mutanda LN (1980) Epidemiology of acute gastroenteritis in early childhood in Kenya. III. Distribution of the aetiological agents. East Afr Med J 57: 317-326
41. Olayiwola B, Tai da Rocha-Afodu J (1985) Experience with acute appendicitis in children in Lagos University. East Afr Med J 62: 679-685
42. Pickering LK, Dupont HL, Olarte J (1978) Single-dose tetracycline therapy for shigellosis in adults. JAMA 239: 853-854
43. Pickering LK, Evans DJ, Muñoz O, DuPont HL, Coella-Ramirez P, Vollet JJ, Conklin RH, Olarte J, Kohl S (1978) Prospective study of enteropathogens in children with diarrhea in Houston and Mexico. J Pediatr 93: 383-388
44. Pickering LK, Woodward WE, Plotkin SA (1982) Acute infectious diarrhea of children in the tropics. Pediatr Ann 11: 974-990
45. Pitarangsi C, Echeverria P, Whitmire R, Tirapat C, Formal S, Dammin GJ, Tingtalapong M (1982) Enteropathogenicity of Aeromonas hydrophila and Plesiomonas shigelloides: prevalence among individuals with and without diarrhea in Thailand. Infect Immun 35: 666-673
46. Pizarro D, Posada G, Mata L (1983) Treatment of 242 neonates with dehydrating diarrhea with an oral glucose-electrolyte solution. J Pediatr 102: 153-156
47. Rajan DP, Mathan VI (1982) Prevalence of Campylobacter fetus subsp. jejuni in healthy populations in southern India. J Clin Microbiol 15: 749-751
48. Sack DA, Islam S, Rabbani H, Islam A (1978) Single-dose doxycycline for cholera. Antimicrob Agents Chemother 14: 462-464
49. Sack DA, Islam S, Brown KH, Islam A, Kabir AKMI, Chowdhury AMAK, Ali MdA (1980) Oral therapy in children with cholera: a comparison of sucrose and glucose electrolyte solutions. J Pediatr 96: 20-25
50. Sack RB (1982) Acute diarrheal diseases in humans caused by bacteria. In: Tyrrell DAJ, Kapikian AZ (eds) Virus infections of the gastrointestinal tract. Dekker, New York, pp 239-290
51. Sakdisiwasdi O, Achananuparp S, Limsuwan A, Nanna P, Barnyen L (1982) Salmonella and shigella carrier rates and environmental sanitation in a rural district central Thailand. Southeast Asian J Trop Med Public Health 13: 380-384
52. Samadi AR, Wahed MA, Islam MR, Ahmed SM (1983) Consequences of hyponatraemia and hypernatraemia in children with acute diarrhea in Bangladesh. Br Med J 286: 671-673
53. Samadi AR, Islam R, Huq MI (1983) Replacement of intravenous therapy by oral rehydration solution in a large treatment centre for diarrhea with dehydration. Bull WHO 61: 471-476
54. Schyns C, Fossa A, Mutombo-Nfendu, Kabuyahiya, Hennart P, Piot P, Colaert A (1979) Cholera in eastern Zaire. Ann Soc Belg Méd Trop 59: 391-400

55. Sen D, Saha MR, Niyogi SK, Balakrish Nair G, De SP, Datta P, Datta D, Pal SC, Bose R, Roychowdhury J (1983) Aetiological studies on hospital in-patients with acute diarrhoea in Calcutta. Trans R Soc Trop Med Hyg 77: 212-214
56. Sen S, Lalitha MK, Fenn AS, Mammen DE (1983) Primary peritonitis in children. Ann Trop Paediatr 3: 53-56
57. Sharma PL, Sharma KB, Prakash K (1984) Incidence of conjugative plasmid amongst multidrug resistant Salmonella typhimurium strains isolated from different parts of India (1979-1982). Indian J Med Res 79: 473-478
58. Snyder JD, Merson MH (1982) The magnitude of the global problem of acute diarrheal disease: a review of active surveillance data. Bull WHO 60: 605-613
59. Soenarto Y, Sebodo T, Suryantoro P, Krisnomurti, Haksohusodo S, Ristanto IK, Romas MA, Noerhajati, Muswiroh S, Rohde JE, Ryan NJ, Luke RK, Barnes GL, Bishop R (1983) Bacteria, parasitic agents and rotaviruses associated with acute diarrhoea in hospital in-patient Indonesian children. Trans R Soc Trop Med Hyg 77: 724-730
60. Stoll BJ, Glass RI, Huq IM, Khan MU, Holt JE, Banu H (1982) Surveillance of patients attending a diarrhoeal disease hospital in Bangladesh. Br Med J 285: 1185-1188
61. Seto WH (1985) The changing bacteriology of recurrent pyogenic cholangitis in Hong Kong. J Trop Med Hyg 88: 359-365
62. Smith G, van den Ende J (1986) Cryptosporidiosis among black children in hospital in South Africa. J Infect 13: 25-30
63. Sundaram SP, Murthy KV (1984) Occurrence of transferable multi-drug resistance in *Vibrio cholerae*-O1 in an endemic area. Indian J Med Res 79: 722-729
64. Taylor DN, Echeverria P, Pál T, Sethabutr O, Saiborisuth S, Sricharmorn S, Rowe B, Cross J (1986) The role of *Shigella* spp., enteroinvasive *Escherichia coli*, and other enteropathogens as causes of childhood dysentery in Thailand. J Infect Dis 153: 1132-1138
65. Werner D (1980) Where there is no doctor - a village health care handbook. Macmillan, London
66. WHO (1981) Surveillance and control of acute diarrhoeal diseases. European reports and studies, vol 44. Regional Office for Europe, WHO, Copenhagen
67. Bogaerts J, Lepage P, Rouvroy D, Vandepitte J (1984) *Cryptosporidium*, a frequent cause of diarrhea in Central Africa. J Clin Microbiol 20: 874-876

15 Infections of the Musculoskeletal System

Osteomyelitis

Types and Incidence

Several types of osteomyelitis can be distinguished: acute hematogenous osteomyelitis, exogenous osteomyelitis (direct inoculation of bacteria by trauma or at operation), and chronic osteomyelitis. In Africa, osteomyelitis and its sequelae continue to be a major medical problem. The majority of patients present late in hospital. Often bone necrosis (sequestrum formation) has already supervened. Other complications, such as pathological fractures, dislocation of the hip, infection of an adjacent joint, and limitation of joint movements, are also common. In cases of chronic osteomyelitis, discharging sinuses are frequent.

Acute hematogenous osteomyelitis accounted for less than 25% of 125 consecutive cases of osteomyelitis in Ibadan, Nigeria [8]. Of 34 infants and children treated for osteomyelitis in Dar Es Salaam, 32 had chronic osteomyelitis [16]. Of 1000 children hospitalized for osteomyelitis in Dakar, 75% suffered from chronic osteomyelitis [26]. Of 56 cases of chronic osteomyelitis in Ile-Ife, Nigeria, 41 were hematogenous in origin, acute or subacute [19]. All these patients had been treated by traditional healers first, before seeking medical help. Most of them were living far from town in rural villages where medical facilities were inadequate or not available [18]. Osteomyelitis due to major trauma played a lesser role in that series, but it is certainly on the increase - due to the rising incidence of road accidents.

An incidence of 1.1 cases per week was noted in a review referring exclusively to chronic cases in Ile-Ife, Nigeria [19]. On average, 24 cases of childhood osteomyelitis per year were encountered in another hospital-based study from Enugu, Nigeria [20].

The increased susceptibility of patients with sickle-cell disease to osteomyelitis is a contributory factor of this high incidence. In Enugu, 12 cases of osteomyelitis were observed in a series of 125 children with sickle-cell anemia followed up for an average period of 33 months (min. 12 months) [13]. Two cases developed as complications of a "hand-foot syndrome", the other 10 cases complicated episodes of vasoocclusive bone crisis.

Staphylococcus aureus ranks as first pathogen in osteomyelitis, being found in 40%-60% of cases (Tables 15.1 and 15.2). This is less than the isolation rate (of 80%-90%) in cases of acute hematogenous osteomyelitis in developed countries. The high percentage of chronic cases - where gram-negative rods, often in mixed infection with *S. aureus*, are more frequently isolated (representing colonization or superinfection, rather than the primary pathogen) - is presumably one of the reasons for this difference. Similarly, enterobacteria *(E. coli, Klebsiella, Proteus)* are

Table 15.1. Etiology of osteomyelitis in Africa

Study location [Reference]	Enugu [20]			Ibadan [24]	Ibadan [8]		Ile-Ife [19]	Dar Es Salaam [16]
Period of review	1976–1980			1970–1975	ND		1976–1977	1980–1981
Cases *(n)*	118			40	125		56	34
Acute osteomyelitis	78			30	29[a]		0	2
Chrônic osteomyelitis	40			10	96		56	32
Age group	Children (2 weeks to 16 years)			Children (1 month to 12 years)	Children and adults (4 months to 26 years)		All ages	Children (Neonates to 12 years)
Sex								
Male	80			28	15		33	14
Female	38			12	14		23	20
Genotype								
Hb AA	58			ND	10		ND	19
Hb AS	15			ND	5		5	8
Hb SS[b]	23			13	7		10	4
Hb SC[b]	0			2	2		ND	ND
Hb AC	0			ND	2		ND	ND
unknown	17			ND	3		ND	3
Multiple bone involvement								
All cases	15			17	11		ND	4
In Hb SS and Hb SC disease	12			ND	6		ND	ND
Bacteria isolated *(n)*								
Genotypes	Hb SS	Other	Un-known	All	Hb SS + Hb Sc	Other	All	All
Staphylococcus aureus	9	34	10	16	0	8	27	23
Staphylococcus epidermidis	0	0	0	0	0	1	1	0
Beta-hemolytic streptococci	1	6	0	0	0	0	0	0
Streptococcus pneumoniae	0	0	0	0	0	1	0	0
Escherichia coli	7	6	2	0	0	0	0	1
Proteus spp.	2	0	0	1	1	1	4	1
Klebsiella spp.	0	0	0	2	0	2	0	1[e]
Pseudomonas spp.	1	1	0	0	1	2	1	2
Salmonella spp.	0	1	0	4 (All Hb SS)	4	0	0	0
Mixed infection		14[c]		7[d]	0	0	13[d]	0
No bacteria isolated		13		1	1	1	6	3
No culture taken (or result lost)		38		9	2	3	3	3

ND, no data.

[a] All data on acute cases only.

[b] Sickle-cell disease has been found to affect approximately 2%–3% of Nigerian children [11].

[c] *S. aureus* + *E. coli* (9), *S. aureus* + β-hemolytic strept. A (3) *S. aureus* + *Pseudomonas aeruginosa* (2) are included in the total numbers of isolates.

[d] No details given.

[e] Enterobacter spp.

Table 15.2. Etiology of 400 consecutive cases of bacterial infection in an orthopedic ward (Kasturba Medical College, Manipal, India, 1971–1975. Modified from [27]

	Osteomyelitis and septic arthritis	Infected compound fractures and crush injuries	Postoperative wound infection	Total
Total cases (*n*)	225	129	46	400
Culture-positive cases (*n*, %)	190 (84.4)	118 (91.5)	46 (100.0)	354 (88.5)
Bacteria isolated (*n*, %)				
Staphylococcus aureus	125 (65.8)	34 (28.8)	18 (39.1)	177 (50.0)
Beta-hemolytic streptococci	13 (6.9)	5 (4.2)	2 (4.4)	20 (5.7)
Pseudomonas aeruginosa	16 (8.4)	23 (10.5)	7 (15.2)	46 (13.0)
Proteus spp.	3 (1.6)	5 (4.2)	1 (2.2)	9 (2.8)
Klebsiella spp.	8 (4.2)	12 (10.2)	3 (6.5)	23 (6.5)
Escherichia coli	4 (2.1)	11 (9.3)	4 (8.7)	19 (5.4)
Salmonella typhi	2 (1.1)	- (-)	- (-)	2 (0.6)
Other	3 (1.6)	5 (4.2)	3 (6.5)	11 (3.0)
Mixed infection, total	16 (8.4)	23 (19.5)	8 (17.4)	47 (13.3)
with *S. aureus*	ND	ND	ND	42 (11.9)
Total	190 (100.0)	118 (100.0)	46 (100.0)	354 (100.0)

ND, no data.

often grown from cases of posttraumatic osteomyelitis (Table 15.2), usually as secondary invaders. *Pseudomonas* osteomyelitis is usually of exogenous origin, but is also a typical complication in heroin addicts. The peculiar susceptibility of patients with sickle-cell disease to *Salmonella* osteomyelitis is well documented in studies from developing and developed countries [23]. The Nigerian studies on the relative frequency of *Salmonella* osteomyelitis compared with staphylococcal osteomyelitis in sickle-cell disease provide conflicting data (Table 15.1).

63 children with *Salmonella* osteomyelitis were reviewed in a study from Nigeria [3]. Fifty-seven patients carried the HbS gene in the homozygous state, and in another four it was associated with other hemoglobins. By the same token, 31 of 41 cases of *Salmonella* osteomyelitis occurred in those with sickle-cell disease as compared with 19 of 248 cases of staphylococcal osteomyelitis [5].

Characteristic features of *Salmonella* osteomyelitis are insidious onset, multiple bone involvement, and the need for prolonged treatment. Due to a comparatively mild course, many patients could be managed entirely as outpatients in the Nigerian series [3].

The differentiation of *Salmonella* osteomyelitis from bone infarction in vasoocclusive bone crisis is often impossible initially, because fever, bone pain, swelling, leukocytosis, periosteal elevation, and cortical destruction occur in both. Radiological features suggesting osteomyelitis rather than bone infarction are longitudinal, intracortical diaphyseal fissuring, multiple symmetrical bone involvement, and involucrum formation.

Diagnosis

The early diagnosis of acute osteomyelitis is difficult. Fever, local symptoms, such as pain and swelling, leukocytosis, and raised ESR are suggestive of osteomyelitis. The differential diagnosis includes pyomyositis and septic arthritis, both of which may coexist with osteomyelitis. Closed greenstick fracture, sickle-cell infarction, abscess, osteosarcoma, and deep vein thrombosis were other initial diagnoses in cases of osteomyelitis [24]. Diagnostic aspiration should always be considered. Treatment for osteomyelitis is justified on the basis of clinical suspicion because the chances of cure diminish with the time elapsed since the onset of disease. Radiological changes of osteomyelitis are usually apparent by 2-3 weeks after the onset of symptoms. This period is reduced to 7-10 days in infants and small children. Isotope scanning of bone reveals diagnostic changes earlier than radiology does. Ultrasonography of adjacent soft tissues may also be of value.

However, X-rays are also needed to establish the diagnosis of chronic osteomyelitis when there are no draining sinuses. The presence of sequestrum worsens the prognosis considerably and renders surgery necessary to achieve cure, if it can be achieved at all.

Isolation of the causative organism should always be attempted, as antibiotic therapy guided by culture and sensitivity testing can avoid treatment failure.

Even microscopic examination (Gram's stain and staining for acid-fast bacilli) of pus influences the choice of antibiotic, by differentiating between staphylococci, streptococci, gram-negative rods, and mycobacteria (see Table 15.7 in the next chapter: suggestions for initial choice of antibiotic in infective arthritis according to Gram's stain).

Blood cultures and aspirates of subperiosteal pus or bone are adequate in acute osteomyelitis. Wound swabs and cultures of intraoperative specimens of bone suffice in postoperative and posttraumatic osteomyelitis. The bacteriological examination of pus from draining sinuses in chronic osteomyelitis may be misleading, except for the finding of *Staphylococcus aureus*. Bacteriological examination of intraoperative specimens is preferable.

Treatment of Acute Hematogenous Osteomyelitis

Parenteral antibiotic treatment is started in acutely ill patients as soon as blood cultures have been taken. Subperiosteal abscesses should be aspirated prior to treatment and specimens submitted for bacteriological examination (Gram's stain, culture, sensitivity testing).

Ambulatory patients presenting with subacute hematogenous osteomyelitis may be treated with oral antibiotics from the onset (for dosage, see Table 15.4).

Cure and prevention of recurrent infection depend upon prolonged high-dose treatment (poor accessibility of bone), even if some African recommendations state rather low doses (Table 15.3).

Repeated aspiration or drainage may be necessary, especially in infants and young children, for relief of joints next to the focus of infection.

Rest and splintage of the affected limb are important because pathological frac-

Table 15.3. Daily dosage for parenteral treatment of osteomyelitis (administered in three to four doses)

Antibiotic	Adults	Children
Benzylpenicillin	8-20 MU	0.5-1.0 MU/kg
Cloxacillin	6-10 g	150-200 mg/kg
Clindamycin	1.2-1.8 (2.4) g	30-40 mg/kg
Lincomycin	1.2-1.8 g	30-40 mg/kg
Chloramphenicol	3-4 g	80-100 mg/kg

tures are common complications. Associated conditions such as severe anemia or malnutrition must be treated accordingly.

The following regimens of parenteral treatment cover the main pathogens, *Staphylococcus aureus* and streptococci. Treatment should be guided by sensitivity testing, whenever possible.

International Standard Therapy

Two alternative treatment regimens are widely used:

1. Cloxacillin + benzylpenicillin. The reasons for adding benzylpenicillin are twofold: (a) pharmacokinetics (the increased serum levels of cloxacillin) and (b) the superior activity against penicillin-sensitive pathogens.
2. Monotherapy with clindamycin (which replaces the former treatment with lincomycin).

Limited Resources

Monotherapy with cloxacillin and benzylpenicillin + chloramphenicol are the two regimens widely used (Table 15.3) under conditions of limited resources. Use of cloxacillin is preferable, since resistance of *Staphylococcus aureus* is more likely to be encountered with the latter combination regimen.

Tetracyclines are no longer considered effective agents in acute staphylococcal osteomyelitis.

Following clinical improvement (defervescence, resolution of local symptoms) - usually after 1-2 weeks - treatment can be changed to oral antibiotics (Table 15.4). In developed countries, clindamycin, penicillinase-resistant penicillins, or oral cephalosporins (cefalexin, cefradine, cefaclor) are used. Under conditions of limited resources either cloxacillin or chloramphenicol remain the drugs to use. Erythromycin is less effective in bone infections.

Total duration of treatment of 4-8 weeks is often recommended; acute staphylococcal osteomyelitis should be treated for a minimum of 3 weeks. Determinations of the ESR are helpful in monitoring treatment progress once resolution of clinical symptoms has been achieved.

In cases where penicillin-sensitive staphylococci or streptococci have been isolated, benzylpenicillin is the drug of choice, followed by oral penicillin later.

Table 15.4. Daily dosage for oral treatment of osteomyelitis (administered in three to four doses)

Antibiotic	Daily dosage	
	Adult (g)	Children (mg/kg)
Phenoxymethylpenicillin	4	100
Ampicillin	4	150
Amoxicillin	4	100
Cloxacillin	4	100
Cefradine	4	100
Cefalexin	4	100
Cefaclor	4	150
Clindamycin	1.2–1.8	30
Chloramphenicol	3	80–100
Fusidic acid	2	30

Hemophilus influenzae, a well-documented cause of osteomyelitis in infants and young children in developed countries, is covered by regimens including chloramphenicol. Alternatively, ampicillin can be used. Modern tetracyclines (doxycycline, minocycline) are further alternatives.

Osteomyelitis due to gram-negative enteric bacilli (except salmonella osteomyelitis) is best treated with a combination of cefotaxime and an aminoglycoside; chloramphenicol or co-trimoxazole may be used instead. Quinolones are apparently useful in adults.

When acute osteomyelitis of unknown etiology has responded to one of the regimens described above, oral treatment is continued with an antibiotic effective against *Staphylococcus aureus,* such as cloxacillin, clindamycin, or chloramphenicol.

Acute hematogenous osteomyelitis usually responds rapidly to adequate antibiotic therapy. If there is *no clinical improvement* within 48–72 h, surgical procedures should be considered for removal of any deep-seated collection of pus; material obtained from that procedure should be examined bacteriologically. Other foci of infection may be present elsewhere in the body.

Antibiotic resistance of S. aureus is another reason for treatment failure. This is the more likely if, in the absence of other reasons of nonresponse, Gram's stain has shown gram-positive cocci or culture has grown *S. aureus*. Chloramphenicol-resistant and even cloxacillin (methicillin)-resistant strains of *S. aureus* can be found in rural areas [12] where facilities for sensitivity testing do not commonly exist.

In children with *sickle cell disease* the antibiotic regimen should cover salmonellae from the onset (benzylpenicillin + chloramphenicol, or amoxicillin + cloxacillin). It is later modified according to bacteriological results. Chloramphenicol is still the drug of choice for *Salmonella osteomyelitis*. High doses of chloramphenicol (100 mg/kg/day) have to be given for up to 6 weeks, contrary to generally accepted recommendations, because of persistence of sensitive organisms in infarcted bone and lack of bactericidal activity. No case of marrow toxicity (reversible reticulocyte depression or irreversible aplastic anemia) was reported in a series of 63 children with *Salmonella* osteomyelitis from Nigeria [3]. Co-trimoxazole or amoxicillin in high dosage are alternatives. Quinolones may prove to be effective alternatives in the future. The safety of these agents in children, however, is not established, since cartilage damage has been reported in experimental animals. Cefotaxime (+gentamicin) may also be an effective, but expensive, alternative in the treatment of systemic *Salmonella* infections.

Surgical management tends to be limited to drainage of an obvious subperiosteal abscess or sequestrectomy, if indicated. The risks of surgery and anesthesia, as well as multiple skeletal involvement, should be borne in mind.

Treatment of Exogenous Osteomyelitis

The principles of treatment of exogenous osteomyelitis (posttraumatic and postoperative) consist in débridement surgery, removal of foreign bodies, and stabilization of the bone by external fixation. Chemotherapy is a necessary adjunct to surgery.

The choice of antibiotics for staphylococcal osteomyelitis is the same as in acute hemotogenous osteomyelitis (clindamycin, lincomycin, cloxacillin, chloramphenicol, in that order of preference).

In cases of exogenous osteomyelitis where cultures of wound swabs yield mixed growth of *Staphylococcus aureus* and enterobacteria, such as *E. coli, Proteus,* or *Klebsiella,* treatment of the staphylococcal component is more important.

Osteomyelitis due to monoinfection with enterobacteria should be treated according to the antibiogram. High-dose cefotaxime+gentamicin, or mezlocillin+gentamicin are recommended in Europe. Under limited conditions, chloramphenicol, co-trimoxazole, or ampicillin (invariably inactive against *Klebsiella*) can be used. First- and second-generation cephalosporins (cefazolin, cefamandole, cefuroxime) provide a more active, but expensive alternative. Monotherapy of osteomyelitis due to gram-negative rods with aminoglycosides has occasionally been recommended, but should not be practiced because of the poor therapeutic results experienced.

Osteomyelitis caused by *Pseudomonas aeruginosa* is difficult to treat. Commonly used antibiotics are ineffective. The use of synergistic combinations like piperacillin, azlocillin, or ticarcillin+gentamicin or tobramycin is necessary. Polymyxins are no therapeutic alternative because of their very poor tissue penetration. Quinolones may prove advantageous in the future.

Treatment of Chronic Osteomyelitis

Chemotherapy of chronic osteomyelitis, according to results of culture and sensitivity tests, is used as an adjunct to adequate operative treatment.

The culture of intraoperative specimens is preferable, since discharging sinuses may become colonized secondarily by bacteria not representing the initial pathogen. The isolation of *Staphylococcus aureus* should, however, not be dismissed as colonization.

Staphylococcus aureus is the commonest pathogen in chronic osteomyelitis, es-

pecially in cases originating from untreated or inadequately treated acute or subacute osteomyelitis. Clindamycin, lincomycin, or cloxacillin are the first-choice drugs, but erythromycin, co-trimoxazole, tetracyclines, or even rifampicin may also be effective. Prolonged treatment, often of several months' duration, is indicated. Insertion of chains of gentamicin-containing plastic beads during surgery, in addition to systemic chemotherapy, leads to increased antibacterial activity at the site of infection. The chains are generally removed several weeks after surgery by simply pulling (one end of the chain is kept loose in the wound at the time of surgery). Repeated instillations, suction drainage, or irrigation of bone cavities - sometimes performed in Europe - carry an even higher risk of contamination and secondary infection under conditions of poor hygiene.

The position of quinolones in the treatment of chronic osteomyelitis has yet to be defined.

Conservative management of chronic osteomyelitis is generally not recommended. Occasionally, exacerbations of chronic osteomyelitis with fever and pain may be controlled by chemotherapy alone, when surgery cannot be carried out.

Brucellosis, tuberculosis, infection with *Salmonella typhi* and systemic mycoses, paracoccidioidomycosis, and histoplasmosis, can also present as chronic osteomyelitis.

Osteomyelitis caused by *Brucella* spp., which is quite frequent in some parts of South America, has a predilection for the lumbar spine. The diagnosis is confirmed by serology, since blood cultures are often negative by the time of presentation. Therapy with tetracycline (+streptomycin), co-trimoxazole, or rifampicin must be continued for 6-8 weeks. The triple drug combination used initially in the treatment of tuberculous osteomyelitis of the spine should include rifampicin, if brucellosis cannot be excluded.

Penicillin (+metronidazole), tetracyclines, clindamycin, or chloramphenicol, which are all active against the mixed aerobic and anaerobic flora of the oral cavity, are used in cases of dental osteomyelitis. Actinomycosis, usually involving perioral tissues, can also affect the bone. Benzylpenicillin, 8-10 MU daily, for at least 2 weeks, is the drug of choice.

Osteomyelitis is common in advanced cases of mycetoma, mostly affecting the foot. Multiple discharging sinuses are usually present. If the infection cannot be controlled medically, amputation may become necessary to prevent the spread of the disease.

Secondary osteomyelitis due to *Staphylococcus aureus, Streptococcus pyogenes,* clostridia, *Bacteroides,* and other anaerobic bacteria complicating atherosclerotic ulcers of the foot is seen frequently in developed countries. Gram-negative enteric bacilli isolated from the ulcer have usually colonized the poorly perfused tissues. Because of vascular insufficiency, treatment with clindamycin or penicillin + metronidazole aims only at limitation of the infection and facilitation of demarcation. Subsequent surgery (amputation of the affected bones) is necessary in almost all instances.

Septic Arthritis

Incidence and Etiology

Septic arthritis, like osteomyelitis, is still a common medical problem in the tropics. Additional difficulties arise from late presentation. Cases of septic arthritis of the hip frequently become complicated by dislocation, avascular necrosis of the femoral head, destruction of the femoral head or of the epiphyseal growth plate, or bony ankylosis by the time of presentation [14, 25]. Data from several studies on the etiology of septic arthritis in developing countries are summarized in Table 15.5.

Septic arthritis occasionally is the result of direct inoculation of bacteria into the joint (trauma, surgery, injection of drugs). Extension of contiguous infection (osteomyelitis) and joint invasion of bacteria from the bloodstream during bacteremia are common causes of septic arthritis in the tropics.

Septic arthritis is mainly a disease of young children. The age distribution as reported in the study from Zaria, Nigeria [22] is shown in Fig. 15.1.

Staphylococcus aureus and *H. influenzae* are the most common pathogens of septic arthritis in childhood in developed countries [31]. *H. influenzae* was only infrequently isolated in cases of septic arthritis reported in developing countries (Table 15.5). The high incidence of *Salmonella* infections reported from Malawi is striking. No association with sickle-cell disease was found. *Salmonella* arthritis occurred only during the rainy season, the time of the peak incidence of diarrheal

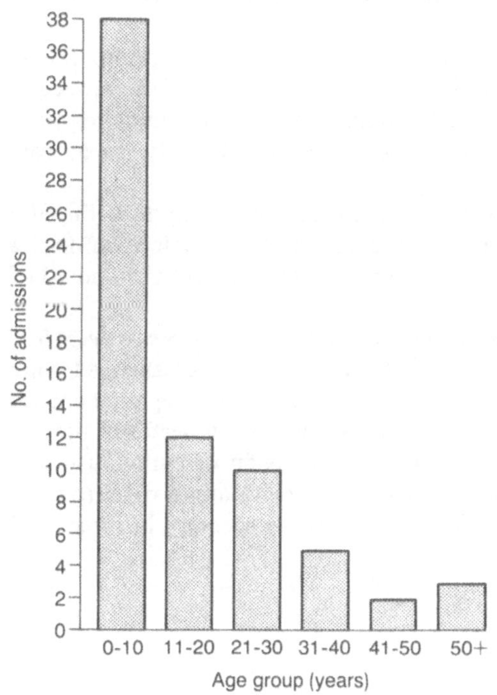

Fig. 15.1. Age distribution of 77 patients with septic arthritis, Zaria, Nigeria. From [22]

Table 15.5. Etiology of septic arthritis

Country [Reference]	Ibadan, Nigeria [14]	Ibadan, Nigeria [25]	Zaria, Nigeria [2]	Blantyre, Malawi [17]	Benin, Nigeria [21]
Type of study	Review of SA of the hip and shoulder	Review of SA of the hip	Review of all cases of SA	Prospective study of all cases of SA	Review of neonatal SA
Period of study	1967-1970	1970-1975	1973-1976	1979-1980 (15 months)	1974-1977
Total patients *(n)*	65	57	77	88	14
Joints affected *(n)*	85[a]	65	112[b]	100	14
Age (range)	ND	1 month to 13 years	Neonatal to 50 years	Neonatal to 11 years	Neonatal
Patients ≤1 year (%)	50	32	33	50	NA
Patients ≤5 years (%)	72	39	ND	85	NA
Sex					
Male	50	35	ND	ND	8
Female	15	22	ND	ND	6
Patients with osteomyelitis in adjacent bone *(n)*	65	11	3[c]	3	11
Bacteria isolated *(n)*					
Staphylococcus aureus	5	20	16	9	6
β-Hemolytic streptococci	1	0	2	10	1
Streptococcus pneumoniae	0	2	0	6	0
Hemophilus influenzae	1	1	0	3	0
Salmonella spp.	3[d]	6[e]	0	20[f]	0
Escherichia coli	1	0	0	1	0
Klebsiella spp.	0	4	3	0	0
Proteus spp.	3	2	0	1	2
Mixed infection	8	7	6	0	0
No pathogen isolated	5	15	15	38	5

SA, septic arthritis; ND, no data; NA, not applicable.

[a] Bacteriological data in 27 cases.

[b] Bacteriological data in 42 cases.

[c] Osteomyelitis developed subsequent to septic arthritis; cases showing radiological evidence of osteomyelitis on initial examination were excluded from the study.

[d] All three cases associated with Hb AS.

[e] All six cases associated with sickle-cell disease.

[f] No case associated with sickle-cell disease.

diseases, measles, malaria, and malnutrition. Children affected by *Salmonella* arthritis were younger than 5 years of age, often undernourished, and significantly more anemic than children with septic arthritis of different etiology [17].

Gonococcal arthritis is a typical and dangerous manifestation of disseminated gonococcal infection. Purulent monarthritis of the knee with recovery of gonococci from synovial fluid used to be the classical description. Presentation with systemic symptoms, skin rash (starting in Europeans as tiny erythematous papules, which may then become vesicular, pustular, or even necrotic), and polyarthralgia involving small joints is, however, more common. Blood cultures are negative, but in many patients, there is evidence of gonococcal infection in the genital tract. In contrast, cultures of synovial fluid are usually sterile. There may be concomitant salpingitis in women and a history of recent urethritis in men. A predisposing factor of pelvic sepsis was found in 4 of 77 patients with septic arthritis [22]. The differential diagnosis of gonococcal arthritis in young adults includes nongonococcal septic arthritis, Reiter's syndrome, rheumatic fever, and, in some cases, brucellosis.

In addition, infection with *Neisseria meningitidis* may mimic disseminated gonococcal infection or lead to septic arthritis.

The salient features of gonococcal and nongonococcal septic arthritis as observed in developed countries, are summarized in Table 15.6.

Table 15.6. Typical features of gonococcal and nongonococcal septic arthritis [31]

Gonococcal arthritis	Nongonococcal septic arthritis
Female preponderance	Males somewhat more often affected
Most common in adolescents, young adults	All ages involved
History of good health usual	Commonly predisposing factors
Polyarticular involvement usual	Monarthritis in most cases
Wrists and small joints often affected	Lower, large, or central joints often affected
Tenosynovitis, rash common	Tenosynovitis, rash rare
Venereal infection often apparent	Nonvenereal focus of infection frequent
Rapid response to antibiotic treatment	Slow response to antibiotics and drainage
Generally good prognosis	Variable prognosis

The common predisposing factors in nongonococcal septic arthritis are:

"1. Extraarticular site of infection, distant or contiguous
2. Infancy
3. Malignancy
4. Diabetes mellitus
5. Alcoholism
6. Immunosuppressive therapy
7. Chronic arthritis, especially rheumatoid arthritis
8. Prior joint trauma, joint surgery, intraarticular injection
9. Drug abuse
10. Presence of joint prosthesis
11. Hemoglobinopathy
12. Dog or cat bite or scratch" [31].

A less acute presentation is typical of tuberculous or fungal *(Sporothrix schenckii, Blastomyces dermatides,* and *Coccidioides immitis)* arthritis.

Arthritis in guinea worm infection *(Dracunculus medinensis)* occurs either as a result of the worm discharging larvae into the joint or as secondary bacterial infection. Aseptic arthritis is a third possibility.

Diagnosis

Clinical symptoms suggestive of septic arthritis (fever, sometimes with chills; warmth; pain and swelling in a single joint or, less commonly, in multiple joints, developing over several days; limited range of motion) will prompt needle aspiration of the joint.

The aspirate is examined microscopically (Gram's stain, acid-fast bacilli) and, when facilities are available, cultured. Blood cultures should also be taken, since they may be positive in cases where cultures of synovial fluid are sterile.

A primary focus of bacteremia (skin sepsis, otitis media, pneumonia, meningitis, salpingitis, urethritis) should be looked for and the appropiate specimens taken for bacteriological investigation.

Changes on X-ray are often not seen before 1 week after the onset of symptoms. They are, however, useful to demonstrate possible contiguous osteomyelitis. Ultrasonography may also be of value (effusion, abscess).

Synovial biopsy may be needed to establish the diagnosis of tuberculous or fungal arthritis.

Treatment

Cure and prevention of sequelae depend on early and vigorous treatment. Repeated aspirations of the joint and adequate systemic chemotherapy are crucial.

The initial antibiotic regimen is chosen according to the clinical setting and Gram's stain of synovial fluid (Table 15.7). This may be modified later according to the results of culture and sensitivity testing. Directed treatment following the isolation of a pathogen is as for septicemia (Table 9.7 of chapter "Septicemia"), since most antimicrobials achieve adequate concentrations in synovial fluid. For this reason, and because of the risk of superinfection and local irritation, instillation of antibiotics into the joint is not recommended.

As in other serious infections, parenteral therapy at full dosage is advisable, at least initially. When synovial fluid has become sterile and symptoms and signs of joint inflammation have diminished (after 3-6 weeks in staphylococcal and enterobacterial infections), treatment may be continued with oral antibiotics. In contrast to nongonococcal septic arthritis, prolonged treatment is usually not necessary in the gonococcal dermatitis-arthritis syndrome. The following regimens are suggested for treatment of gonococcal arthritis due to beta-lactamase-producing strains (PPNG): spectinomycin (2 g i.m.), cefotaxime (2 g i.m. or i.v.), or cefuroxime (1.5 g i.m. or i.v.) twice daily for 5-7 days. Alternatively, erythromycin, chloramphenicol, or tetracyclines may be employed (for 7-10 days).

Because of the high percentage of penicillin-resistant isolates of *Neisseria gonorrhoeae* in Southeast Asia and parts of Africa, penicillin should no longer be used in the initial treatment of known or suspected gonococcal infections in these areas. Where the prevalence of PPNG is still low, treatment with penicillin is appropriate: 5 MU benzylpenicillin i.m. + 1 g probenecid orally is recommended initially, followed by 1-2 MU benzylpenicillin every 6-8 h for 7 days, or for 2 days only followed by 3 MU procaine penicillin i.m. + 1 g probenecid orally daily for 5 days. Tetracyclines are a more convenient alternative in outpatient departments.

Table 15.7. Suggestions for initial choice of antimicrobial agents in the treatment of infective arthritis. Modified from [31]

Gram's stain of synovial fluid	Likely organisms[a]	Drugs of choice under conditions of limited resources	Alternatives	Comments
Gram-positive cocci				
Neonate	*S. aureus*, beta-hemolytic streptococci, *Strept. pneumoniae*	Cloxacillin	Clindamycin, cefazolin	Benzylpenicillin is the drug of choice for all streptococci
Child or adult	*S. aureus*, streptococci	Cloxacillin, chloramphenicol	Vancomycin, clindamycin, cefazolin	Resistance of *S. aureus* to cloxacillin is less likely than to chloramphenicol
Gram-negative cocci, tender rods				
Child, postneonatal to 3 years	*H. influenzae*, occ. *N. meningitidis*	Chloramphenicol, ampicillin	Cefotaxime	
Adolescent or adult	*N. gonorrhoeae*, occ. *N. meningitidis*	Spectinomycin, chloramphenicol, tetracycline, erythromycin	Cefotaxime, cefuroxime	Quinolones may be of value
Gram-negative bacilli				
Neonate	Enterobacteria, *Ps. aeruginosa*	Gentamicin (+ampicillin)	Cefotaxime + aminoglycoside	
Child, especially up to 5 years	*Salmonella* spp., occ. *H. influenzae* or enterobacteria	Chloramphenicol, ampicillin	Co-trimoxazole, cefotaxime, gentamicin	
Adult	Enterobacteria, *Ps. aeruginosa*	Gentamicin (+ampicillin)	Co-trimoxazole, chloramphenicol, cefotaxime + aminoglycoside	Quinolones may be of value
No organisms seen				
Neonate	*S. aureus*, enterobacteria, *Ps. aeruginosa*, streptococci	Cloxacillin + gentamicin	Clindamycin + cefotaxime + aminoglycoside	
Child, postneonatal to 5 years	*S. aureus*, *Salmonella* spp., enterobacteria, streptococci, occ. *H. influenzae*	Chloramphenicol (+cloxacillin), cloxacillin + ampicillin	Vancomycin, clindamycin, cefazolin + ampicillin, co-trimoxazole, cefotaxime	
Child, 5-15 years	*S. aureus*, streptococci, occ. enterobacteria	Cloxacillin + ampicillin	Vancomycin, clindamycin + chloramphenicol, cefotaxime	*N. meningitidis*, *N. gonorrhoeae*, *H. influenzae* rarely involved
Sexually active, healthy adult	*N. gonorrhoeae*, occ. *S. aureus*	Chloramphenicol, spectinomycin, tetracycline, (+cloxacillin)	Cefuroxime, cefotaxime	Quinolones may be of value
Chronic arthritis	*M. tuberculosis*	Triple drug combination		Rifampicin acts on both *M. tuberculosis* and *Brucella* spp.
	Brucella spp.	Tetracycline (+streptomycin), co-trimoxazole	Quinolones	

occ., occasional.

[a] Sensitivity testing is required for *S. aureus*, *N. gonorrhoeae*, enterobacteria, *Ps. aeruginosa*, and *Salmonella* spp.

(Where PPNG are prevalent, resistance of *N. gonorrhoeae* to tetracyclines is also common!)

Immobilization and rest of the inflamed joint is generally recommended only for the first few days of treatment.

Surgical drainage rather than needle aspiration is usually advised for septic arthritis of the hip in infancy (risk of dislocation, destruction of the epiphyseal growth plate, etc.). Surgical drainage should also strongly be considered in cases responding poorly to treatment (except tuberculosis). Substantial improvement should occur by 4-7 days of treatment. Patients who present late after several weeks often require surgery for removal of necrotic material, in addition to chemotherapy. Chronic arthritis in brucellosis is treated with tetracycline (+streptomycin) or co-trimoxazole. Tuberculous arthritis requires long-term triple-drug therapy.

Pyomyositis

Incidence and Etiology

Pyomyositis (synonyms: tropical myositis, tropical pyomyositis) is a disease characterized by the development of one or more abscesses in the skeletal muscles together with the symptoms and signs of acute infection. The disease is relatively common in some parts of the tropics (Africa, Papua New Guinea), but is only rarely seen in temperate climate [10, 29, 30].

The initial pathogenesis of pyomyositis is unknown [29]. Apparently, myositis or a myositis-like state (resulting from viral infection, migrating filarial larvae, trauma, or bacterial toxin) predisposes to secondary bacterial infection. Pyogenic organisms, mostly ***Staphylococcus aureus,*** invade the damaged muscle during an episode of bacteremia leading to the formation of single or multiple intramuscular abscesses. The large muscle groups of the thigh, calf, buttocks, back, abdominal wall, arm and shoulder, are usually affected, but no muscle is exempted.

The majority of cases occur in children and young adults, with a preponderance of males; mortality is low, in the range of 1%-4% (Table 15.8).

A small proportion of patients, however, present with severe fulminant disease, the clinical picture being dominated by septicemia and metastatic abscesses. Pneumonia, lung abscesses, pericarditis, and myocarditis are the commonest complications, but metastatic abscesses of the kidney, liver, and brain have all been reported [1, 28, 32]. In one pediatric series of pyomyositis, about 25% of the children were found to develop concomitant osteomyelitis [2].

Diagnosis

Most patients present with the clinical features of an acute abscess, fever, and a hot, painful swelling. As a rule, a tender, tense mass, which may be fluctuant, is palpable beneath the skin. Large-bore needle aspiration will yield thick, yellow pus (though not in the early nonsuppurative stage). At that stage, signs of inflam-

Table 15.8. Etiology of pyomyositis

Study location [Reference]	Kampala, Uganda [9]	Kampala, Uganda [26 a]	Harare, Zimbabwe [7]	Zaria, Nigeria [15]	Benin, Nigeria [6]	Ibadan, Nigeria [2]	
Period of review	1964	ND	1970–1974	1971–1974	1973–1977	1972–1978	
Cases *(n)*	79	106	100	90	112	63	
Age							
⩽10 years	11[a]	32	50	29	35.7%[b]	Children only[c]	
⩽20 years	14[a]	30	27	20			
⩽30 years	11[a]	24	8	20			
Sex							
Male	36[a]	72	58	66	87	37	
Female	17[a]	34	42	24	25	26	
						With osteomyelitis[d]	Others[e]
Multiple abscesses (%)	ND	12.0	6.0	37.8	42.8	56.3	55.3
Mortality (%)	ND	0.9	4.0	3.3	0.9	6.3	2.1
Bacteria isolated from pus *(n)*							
Staphylococcus aureus	74 (93.7%)	94 (88.7%)	75 (75.0%)	35[f] (75.5%)	101 (90.2%)	15[g] (100%)	39 (83.0%)
Streptococcus (beta-hemolytic)	1	1	2	0	4	0	0
Streptococcus pneumoniae	0	0	1	0	0	0	0
Other streptococci	3	0	0	0	1	0	0
Escherichia coli	0	4	1	1	1	0	3
Klebsiella spp.	0	0	0	0	0	0	3
Mixed infection	1	0	0	2[j]	0	0	0
No pathogen isolated	0	2	21	9	5	0	2
Results lost	0	5	0	0	0	0	0
Cases with positive blood cultures (%)	ND	7.5	3.0	10[h]	ND	100	47.1[i]

ND, no data. [a] Out of 53. [b] Pediatric patients. [c] 6 months–12 years, peak at 6 years. [d] n=16. Cases with radiological evidence of osteomyelitis at presentation were excluded. Nine cases were detected radiologically within 2 weeks of admission; the other seven by 3 weeks after admission. [e] $n=47$. [f] With regard to cases of bacteria isolated from pus, $n=47$ in this study. [g] With regard to cases of bacteria isolated from pus, $n=15$ in this study. [h] Out of 30. [i] Out of 34. [j] Contamination.

mation are less prominent and a hard indurated mass is felt upon palpation of the muscle. The differential diagnosis includes sarcoma and tuberculosis in these cases.

Septic arthritis and osteomyelitis are the main differential diagnoses of an abscess occurring in a limb. Tenderness is diffuse and not well localized, as in osteomyelitis. Pyomyositis involving the cervical muscles may mimic meningitis because of neck stiffness. An abscess located in the psoas muscle is difficult to distinguish from appendicitis.

Laboratory signs of acute pyogenic infection, neutrophil leukocytosis, and raised ESR are usually present; anemia is another common finding. Radiology is helpful to exclude osteomyelitis. Ultrasonography may localize any abscess(es). Where facilities are available, pus should be stained and cultured and sensitivity testing performed. Blood cultures may show septicemia.

Patients with the septicemic type of illness should be carefully examined for the presence of murmurs (endocarditis), cardiac enlargement (pericarditis, myocarditis), and metastatic abscesses.

Treatment

Nearly all patients presenting in the early nonsuppurative stage can be expected to respond to systemic chemotherapy.

All abscesses require incision and adequate drainage. In addition, systemic chemotherapy is suggested in order to prevent or abort further abscess formation. Chemotherapy is always indicated in patients with septicemia.

Since most strains of *Staphylococcus aureus* are resistant to benzylpenicillin [4, 6], cloxacillin or, alternatively, chloramphenicol should be employed. Erythromycin or co-trimoxazole may be used according to the sensitivity pattern, but are less valuable, especially in septicemia.

Experience in pyomyositis with international standard regimens for severe staphylococcal disease (clindamycin, vancomycin, cephalosporins, fusidic acid, and rifampicin) has not been reported. Regimens used in staphylococcal osteomyelitis can, however, be expected to be of value in the treatment of pyomyositis.

References

1. Adeloye A (1982) Intracranial suppuration complicating tropical myositis. Report of 2 cases. Trans R Soc Trop Med Hyg 76: 463–464
2. Aderele WI, Osinusi K (1980) Pyomyositis in childhood. J Trop Med Hyg 83: 99–104
3. Adeyokunnu AA, Hendrickse RG (1980) Salmonella osteomyelitis in childhood. Arch Dis Child 55: 175–184
4. Buttner DW, Westhoff H (1973) Antibiotic sensitivity of *Staphylococcus aureus* in Uganda with special reference to pyomyositis and osteomyelitis. East Afr Med J 50: 74–85
5. Carayon A, Courbil J, Ferro R-M, Merrien Y (1978) Les infections osseuses drepanocytaires. Med Trop 38: 213–221
6. Chiedozi LC (1979) Pyomyositis. Review of 205 cases in 112 patients, Am J Surg 137: 255–259
7. Coots D, Gelfand M (1977) Tropical myositis in Rhodesian Africans. J Trop Med Hyg 80: 9–10

8. Ebong WW, Oyemade GAA (1978) Acute haematogenous osteomyelitis in Nigeria. Trop Geogr Med 30: 451-461
9. Foster WD (1965) The bacteriology of tropical myositis in Uganda. J Hyg 63: 517-524
10. Horn CV, Master S (1968) Pyomyositis tropicans in Uganda. East Afr Med J 45: 463-471
11. Isaacs-Sodeye WA (1975) Sickle cell disease: a handbook for general clinicians. In: Onwubalili JK (1983) Sickle cell disease and infection, special review. J Infect 7: 2-20
12. Jayaguru AS, Martin KS (1983) Acute osteomyelitis in Marondera: cause for concern. Cent Afr J Med 29: 191-193
13. Kaine WN (1983) Morbidity of homozygous sickle cell anaemia in Nigerian children. J Trop Pediatr 29: 104-111
14. Kolawole TM, Bohrer SP (1972) Acute septic arthritis in Nigeria; A review of 65 cases involving the hip and shoulder joints. Trop Geogr Med 24: 327-338
15. Ladipo GOA, Fakunle YF (1977) Tropical pyomyositis in the Nigerian Savanna. Trop Geogr Med 29: 223-228
16. Mlay SM, Shija JK (1985) The incidence and pattern of osteomyelitis in infancy and childhood and the role of sickle cell disease in Dar Es Salaam. J Trop Pediatr 31: 292-294
17. Molyneux E, French G (1982) Salmonella joint infection in Malawian children. J Infect 4: 131-138
18. Ogunjumo DO (1981) Aetiology and control of chronic pyogenic osteomyelitis in Ile-Ife, Nigeria. Trop Doct 11: 155-159
19. Ogunjumo DO (1982) The clinical pattern of chronic pyogenic osteomyelitis in a Nigerian community. J Trop Med Hyg 85: 187-194
20. Okoroma EO, Agbo DC (1984) Childhood osteomyelitis, a five-year analysis of 118 cases in Nigerian children. Clin Pediatr 23: 548-552
21. Omene IA, Odita IC (1979) Clinical and radiological features of neonatal septic arthritis. Trop Geogr Med 31: 207-212
22. Onyemelukwe G, Sturrock RD (1979) Septic arthritis in Northern Nigeria. Rheumatol Rehabil 18: 13-17
23. Onwubalili JK (1983) Sickle cell disease and infection, special review. J Infect 7: 2-20
24. Oyemade GA, Dawodu AH, Olusanya AO (1977) Osteomyelitis in Nigerian children. J Trop Med Hyg 80: 183-186
25. Oyemade GAA (1980) Presenting features of septic arthritis of the hip in Nigerian children. Trop Geogr Med 32: 145-150
26. Padonou N, Diagne BA, N'Doye M, Díop A (1979) Ostéomyélitis à salmonella de l'enfant (à propos de 20 observations). Dakar Méd 24: 29-35
26a. Pinto DJ (1972) Treatment of pyomyositis. East Afr Med J 49: 643-650
27. Raghava Rao LV, Chacko V, Shivananda PG, Rao KNA (1977) Changing trends in the bacteriology of bone and joint infections. J Indian Med Assoc 69: 190-192
28. Scrimgeour EM, Kaven J (1982) Severe staphylococcal pneumonia complicating pyomyositis. Am J Trop Med Hyg 31: 822-826
29. Shepherd JJ (1983) Tropical myositis: Is it an entity and what is its cause. Lancet II: 1240-1242
30. Smith PG, Pike MC, Taylor E, Taylor JF (1978) The epidemiology of tropical myositis in the Mengo District of Uganda. Trans R Soc Trop Med Hyg 72: 46-53
31. Steigbigel NH (1983) Diagnosis and management of septic arthritis. In: Remington JS, Swartz M (eds) Current clinical topics in infections diseases, vol 4. McGraw-Hill, New York, pp 1-29
32. Taylor JF, Templeton AC, Henderson B (1970) Pyomyositis. A clinico-pathological study based on 19 autopsy cases, Mulago Hospital 1964-1968. East Afr Med J 47: 493-501

16 Skin and Wound Infections

Common Bacterial Infections

There are many similarities between skin and wound infections, including several intermediate forms. Skin and wound infections are often caused by the same spectrum of pathogens.

Incidence

For convenience, bacterial skin infections may be divided into four groups:

1. Superficial skin infections
2. Deep skin infections
3. Cutaneous manifestations of systemic infections
4. Bacterial superinfections of other infectious or noninfectious conditions

All these types of skin infections are very common in developing countries. Malnutrition, overcrowding, lack of water, poor hygiene, and also microlesions and parasitic diseases contribute to the high frequency [32]. In particular, streptococcal pyoderma, either primary or as superinfection of scabies or eczema, is far more frequent than in developed countries. Some data on its prevalence are given in Table 16.1. Streptococcal pyoderma is commonest in children under 10 years of age (Figs. 16.1 and 16.2). The warm, humid climate is an additional factor accounting for its high prevalence in developing countries (Figs. 16.2 and 16.3). Strepto-

Table 16.1. Prevalence of pyoderma: data from four studies from tropical countries

Country	Reference	Year of study	Study population		Prevalence of pyoderma (%)
Nigeria	[30]	1972	5300	6-year-old schoolchildren	15.0
Panama	[2]	1972	1084	villagers of all ages	15.2
India	[19]	1973-1974	6651	children below 15 years of age in 23 rural villages	4.3[a]
		1974	374	rural schoolchildren	15.0
		1974	664	urban schoolchildren	14.9
Tanzania	[23]	1973	532	preschool children in five villages	7.0[b]

[a] Case finding by trained field workers.
[b] Only cases of primary pyoderma were included - if, in addition, secondarily infected scabies and eczema had been included, the total prevalence rate would have risen to 38%!

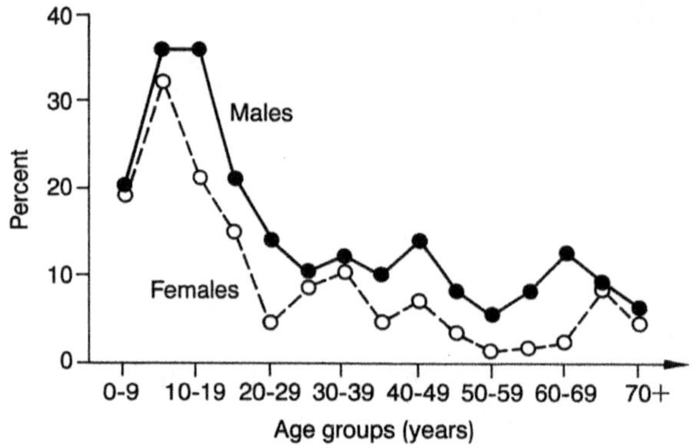

Fig. 16.1. Age-specific distribution of pyoderma. Survey of 3770 villagers in rural Ghana, 1975. From [4]

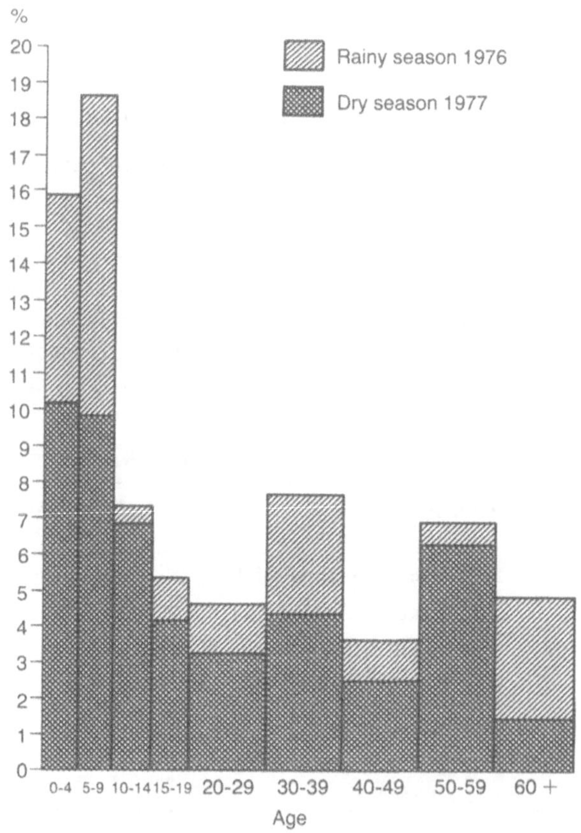

Fig. 16.2. Age-specific distribution of pyoderma during the rainy and dry seasons in a Gambian village. From [33]

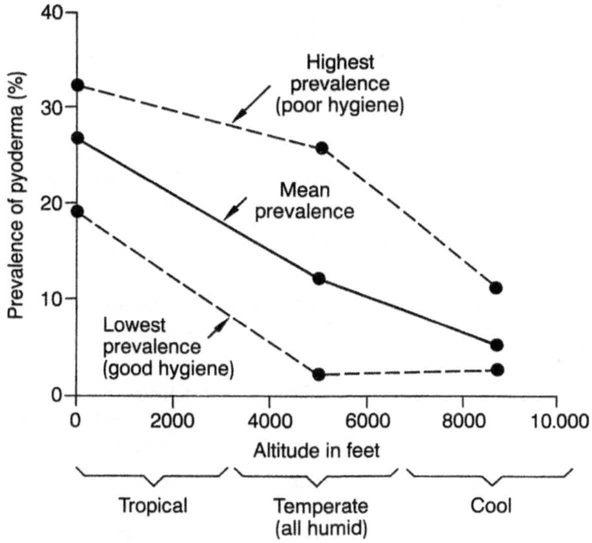

Fig. 16.3. Prevalence of pyoderma during September, 1971, in lower socio-economic class Colombian children in relation to altitude (climate) and level of hygiene. From [44]

coccal superinfection of scabies, also frequently encountered in developing countries, has been associated with acute glomerulonephritis [16, 24, 48].

Wound infections play an important role in tropical traumatology. Highly contaminated wounds are frequently seen. Traditional forms of nonsterile treatment are sometimes involved in the pathogenesis of wound infections, e.g., in the development of neonatal tetanus. Often medical help is sought only when the infection is already far advanced. Postoperative wound infections are also seen with remarkable frequency (Table 16.2). Extensive cross-infection due to overcrowding, poor hygiene, inadequate supplies, and nonsterile dressings - but also to a remarkable abuse of antibiotics - may all contribute to this situation.

Treatment of wound infections can be grouped according to four major situations:

1. Treatment of established wound infections
2. Prophylactic treatment of wounds
3. Perioperative prophylaxis
4. Use of antibiotics in septic surgery

Etiology

Group A streptococci play a far more important role in developing countries, but beta-hemolytic streptococci of other serological groups are also occasionally isolated from pyoderma lesions (Table 16.3). Peptostreptococci and even pneumococci are sometimes isolated in skin and wound infections.

Table 16.2. Incidence of postoperative wound infections

Study location	[Reference]	Year of study	Classification of operations	No. of operations	Incidence of postoperative wound infections (%)
Ibadan, Nigeria	[38]	ND	Caesarean section, gynecological laparatomy	76 24	27.0
New Delhi, India	[43]	ND	All types	244	21.7
Port Moresby, Papua New Guinea	[14]	1972–1973	"Infection was not likely to occur"	100	10.0
Pune, India	[21]	ND	Clean	230	10.4
			Clean-contaminated	107	21.5
			Contaminated	23	56.5
Hyderabad, India	[36]	ND	Clean	22	40.9
			Potentially contaminated	68	35.3
			Contaminated	10	90.0
Ibadan, Nigeria	[10]	1978	Clean	2050	8.6
			Potentially contaminated	1313	10.1
			Contaminated	432	18.1

ND, no data.

Table 16.3. Distribution of beta-hemolytic streptococcal groups in pyoderma studies

Country	[Reference]	Year of study	Total no. of strains	Distribution of groups (%)				
				A	B	C	G	Other[a]
Tanzania	[25]	1972	50	92.0	-	2.0	-	6.0
Uganda	[29]	ND	71	96.0	ND	ND	ND	ND
Panama	[2]	1972	133	89.0	ND	ND	ND	ND
Nigeria	[30]	1972	402	89.5	-	2.9	6.4	9.0
Ghana	[4]	1975	56	-	-	17.9	82.1	-
India	[19]	1973–1974	310	91.6	1.3	1.6	5.5	-
India	[24]	1976	67	86.6	-	4.5	6.0	3.0
India	[17]	ND	56	87.5	-	5.4	5.4	1.8
Papua New Guinea	[8]	1981	45	46.7	6.7	22.2	24.4	-

ND, no data.
[a] Includes untypable strains.

Staphylococcus aureus is the other main pathogen of skin and wound infections. Staphylococcal skin sepsis should never be disregarded in neonates, where it may give rise to septicemia. In pyoderma, however, *S. aureus* is generally considered a secondary invader, although staphylococci are recovered from pyoderma lesions with considerable frequency (Table 16.4). These lesions usually respond to treat-

Table 16.4. Types of bacteria isolated from pyoderma

Country	[Reference]	Year of study	Study population	Bacteria isolated (%)				
				Beta-hemolytic streptococci only	*Staphylococcus aureus* only	Both	Neither	Total beta-hemolytic streptococci
Vietnam	[1]	1968–1969	122 soldiers	6.0	20.0	47.0	19.0	53.0
Uganda	[29]	ND	89 pediatric clinic patients	28.7	11.7	ND	ND	86.0
Tanzania	[25]	1972	96 pediatric clinic patients	16.7	15.6	31.3	28.1	52.1
Panama	[2]	1972	160 villagers of all ages	ND	ND	ND	ND	83.1
Ghana	[4]	1975	76 villagers of all ages	27.6	17.1	46.1	18.4	73.7
India	[17]	ND	101 mostly pediatric clinic patients	31.7	29.7	20.8	ND	55.4
Papua New Guinea	[8]	1981	39 pediatric clinic[a] patients	ND	ND	ND	ND	94.9

ND, no data.
[a] Includes 12 patients with typical ulcers.

Table 16.5. Bacteriology of postoperative wound infection

Bacteria isolated (%)	New Delhi, India [43]	Hyderabad, India [36]	Ibadan, Nigeria [10]	Addis Ababa, Ethiopia [13][a]
Staphylococcus aureus	30.9	23.8	38.2	19.5
Beta-hemolytic streptococci	–	1.2	1.3	3.2
E. coli	25.0	19.4	15.1	13.9
Klebsiella spp.	13.2	10.0	17.4	9.0
Proteus spp.	11.8	11.3	15.5	27.9
Pseudomonas spp.	19.1	27.5	9.3	15.0
Clostridium tetani	–	0.6	–	–[b]
Bacteroides spp.	–	–	0.6	–[b]
Peptostreptococci	–	–	0.4	–[b]
Others	–	6.2	2.1	11.5

All data expressed in percent.
[a] Some of the isolates were obtained preoperatively.
[b] Only aerobic culture was performed.

Table 16.6. Isolation rates (%) from burn wounds of various pathogens in relation to time. Kampala, Uganda [18]

Time after burn	1st and 2nd day	3rd day	4th and 5th day	6th and 7th day
No. of patients	63	53	9	32
Bacteria isolated				
Beta-hemolytic streptococci	14.3	5.7	11.1	–
Staphylococcus aureus	12.7	17.0	11.1	46.9
Pseudomonas aeruginosa	1.6	5.7	11.1	12.5
Gram-negative enteric bacilli	7.9	47.2	55.5	50.0
Sterile	44.4	45.3	33.3	28.1

Some totals are greater than 100% because of mixed infections.

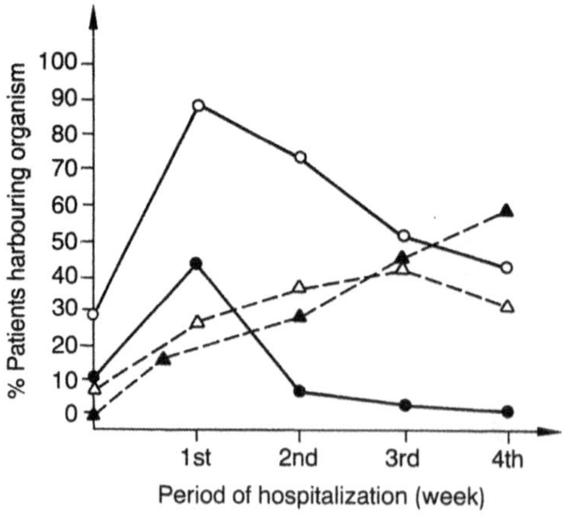

Fig. 16.4. Isolation rates (%) from burn wounds of various pathogens at the time of admission and during the course of hospitalization. ○——○ *Staphylococcus aureus;* ●——● beta-hemolytic streptococci; ▲——▲ *Pseudomonas aeruginosa;* △——△ gram-negative bacilli. India [39]

ment with benzylpenicillin despite colonization with penicillin-resistant staphylococci. *S. aureus* is also the causative agent of a considerable proportion of postoperative wound infections (Table 16.5). Coagulase-negative staphylococci *(S. epidermidis)* are the commonest pathogens in infections of intravascular devices and other foreign bodies. Their isolation in other situations does not usually represent true infection.

Pseudomonas aeruginosa plays an important role worldwide as secondary invader of preexisting ulcers. It is a main pathogen in burns, typically occurring 2-4 weeks after the injury. The bacteriology of burns shows worldwide a typical time course (Table 16.6, Fig. 16.4); group A streptococci are isolated predominantly in the first few days, followed by *S. aureus*, enterobacteria, and *Pseudomonas*.

Enterobacteria (E. coli, Proteus spp., *Klebsiella* spp.) are frequently isolated in superficial mixed infection from burns and from postoperative and posttraumatic wounds. Their etiological role is always difficult to evaluate. Colonization of wounds by enterobacteria often hinders the isolation of far more important pathogens (*Bacteroides* spp., peptostreptococci, clostridia). These anaerobes are common pathogens of postoperative and posttraumatic wound infections, but not of skin infections. Group D streptococci (enterococci) are frequently isolated from the skin and superficial wounds, but their clinical importance is not very pronounced.

Specific wound infections are now very rare in developed countries. Cutaneous diphtheria (desert sore, veld sore [22]) and wound diphtheria are still encountered in some developing countries, but several studies indicate that colonization of preexisting skin lesions by *Corynebacterium diphtheriae* is much more common (Table 16.7). Other corynebacteria, like *Corynebacterium pyogenes*, have also been associated with skin infections [20]. Tuberculosis, leprosy, and other mycobacterial infections *(M. ulcerans, M. balnei)* are important in the differential diagnosis of chronic skin infections. Tetanus is still common in many developing countries, usually originating from minor trauma or superficial infections.

Classical infectious diseases with cutaneous manifestations (plague, anthrax, tularemia, glanders, melioidosis, lymphogranuloma inguinale) may be of regional importance. Nonbacterial infections (leishmaniosis, histoplasmosis) may mimic bacterial infections.

Certain agents which are confined to warm, humid climates can be isolated as pathogens of severe skin infections only in the tropics and subtropics: *Chromobacterium violaceum, Vibrio vulnificus, Aeromonas hydrophila, Mycobacterium ulcerans, Mycobacterium balnei.*

Tropical ulcer, a frequent skin condition in tropical climates, is caused by the same pathogens as cancrum oris (noma), namely *Fusobacterium fusiforme* and *Borrelia vincenti*. Why common and widely distributed pathogens lead to chronic necrotizing ulcers, mainly of the lower leg, is poorly understood, but tropical ulcers are more common in malnourished and debilitated children.

There are several other skin conditions in the tropics which frequently require the use of antibiotics. These tropical skin infections are listed in Table 16.8.

Similarly, there are several wound infections characteristic for developing countries which, because of frequency or type, also commonly require the use of antibiotics. This group of typical tropical wound infections is presented in Table 16.9.

Table 16.7. Isolation rates, biotypes, and toxigenicity of *Corynebacterium diphtheriae* from skin lesions: data of five studies from tropical countries

Country	[References]	Year of study	Study population	Type of skin lesion *(n)*	Isolation rate of *C. diphtheriae* (%)	No. of strains typed	Biotypes (%)			No. of strains tested for toxigenicity	Toxigenicity (%)
							Gravis	Mitis	Intermedius		
Uganda	[5]	1968-1969	139 preschool and schoolchildren	Impetigo and ulcers (50)	32.0	17	11.8	76.5	ND	47	4.3
				Ulcers only (89)	59.6	57	3.5	84.2	ND		
Tanzania	[26]	ND	94 children	Pustules and vesicles (33)	51.5	17	-	94.1	5.9	17	41.2
			9 adults	Ulcers (44)	56.8	25	20.0	76.0	4.0	25	40.0
				Crusted lesions (9)	88.9	8	12.5	75.0	12.5	8	50.0
				Others (17)	82.4	14	28.6	71.4	-	14	35.7
Burma	[45]	1971	493 patients at the VD and dermatology clinic	Scabetic eczema (21)	42.8	ND	ND	ND	ND	4	50.0
				Impetiginous scabies (358)	71.2	ND	ND	ND	ND	85	17.6
				Impetigo (56)	41.0	ND	ND	ND	ND	13	15.3
				Impetiginous eczema (58)	50.0	ND	ND	ND	ND	6	16.6
India	[9]	ND	200 surgical patients	chronic skin ulcers (120), posttraumatic wounds (50), postoperative wounds (30)	12.0	24	12.5	83.3	4.2	24	20.8
			50 patients with burns	Burn (50)	0.0	-	-	-	-	-	-
Papua New Guinea	[28]	1983	99 pediatric clinic patients	Infected skin sores (99)	87.9	92[a]	21.7	76.1	2.2	92[a]	5.4

ND, no data.
[a] Five double populations of *C. diphtheriae* gravis and mitis occurred.

Table 16.8. Characteristic tropical skin infections

Streptococcal superinfection of scabies
Secondary infections in tungiasis
Secondary infections in dracontiasis
Secondary infections of trophic ulcers in leprosy
Foot infections in people walking barefoot
Burns in epileptics
Secondary infections in snakebites, spiderbites, scorpion stings, and other injuries associated with local necrosis by other poisonous agents
Infections in animal bites, mainly inflicted by carnivores (dogs, cats, lions, etc.)
Secondary infections as a result of superstitious or traditional procedures (incisions, circumcision, tattoes, etc.)
Umbilical infections in neonates due to nonsterile procedures

Table 16.9. Characteristic events leading to tropical wound infections

Traffic accidents in the bush or jungle
Emergency operations under nonsterile conditions
Surgery under warlike conditions
Farm accidents
Attacks using traditional weapons (bush knife, spear, arrow, etc.)

Diagnosis

Treatment of skin and wound infections in developing countries cannot usually rely on bacteriological services. It must, however, be taken into consideration that not even sophisticated bacteriological techniques permit easy distinction between superficial colonization and true infection. The correlation between isolated bacteria and real pathogens is often quite poor. In severe wound and tissue infections, the underlying pathogens can be isolated in a remarkable percentage of cases by blood culture. In less severe cases, important information can be gained by Gram's stain of pus or wound secretions. It is comparatively easy to differentiate between streptococci, staphylococci, gram-negative rods, and the mixed bacteria usually seen in anaerobic infections. Absence of bacteria on Gram's stain may be a sign of mycobacterial infection (acid-fast stain!), some rare pathogens *(Brucella)*, or a nonbacterial condition.

In general, clinical signs of infection (aspect, swelling, pain, lymphangitis, lymphadenitis, exudate, suppuration, odor, fever, and prostration) are far more important than laboratory results. For several pathogens, there is a typical appearance of pus in skin and wound infections (Table 16.10). Meticulous anamnesis concerning trauma, operation, foreign bodies, course and duration of illness, pain, fever, and underlying diseases may give important information on the nature of the skin or wound infection.

If bacteriology is available, certain bacteria, like *Acinetobacter,* some *Pseudomonas* species (except *Ps. aeruginosa*), enterococci, *Staphylococcus epidermidis,* and *Propionibacterium acnes,* are frequently isolated. They represent common external

Table 16.10. Appearance of pus and related pathogens

Appearance of pus	Pathogen(s)
Yellow, sticky	*Staphylococcus aureus*
Fluid, sanguinolent	Group A streptococci
Fluid, sanguinolent, sometimes green or with typical odor	*Pseudomonas aeruginosa*
Fluid, feculent, malodorous	Anaerobes (*Bacteroides*, peptostreptococci)
Chronic, friable, whitish gray	*Mycobacterium tuberculosis*

colonizations and not true infections. Even though enterobacteria (*E. coli, Klebsiella* spp., *Proteus* spp., *Serratia* spp.) may give rise to skin and wound infections (usually together with anaerobes), far more often they are isolated as superficial colonizations. This is particular common in lesions of the genital and perianal region. Sometimes isolation of *E. coli* in wounds is only a sign of fecal contamination of the wound. Even culture of *Staphylococcus aureus* does not prove true infection; superficial growth without clinical importance is not infrequent.

Treatment

Initially, a decision has to be made whether treatment with antibiotics is needed at all. Many minor local infections have a tendency to heal spontaneously.

Topical Treatment

Topical treatment is sufficient in many skin and wound infections [6, 31]. Local antibiotic treatment is only justified when the infection is very superficial. Antibiotics do not diffuse into intact skin, thus topical antibiotics are unlikely to influence deep skin infections. Superficial infections susceptible to local treatment include superficial pyoderma, impetigo, some purulent wounds, second- or third-degree burns, secondarily infected ulcers, and eczema. Local antibiotics are ineffective in cutaneous and subcutaneous infections such as erysipelas, boils, cellulitis, erysipeloid, anthrax, and cutaneous tuberculosis.

Local antibiotics for skin infections should fulfill the following criteria:
1. Slow development of bacterial resistance during treatment
2. Low risk of hypersensitivity
3. No systemic use (risk of resistance or allergy)

Topical antibacterial therapy is not yet really developed. Classical, but highly toxic antibiotics like neomycin, bacitracin and tyrothricin are often mixed with several other agents in crude commercial preparations. In general it can be recommended that only single antibiotics or topical antibacterial agents be used. Well-established antibiotic combinations (neomycin + bacitracin; neomycin + bacitracin + polymyxin) are expensive and do not fulfill the criteria of modern antibiotic therapy. The topical pharmacology, even the galenic formulas of topical antibiotics, is not yet fully understood. Cheap disinfectants, which usually display 1/1000 to 1/10000 of the efficacy of an antibiotic, are fairly effective in many superficial infections.

Modern disinfectants like povidone-iodine and chlorhexidine are apparently superior to clioquinol, ethacridine, benzalkonium, and other traditional disinfectants. Silver sulfadiazine and mafenide are valuable in the treatment of burns.

The WHO list of essential drugs contains two disinfectants, chlorhexidine and iodine, both as solution, and a combination of neomycin and bacitracin prepared as an ointment for topical antibacterial treatment [49].

The advantages of topical antibiotics in the treatment of skin infections are controversial. High concentrations on the skin may achieve a greater bactericidal activity than usual systemic treatment. Even resistant strains are usually covered by topical concentration. Antibiotic sensitivity testing therefore has little or no application for topical treatment. Traditional antibiotics like tetracycline or chloramphenicol are of limited value for topical use. The risk of hypersensitivity renders all penicillins and cephalosporins unsuitable for any superficial topical application.

For topical treatment of skin infections, the appropriate preparation must be chosen (ointment, cream, spray, powder, or solution). Sprays and solutions are generally more effective than creams or ointments; a cream, as an emulsion of oil in water, is usually better than an ointment, which does not contain water. When the skin is dry, ointments are preferable, especially for longer courses of treatment. The removal of exudate, crusts, or areas of thick skin with a keratolytic ointment improves the conditions for topical antibiotic treatment.

Systemic Treatment

The range of activity of available antibiotics does not match perfectly with the spectrum of common pathogens of skin and wound infections. The main agents, staphylococci, streptococci, and anaerobes, are only partially covered by antibiotics of the WHO list. Cloxacillin is active only against streptococci and staphylococci; it is inactive against many anaerobic bacteria. Many staphylococci are now tetracycline-resistant. Co-trimoxazole is generally not recommended for cutaneous and wound infections. Chloramphenicol is quite effective but carries considerable risks in minor infections. Treatment with benzylpenicillin, procaine penicillin, or phenoxymethylpenicillin covers only the very dangerous beta-hemolytic streptococci. Treatment with procaine penicillin was found to be associated with an improved healing rate compared to placebo in pediatric pyoderma, despite the fact that penicillin-resistant staphylococci are frequently isolated from these lesions [7] (see Table 16.4). Aminopenicillins such as ampicillin offer no advantage.

Certain agents (not in the WHO list) are quite suitable for treatment of skin and wound infections. Cefazolin, a first-generation cephalosporin, is advantageous in the treatment of wound infections. Clindamycin and, to a lesser extent, lincomycin cover the three major pathogens - streptococci, staphylococci, and anaerobes. The use of erythromycin in wound infections cannot be advocated because of the high risk of secondary resistance. Some staphylococcal wound infections that are difficult to treat require the use of expensive special antistaphylococcal antibiotics such as vancomycin or fusidic acid. Aminoglycosides should not be used for non-directed monotherapy of skin or wound infections because of high failure rates.

Duration of treatment depends on the location and type of infection. A treatment course of 10 days is considered necessary in streptococcal infections to prevent acute nephritis. With other pathogens, shorter treatment courses are apparently sufficient. Deep-seated infection of longer duration with fistulas and suppuration sometimes requires treatment for many weeks. Antibacterial treatment of skin and wound infections is generally ineffective as long as infected foreign bodies are not removed.

In general, when treating skin and wound infections, the actual type of infection should be taken into account and the antibiotic, if deemed necessary, chosen accordingly (Tables 16.11-16.14).

Prophylaxis of Wound Infections

No antibiotics of any kind replace surgical treatment of wounds: thorough cleaning, débridement, excision, removal of foreign bodies, drainage, immobilization. Antibiotics do not protect a patient from unsterile dressing procedures - on the contrary, antibiotics favor the selection of resistant strains in these circumstances. Nevertheless, in a number of instances routine chemoprophylaxis minimizes the incidence either of high-risk common infections (e.g., in vaginal hysterectomy) or of low-risk, rare but very dangerous infections (e.g., gas gangrene, endocarditis).

Decisions regarding the prophylactic use of antibiotics in surgery should be based upon a classification of operations according to the likely degree of bacterial contamination of the operation site. Surgical operations are, therefore, often classified into four categories:

1. Clean
2. Clean-contaminated
3. Contaminated
4. Dirty

The routine use of antibiotics after aseptic (i.e., "clean") operations should be discouraged, because it is unlikely to be of benefit yet exposes a patient to possibly harmful side effects and carries a small risk of selection of resistant strains. Open heart surgery and neurosurgical operations are the main indications for antibiotic prophylaxis in clean surgery. The extended duration of these operations and the very high risk of even rare infections contribute to the indication. Usually, chemoprophylaxis is indicated in "clean-contaminated" or "contaminated" operations. Use of antibiotics in "dirty" operations cannot be considered true prophylaxis, since their use is part of the therapy in septic surgery.

Strict asepsis during operation, débridement, and drainage are still the best means of prevention of wound infections; even in trivial injuries, antibiotic therapy is no substitute for meticulous wound-dressing technique. In general, antibiotic prophylaxis should be given only in exceptional cases, where the risk of infection is particularly high or rare but dangerous infections may arise.

Table 16.11. Treatment of skin infections

Type of skin infection	Topical antibacterial treatment	Systemic antibiotics			Comments
		Drug of choice	Alternatives	Not on WHO list	
Pyoderma (impetigo)	Yes	Phenoxymethyl-penicillin	Erythromycin, cloxacillin	Cephalosporins (oral)	Risk of nephritis!
Erysipelas	No	Benzylpenicillin, phenoxymethyl-penicillin	Erythromycin	Cefazolin	Long-term treatment in recurrent erysipelas with benzathine penicillin
Lymphangitis, Lymphadenitis	No	Benzylpenicillin, phenoxymethyl-penicillin	Erythromycin	Cefazolin	Risk of suppurative lymphadenitis and septicemia
Small furuncle	No	No treatment	Cloxacillin	Cefazolin	Risk of metastatic abscesses
Large furuncle and furuncle of the face	No	Cloxacillin	Erythromycin, chloramphenicol	Clindamycin	Septicemia, risk of septic thrombosis in furuncles of the face
Finger abscess	No	Cloxacillin	Benzylpenicillin, phenoxymethylpenicillin, tetracycline	Clindamycin, cefotaxime	Risk of involvement of joints, bones, tendons
Foot abscess	No	Ampicillin	Benzylpenicillin, metronidazole		Risk of involvement of bones
Cellulitis	No	Benzylpenicillin	Chloramphenicol, cloxacillin	Cefotaxime, clindamycin	Risk of septicemia, Gram's stain
Ecthyma (nonstreptococcal)	Yes	Gentamicin	Amikacin	Azlocillin, piperacillin, tobramycin	Often underlying disease
Tropical ulcer	Possibly	Benzylpenicillin, phenoxymethyl-penicillin	Tetracycline, metronidazole	Clindamycin	Immobilization, surgery may be necessary
Cutaneous diphtheria	Possibly	Benzylpenicillin, phenoxymethyl-penicillin	Erythromycin, tetracycline	Cephalosporins	
Neonatal skin infection	Yes	Cloxacillin	Erythromycin	Cephalosporins	Risk of septicemia

Table 16.12. Treatment of secondary bacterial infections

	Topical antibacterial treatment	Systemic antibiotics			Comments
		Drug of choice	Alternatives	Not on WHO list	
Eczema	Yes	Benzylpenicillin, cloxacillin	Erythromycin	Clindamycin	Consider *Candida* infection
Blister-forming skin diseases	Possibly	Cloxacillin	Benzylpenicillin, erythromycin	Clindamycin	Treatment of underlying disease
Trophic ulcer of leprosy	Yes	Cloxacillin	Benzylpenicillin, tetracycline, metronidazole	Clindamycin	Immobilization of limb
Scabies	Yes	Phenoxymethyl-penicillin	Erythromycin	Cefazolin	
Tungiasis	Possibly	Phenoxymethyl-penicillin	Cloxacillin	Cefazolin	Adequate treatment of parasitic infections
Dracontiasis	Possibly	Benzylpenicillin	Ampicillin + metronidazole, chloramphenicol	Cephalosporins, clindamycin	
Infected ulcer	Yes	Benzylpenicillin	Cloxacillin, gentamicin	Cefotaxime, azlocillin, piperacillin, clindamycin	Treatment of underlying disease

Table 16.13. Treatment of posttraumatic wound infections

Type of post-traumatic wound infection	Topical antibacterial treatment	Systemic antibiotics			Comments
		Drug of choice	Alternatives	Not on WHO list	
Stab or bullet wound	No	Benzylpenicillin	Cloxacillin, chloramphenicol	Cephalosporins	Maximal treatment in severe forms
Perforation of thorax	No	Benzylpenicillin	Chloramphenicol	Cephalosporins	Maximal treatment in severe forms
Perforation of abdomen (or similar)	Possibly	Ampicillin + metronidazole	Chloramphenicol, gentamicin	Cefotaxime + piperacillin, imipenem, clindamycin	Maximal treatment in severe forms
Traffic accident					Maximal treatment in severe forms
minor	Possibly	Benzylpenicillin	Ampicillin + cloxacillin	Cephalosporins	
major	Possibly	Chloramphenicol	Ampicillin + gentamicin	Quinolones	
Delay in initial wound toilet	Possibly	Benzylpenicillin, phenoxymethylpenicillin	Ampicillin, chloramphenicol	Cephalosporins	Risk of clostridial/streptococcal infection
Open fracture	Possibly	Benzylpenicillin, ampicillin	Chloramphenicol	Cephalosporins, clindamycin	Risk of clostridial/streptococcal infection
Human/animal bites	Possibly	Benzylpenicillin	Tetracycline	Cephalosporins	Risk of streptococcal, *Bacteroides, Pasteurella multocida* infection
Burns, minor	Possibly	Benzylpenicillin	Erythromycin	Cephalosporins	Risk of streptococcal infection
Burns, major	Obligatory	Benzylpenicillin (first week)	Chloramphenicol, gentamicin (later)	Cefotaxime, piperacillin, azlocillin	High risk of *Pseudomonas*/staphylococcal infection; in the late phase Gram's stain!

Table 16.14. Treatment of postoperative wound infections

Type of postoperative wound infection	Topical antibacterial treatment	Systemic antibiotics			Comments
		Drug of choice	Alternatives	Not on WHO list	
Orofacial	No	Benzylpenicillin, phenoxymethyl-penicillin	Erythromycin, tetracyclines, metronidazole	Cephalosporins etc.	Risk of streptococcal infection
Thoracic	No	Benzylpenicillin, cloxacillin	Chloramphenicol, gentamicin	Cephalosporins etc.	Maximal treatment in risk of mediastinitis
Abdominal	No	Ampicillin + metronidazole	Chloramphenicol, gentamicin	Cefotaxime, clindamycin	Maximal treatment in risk of peritonitis!
Perianal	Possibly	Metronidazole	Chloramphenicol, tetracycline	Cefotaxime, clindamycin, piperacillin, etc.	
Bone and joint surgery	No	Cloxacillin	Erythromycin, chloramphenicol	Cephalosporins, clindamycin	Risk of chronic infection
Obstetrics/gynecology	No	Benzylpenicillin, ampicillin	Metronidazole, chloramphenicol	Cephalosporins, piperacillin, clindamycin	Risk of severe peritonitis
Neurosurgery	No	Chloramphenicol	Benzylpenicillin, cloxacillin, metronidazole	Cefotaxime, fosfomycin	High risk of permanent sequelae
Amputation, leg	Possibly	Benzylpenicillin + metronidazole	Cloxacillin, chloramphenicol	Cefotaxime, clindamycin, etc.	Risk of clostridial infection
Mastectomy	Possibly	Cloxacillin	Chloramphenicol	Cephalosporins, clindamycin	

The frequent necessity of operating under suboptimal conditions may be an additional argument in favor of prophylaxis in developing countries. The borderline between true prophylaxis and early therapy is not well defined. Use of antibiotics for highly contaminated wounds (gunshot wounds of the abdomen) could be regarded as early treatment since the criteria of surgical prophylaxis are not fulfilled.

Indications of Prophylaxis

Any hospital where surgery is carried out should set up its own catalogue of main indications for chemoprophylaxis in surgery. Suggestions concerning indications and choice of agents are made in Table 16.15. The main indications for surgical prophylaxis should not be easily changeable - they should be part of the general hospital policy.

The antibiotics chosen for chemoprophylaxis may be changed at routine intervals, since there are arguments against long-term continuous use of the same antibiotic for surgical prophylaxis (continuous directed selection possible).

Modern Rules of Prophylaxis

Surgical prophylaxis is often started too late and continued too long. It is very important to achieve effective tissue levels of the antibiotic during the operation. Usually, the first dose of the antibiotic is given at the time of induction of anesthesia (30-60 min prior to the operation). Routine administration of more than one dose of antibiotic prior to the operation is not necessary and carries the risk of selection of highly resistant hospital strains. If circumstances arise during the course of operation which necessitate antibiotic prophylaxis (such as opening of the colon), a suitable agent should be given immediately. The administration of antibiotic prophylaxis in surgery is thus largely the task of the anesthetist. The duration of prophylaxis should be short. Opinions as to the optimal duration vary between one single dose and 3 days of treatment. The main effect of prophylaxis is without doubt achieved by the first dose. There are good arguments for prophylaxis with three doses of antibiotic. The first dose is given, as described above, at the beginning of anesthesia, followed by the second and third dose 4-6 h and 24 h after the operation. The dosage for prophylaxis should be as for treatment. Although it is not uncommon in some units to treat routinely with antibiotics for 10-14 days, this practice is certainly not necessary and may be harmful.

General caution in the prophylactic use of antibiotics in surgery should never lead to omission of the prophylaxis of clostridial infections (gas gangrene), which is essential in contaminated wounds and in amputation of ischemic legs!

The modern rules of short-term prophylaxis are not applicable in case of highly contaminated wounds, delay of wound toilet, or when a wound infection is already present. Established, severe wound infections require treatment of 2-3 weeks' duration. Single-dose or short-term prophylaxis does not suffice in these cases.

Table 16.15. Indications for chemoprophylaxis/early treatment in surgery

Indication	Antibiotics	Alternatives	Reason
Heavily contaminated wounds, delay in wound closure	Benzylpenicillin	Many other antibiotics (except aminoglycosides)	Prevention of tetanus, gangrene, streptococcal infection
Compound fractures, traumatic penetration of joints, thorax	Benzylpenicillin	In Europe cephalosporins, many other antibiotics possible	Mixed infections common, often including streptococci, anaerobes, clostridia
Gunshot wounds, stab wounds	Benzylpenicillin	In Europe cephalosporins, many other antibiotics possible	Mixed infections common, often including streptococci, anaerobes, clostridia
Perforating wound of abdomen	Ampicillin + metronidazole for several weeks	Chloramphenicol; in Europe, combinations like cefotaxime + metronidazole	Mixed infections, mainly *Bacteroides*, enterobacteria inevitable
Animal/human bites	Benzylpenicillin	Many other antibiotics	High risk of infection by anaerobes, streptococci (human), or *Pasteurella multocida* (animal)
Severe burns (Ist week)	Benzylpenicillin	Cephalosporins, many other antibiotics	Prophylaxis of initial streptococcal infection
Neurosurgery	Chloramphenicol	Cefotaxime	Low but deleterious risk of infection, prolonged duration of operation
Implantation of plastics/metals	Cloxacillin	Cephalosporins	Risk of staphylococcal infection
Thoracic surgery/open heart surgery	Cloxacillin	Cephalosporins	Risk of postoperative mediastinitis/endocarditis mainly by staphylococci/streptococci
Amputation of ischemic leg	Benzylpenicillin	Many other antibiotics	Prevention of clostridial infection
Surgery in heavily contaminated areas (esophagus, colon, rectum)	Ampicillin + metronidazole	Chloramphenicol, cephalosporins	High risk of secondary infection
Vaginal hysterectomy, operations in pelvic inflammatory disease	Ampicillin	Metronidazole, Benzylpenicillin, cephalosporins	Prevention of frequent anaerobic infection
Cesarean section with amnion infection	Ampicillin	Benzylpenicillin, cephalosporins	Prevention of severe streptococcal or anaerobic infection (gas gangrene)
Operation under nonsterile conditions	Benzylpenicillin, ampicillin	Cloxacillin, chloramphenicol	Prophylaxis of infections by streptococci, staphylococci, clostridia

Tetanus

Tetanus, in particular neonatal tetanus, is still an important health problem in developing countries [37, 42]. It has been estimated that neonatal tetanus accounts for 23% to more than 70% of all neonatal deaths in different developing countries [42]. Of all tetanus deaths worldwide, 25%-50% have been attributed to tetanus occurring in the first months of life [27].

The spores of *Clostridium tetani,* a strictly anaerobic organism, are ubiquitous in soil, where they can survive for years if not exposed to sunlight. Being highly resistant, their killing requires autoclaving at 121 °C for 10 min. Following inoculation of spores in wounds, vegetative forms develop under anaerobic conditions to produce tetanospasmin, the neurotoxin responsible for tetanus.

In neonatal tetanus, the umbilical stump is contaminated by nonhygienic cutting of the cord or by application of nonsterile dressings, e.g., cow dung. Apart from wounds caused by injuries, tetanus may follow ritual scarifications, circumcisions, ear-piercing, or extractions of *Dracunculus medinensis.* Cases of tetanus following intramuscular injections, surgery, abortion, and childbirth are also well known. Chronic otitis, leprotic ulcers, and tropical ulcers are further portals of entry.

The diagnosis of tetanus is made clinically. Isolation of *C. tetani* from wounds is difficult, and is successful in less than half the cases.

Treatment

The principles of treatment are:
1. Surgical débridement of any wounds.
2. Use of benzylpenicillin, 4(-10) megaunits (MU) daily in divided doses in adults, 0.2-0.4 MU daily in neonates, for 7 days to stop further production of the neurotoxin by *C. tetani.* Intercurrent infections, pneumonia, and urinary tract infections associated with indwelling urinary catheters are treated with appropriate antibiotics.
3. Administration of tetanus antitoxin, either 10000 units of equine antitetanus serum (ATS) i.m. or i.v. (the same dose can be given to neonates) or 2000-4000 (-10000) units of human tetanus immunoglobulin (TIG) i.m. Use of human TIG avoids the serum sickness and possible anaphylactic reactions associated with horse serum, but is usually too expensive for developing countries. Prior administration of a test dose (0.05 ml) ATS is recommended.

   There are conflicting data regarding the reduction in mortality achieved by concurrent intrathecal administration of tetanus immunglobulin. Quantities of 50-100 units (with 2 mg dexamethasone to prevent neural irritation) and 250 units of equine serum have been found beneficial in neonatal tetanus in two different studies [40, 41]. A quantity of 250 units of human TIG given early in the course of the disease was found to be beneficial in a further study [15], but others did not observe a reduction of mortality on giving human TIG or equine ATS intrathecally in adult-type tetanus [46, 47]. Adding 200 units intrathecal ATS to a regimen including systemic corticosteroids and low-dose ATS,

750 units once daily for 3 days, was associated with definite improvement of mortality [35, 46].

4. Sedation and control of spasms with short-acting barbiturates, chlorpromazine, or diazepam. Sedation is carried up to the point of respiratory depression where facilities for artificial ventilation are not available. Muscle relaxation with intermittent positive pressure ventilation via a tracheostomy in an intensive care unit is now standard treatment for severe tetanus in industrialized countries [11] but is usually beyond the means of district hospitals in developing countries [12, 41].
5. Removal of external stimuli.
6. Airway clearance with removal of secretions to maintain adequate ventilation. If tracheostomy is considered, skilled nursing staff must be available.
7. Meticulous nursing care. Attention must be paid to maintenance of nutrition and hydration. Sympathetic overactivity is common in severe tetanus.
8. Active immunization after recovery.

Prevention

Tetanus is preventable by primary immunization with three doses of tetanus toxoid, the second dose given 4–6 weeks after the first, the third 6–12 months thereafter. Correct surgical débridement of wounds also reduces the incidence of tetanus. Administration of tetanus toxoid and tetanus immunglobulin according to the immunization history is summarized in Table 16.6. The role of antibiotics in the prophylaxis of posttraumatic tetanus is controversial.

Neonatal tetanus is preventable by immunization of the mother with two doses of tetanus toxoid and by training of traditional birth attendants [34]. Several strategies for prevention of tetanus have recently been reviewed with regard to feasibility and cost-effectiveness [37].

Gas Gangrene

Clostridium perfringens (welchii) is the most common causative agent of gas gangrene. Other clostridia involved include *C. novyi, C. septicum,* and *C. histolyticum,* often in mixed infection. *C. perfringens* is ubiquitous in nature. It belongs to the normal colonic flora of humans and animals and is isolated from almost every soil sample.

Gas gangrene may arise (1) from traumatic injuries and penetrating wounds (e.g., gunshot wounds), (2) postoperatively following surgery of the colon or the biliary tract, (3) as uterine gas gangrene postpartum or, more commonly, following septic abortion. The latter two, postoperative and uterine gas gangrene, are more common in developing countries. A number of cases of gas gangrene have been associated with vascular insufficiency (diabetes mellitus, atherosclerosis). Clostridial septicemia is quite common in patients with leukemia.

The diagnosis of wound gas gangrene is made clinically (onset of severe pain in the wound, crepitation, edematous, tense skin which is initially pale in color, tachycardia). The exudate from the wound contains large gram-positive rods in

Table 16.16. Recommendations for wound management for prevention of tetanus (based on immunization history). From [3]

Immunization record	Wound[a]	Recommendation
Unimmunized, incomplete, or unknown	Low risk	Toxoid[b] followed by complete immunization
	Tetanus-prone	Toxoid plus TIG[c] 250-500 units using separate syringe and injection sites; complete active immunization
Primary immunization but no booster in 10 years	Low risk	Toxoid
	Tetanus-prone	Toxoid; use of TIG is arbitrary
Primary immunization with booster within 10 years	Low risk	-
	Tetanus-prone	Toxoid if 5 years since booster

TIG, tetanus immune globulin.
[a] "Tetanus-prone" indicates severe, neglected, or more than 24 h old.
[b] Toxoid as DPT for children under 5 years old; DT for persons over 6 years old.
[c] If equine ATS is used instead of human TIG, a dose of 1500 units is suggested after careful screening for sensitivity to horse serum.

great numbers with only few polymorph leukocytes. Prompt surgical exploration of the muscle revealing myonecrosis establishes the diagnosis. Uterine gas gangrene frequently presents with jaundice due to hemolysis, hypotension, and hemoglobinuria resulting in renal failure while the local signs are still minimal.

Treatment consists in immediate administration of high-dose benzylpenicillin, 10-20 megaunits a day in adults, and in prompt débridement of all necrotic muscle. The usefulness of hyperbaric oxygenation remains controversial. Even with modern intensive care a high percentage of these critically ill patients will die.

Thorough débridement and prophylaxis with penicillins is necessary to prevent clostridial infections of contaminated wounds.

References

1. Allen MA, Taplin D, Twigg L (1971) Cutaneous streptococcal infection in Vietnam. Arch Dermatol 104: 271-280
2. Allen MA, Taplin D (1974) Skin infections in Eastern Panama: survey of two representative communities. Am J Trop Med Hyg 23: 950-956
3. Bartlett JG (1984) Tetanus. In: Strickland GT (ed) Hunter's tropical medicine, 6th edn. Saunders, Philadelphia, pp 361-363
4. Belcher DW, Afoakwa SN, Osei-Tutu E, Wurapa FK, Osei L (1977) Endemic pyoderma in Ghana: a survey in rural villages. Trans R Soc Trop Med Hyg 71: 204-209
5. Bezjak V, Farsey SJ (1970) *Corynebacterium diphtheriae* skin lesions in Ugandan children. Bull WHO 43: 643-650
6. Bourlard C, Sghir M, Lahouague A (1979) Standardized treatment of superficial skin infections in children aged 0-4 years in Cap Bon, Tunisia. J Trop Pediatr 25: 146-148
7. Bower M, Bowness P, Shann F (1984) Penicillin in the treatment of skin sores in children. Med J Aust 140 (9): 528-530
8. Bowness P, Bower M, Montgomery J, Lupiwa T, Gratten M, Shann F (1984) The bacteriology of skin sores in Goroka children. Papua New Guinea Med J 27: 83-87

9. Chhabra HL, Sharma RG, Chugh TD, Saini SK, Midha NK (1981) *Corynebacterium diphtheriae* in surgical wounds. Indian J Med Res 73: 179-180
10. Control of Infection Sub-Committee University College Hospital, Ibadan (1979) Epidemiological surveillance of hospital-acquired wound infection - a report. Niger Med J 9: 289-293
11. Edmondson RS, Flowers MW (1979) Intensive care in tetanus: management, complications, and mortality in 100 cases. Br Med J 1: 1401-1404
12. Garnier MJ, Marshall FN, Davison KJ, Lepreau FJ (1975) Tetanus in Haiti. Lancet I: 383-386
13. Gedebou M, Tassew A, Azene G (1983) Frequency and resistance patterns of bacterial isolates from surgical patients in a teaching hospital in Addis Ababa. Trop Geogr Med 35: 133-138
14. Gray FJ (1974) Surgical wound infection in a developing country - Papua New Guinea. Papua New Guinea Med J 17: 373-378
15. Gupta PS, Goyal S, Kapoor R, Batra VK, Jain BK (1980) Intrathecal human tetanus immunglobulin in early tetanus. Lancet II: 439-440
16. Hendrickse RG (1980) Epidemiology and prevention of kidney disease in Africa. R Soc Trop Med Hyg 74: 8-16
17. Ichhpujani RL, Arora DR, Chaudhary U, Arora B, Mohan H, Chugh TD (1981) Bacteriology and serology of pyoderma. J Com Dis 13: 253-256
18. Kakande J (1978) Pattern and sensitivity of bacterial organisms infecting burns in Mulago Hospital, Kampala. East Afr Med J 55: 223-227
19. Koshi G, Benjamin V (1977) Surveillance of streptococcal infections in children in a south Indian community - a pilot survey. Indian J Med Res 66: 379-388
20. Kotrajaras R, Buddhavudhikrai P, Sukroongreung S, Chanthravimol P (1982) Endemic leg ulcers caused by *Corynebacterium pyogenes* in Thailand. Int J Dermatol 21: 407-409
21. Kulshreshtha B (1975) Postoperative wound sepsis - a clinical study. J Armed Forces India 34: 37-42
22. Liebow AA, Maclean PD, Bumstead JH, Welt LG (1946) Tropical ulcers and cutaneous diphtheria. Arch Intern Med 78: 255-295
23. Masawe AEJ, Nsanzumuhire H (1975) Scabies and other skin diseases in pre-school children in Ujamaa villages in Tanzania. Trop Geogr Med 27: 288-295
24. Mehta G, Prakash K, Sharma KB (1980) Streptococcal pyoderma and acute glomerulonephritis in children. Indian J Med Res 71: 692-700
25. Mhalu FS (1973) Bacteriological study of superficial skin infections in Tanzanian children - a preliminary report. East Afr Med J 50: 273-276
26. Mhalu FS, Lindquist KJ (1974) *Corynebacterium diphtheriae* infections in Tanzania. East Afr Med J 51: 44-50
27. Miller JK (1972) The prevention of neonatal tetanus by maternal immunization. J Trop Pediatr 18: 159-167
28. Montgomery J (1984) Carriage of *Corynebacterium diphtheriae* in children of the Eastern Highlands Province. Papua New Guinea Med J 27: 20-23
29. Nsanzumuhire H, Taplin D, Lansdell L (1972) Pyoderma among Ugandan children. East Afr Med J 49: 84-88
30. Ogunbi LOO, Lawal A, Lasi Q, Ogunbi O (1974) Streptococcal pyoderma in a Lagos school population. Niger Med J 4: 178-180
31. Pettit JHS (1977) A simple pharmacopoeia for out-station skin clinics. Trop Doct 7: 107-110
32. Porter MJ (1977) An epidemiological approach to skin disease in the tropics. Trop Doct 7: 59-66
33. Porter MJ (1979) Seasonal change and its effect on the prevalence of infectious skin disease in a Gambian village. Trans R Soc Trop Med Hyg 74: 162-168
34. Rahman S (1982) The effect of traditional birth attendants and tetanus toxoid in reduction of neonatal mortality. J Trop Pediatr 28: 163-165
35. Sanders RKM, Martyn B, Joseph R, Peacock ML (1977) Intrathecal antitetanus serum (horse) in the treatment of tetanus. Lancet I: 974-977
36. Satyavani U, Koteswarakao M, Bhaskaran CS (1978) Bacteriological study of postoperative wound infections. J Indian Med Assoc 70: 121-126
37. Schofield F (1986) Selective primary health care: strategies for control of disease in the developing world. XXII. Tetanus: A preventable problem. Rev Infect Dis 8: 144-156

38. Scott-Emuakpor MB (1970) The problem of post-operative wound sepsis in a city hospital. 1. Bacteriology of post-operative infections. J Trop Med Hyg 73: 39–44
39. Sharma VK, Agarwal DS, Satyanand (1981) Bacteriological study of infection in patients with burns. Indian J Med Res 73: 697–709
40. Singh AK, Bansal A, Goel SP, Agarwal VK (1980) Intrathecal antitetanus serum (horse) with steroid in the treatment of neonatal tetanus. Arch Dis Child 55: 527–531
41. Sow A (1983) Traitement du tétanos néo-natal en Afrique. Dakar Méd 28: 483–496
42. Stanfield JP, Galazka A (1984) Neonatal tetanus in the world today. Bull WHO 62: 647–669
43. Subramanian KA, Prakash A, Shriniwas A, Bhujwala RA (1973) Post-operative wound infection. Indian J Surg 35: 57–64
44. Taplin D, Lansdell L, Allen AM, Rodriguez R, Cortes A (1973) Prevalence of streptococcal pyoderma in relation to climate and hygiene. Lancet I: 501–503
45. Thaung U, Than Naung, Khin Saw Khine, Khai Ming C (1978) Epidemiological features of skin diphtheria infection in Rangoon, Burma. Southeast Asian J Trop Med Public Health 9: 4–10
46. Thomas PP, Crowell EB, Matthew M (1982) Intrathecal anti-tetanus serum (ATS) and parenteral betamethasone in the treatment of tetanus. Trans R Soc Trop Med Hyg 76: 620–623
47. Vakil BJ, Armitage P, Clifford RE, Laurence DR (1979) Therapeutic trial of intracisternal human tetanus immunoglobulin in clinical tetanus. Trans R Soc Trop Med Hyg 73: 579–583
48. Wittle HC, Abdullahi MT, Fakunle F, Parry EHO, Rajkovic AD (1973) Scabies, pyoderma and nephritis in Zaria, Nigeria - a clinical and epidemiological study. Trans R Soc Trop Med Hyg 67: 349–363
49. WHO Expert Committee (1983) The use of essential drugs. Tech Rep Ser No. 685. WHO, Geneva

17 Sexually Transmitted Diseases

Occurrence and Diagnosis

Reliable statistics on the prevalence of sexually transmitted diseases (STDs) in developing countries are not available. However, the general impression is that STDs, particularly gonorrhea, are very common in the urban areas of most of these countries. Secondary complications of gonorrhea, such as urethral strictures, pelvic inflammatory disease (PID), and infertility, account for a major proportion of urological and gynecological consultations in Africa [13, 16]. The high incidence of gonococcal ophthalmia neonatorum is a further cause of concern [13].

Data on the prevalence of nongonococcal urethritis (NGU) range from 6% to more than 30% of all cases of STD seen in several African series [16]. Chlamydia trachomatis, serotypes D through K, has been recognized as the major cause of NGU, PID, and infertility in industrialized countries during the past decade. Neonatal inclusion conjunctivitis is another disease caused by this agent. However, in Africa, the isolation rate of *C. trachomatis* is low compared to the high prevalence of gonorrhea. Association of *C. trachomatis* infection with gonorrhea appears to be much less common in Africa than in industrialized countries [14].

Genital herpes simplex infection is now the most common STD in some industrialized countries. Genital herpes appears to be less common in Africa, where chancroid and syphilis are both diagnosed more frequently as causes of genital ulcer [4, 16]. The prevalence of syphilis in Africa is estimated to be considerably higher than in Europa or USA [16]. As a consequence, congenital syphilis appears to be a common clinical problem in several countries of Africa [4, 9]. Chancroid accounts for 30%-80% of etiologically proven genital ulcers in eastern, central, and southern Africa. Chancroid is also common in South America and Southeast Asia. While granuloma inguinale (donovanosis) is almost nonexistent in temperate climates, the disease is endemic in southern India, Vietnam, Papua New Guinea, Africa, and the Caribbean. Lymphogranuloma venereum (inguinal lymphadenopathy is usually the presenting complaint, rather than a genital ulcer) is worldwide in distribution, but more common in subtropical and tropical climates.

Trichomoniasis, candidiasis, and veneral warts (condyloma acuminata) are other conditions frequently seen in STD clinics in developing countries. In some areas this is also true for sexually transmitted scabies and pediculosis, both of which are not uncommonly complicated by bacterial superinfection.

Extended laboratory support for the diagnosis of syphilis, gonorrhea, chancroid, granuloma inguinale (donovaniosis), lymphogranuloma venereum, genital infections caused by other serotypes of *Chlamydia trachomatis,* and genital herpes is only available at reference centers.

Even minimal support, i.e., Gram's stain for *Neisseria gonorrhoeae* and dark-field microscopy and the rapid plasma reagin (RPR)-card test for syphilis, is usually not available at the primary care level in the absence of laboratory facilities. Management of STDs has to be guided by the clinical picture, epidemiological considerations, and response to therapy.

A symptom-oriented approach directed at the management of genital ulcers and urethral or vaginal discharge therefore appears to be the only practical solution. Several symptom-oriented strategies have been developed for this purpose [2, 13, 30]. Examples of flowcharts are shown in Figs. 17.1 and 17.2.

The strategy depicted in Fig. 17.1 proposed treating all patients with urethral discharge for gonorrhea. This is based on African studies which found a gonococcal etiology in approximately 80% of patients with urethral discharge [13]. Alternatively, a distinction is made clinically between spontaneous, thick, profuse, and purulent discharge and scanty, mucoid, or clear discharge. Treatment for gonorrhea and nongonococcal urethritis, respectively, is then given accordingly [2].

The rationale of treating all women presenting with vaginal discharge or other genital complaints first for uncomplicated gonorrhea (see Fig. 1 of chapter "Obstetric and Gynecological Infections") after clinical exclusion of acute PID - is similar: high prevalence of gonorrhea and prevention of PID on the one hand, and low validity of symptoms for gonorrhea or other STDs in women attending STD clinics, on the other [13].

Purely clinical distinction between genital ulcers due to syphilis and those due to chancroid (Fig. 17.2) may be difficult. Combined treatment of both (single-dose benzathine penicillin and co-trimoxazole, two tablets twice a day for 7 days) may then be preferable [13, 37]. Similar symptom-oriented strategies for use in the presence of laboratory facilities have also been designed [2, 13, 30].

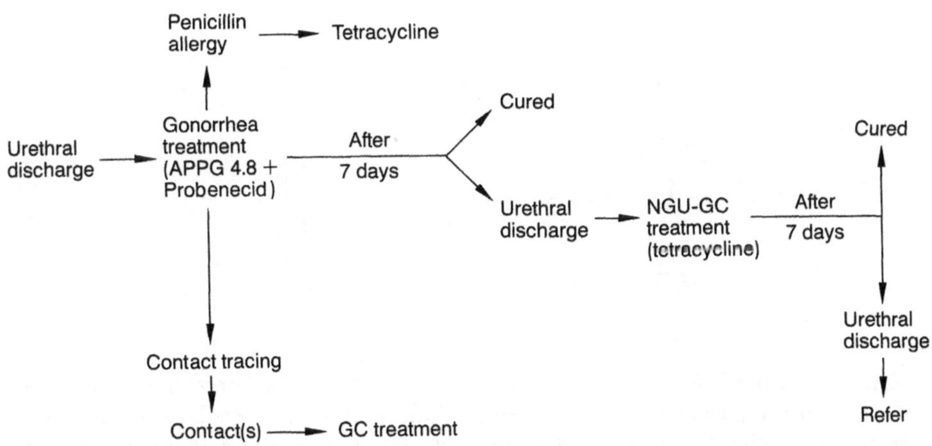

Fig. 17.1. Example of a strategy for the management of urethral discharge when no laboratory facilities are available. Dosage of tetracycline: 500 mg p.o. four times daily for 7 days. Where penicillin-producing strains of *Neisseria gonorrhoeae* are prevalent, one of the regimens for treatment of PPNG infections suggested in Table 1 should be administered. *APPG 4.8+Prob*, aqueous procaine penicillin G 4.8 MU+probenecid 1 g p.o.; *NGU*, nongonococcal urethritis; *GC*, gonorrhea. From [13]

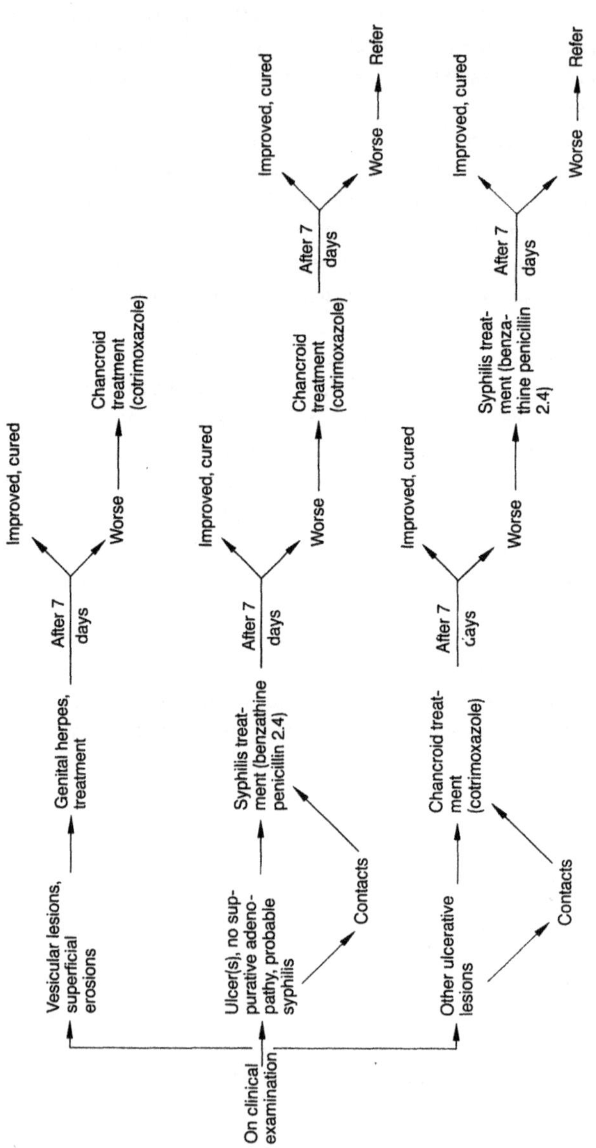

Fig. 17.2. One strategy for the management of "genital ulcer(s)" when no laboratory facilities are available. Dosage of co-trimoxazole: two tablets twice daily for 7 days. From [13]

Gram's stain of urethral smears allows definition of the presence of urethritis [five or more polymorphonuclear neutrophilic leukocytes (PMNs) on the smear] and to distinguish between gonococcal and nongonococcal urethritis by the presence of gram-negative intracellular diplococci.

Wet mounts of vaginal secretions and Gram's stain of cervical smears allow the diagnosis of trichomoniasis (motile trichomonads), candidiasis (yeast cells), nonspecific vaginitis ("clue cells"), or gonorrhea. Availability of dark-field microscopy is of great help in the management of genital ulcers.

Further difficulties in the management of STD in developing countries arise from the emergence of resistance.

Penicillinase-producing strains of *Neisseria gonorrhoeae* (PPNG) are now widespread in Southeast Asia and at least in some countries of eastern, western, and southern Africa [4, 17, 19, 23, 25, 26, 29]. Prevalence rates of 20%-40% of PPNG are associated with ever higher failure rates of penicillin therapy. Effective treatment of gonorrhea caused by PPNG is more expensive than the penicillin regimens used previously. Furthermore, cost-effective regimens for treatment of gonococcal PID or disseminated gonococcal infection due to PPNG have yet to be evaluated in controlled trials. The observed increase in chromosomal resistance of *N. gonorrhoeae* to several antibiotics appears to be of lesser importance at the moment.

Similarly, tetracycline resistance of *Hemophilus ducreyi* has been reported from several areas of the world, and sulfonamide resistance has also been noted, so that more expensive regimens of trimethoprim-sulfonamide combinations often have to be used. Failure rates of more than 30% with co-trimoxazole, associated with resistance to trimethoprim in addition to sulfonamide resistance, have recently been reported from Bangkok [28].

Because of the variable pattern of resistance to, availability, and cost of antimicrobial agents, several regimens have been suggested for treatment of gonorrhea and chancroid (Tables 17.1-17.3). Data on the optimal treatment of lymphogranuloma venereum and granuloma inguinale are limited; regimens commonly suggested have been included in Tables 17.4 and 17.5).

Single-dose treatment and short-course regimens are usually preferable to prolonged courses of oral chemotherapy, since patients may save tablets for the next episode of STD or for their partners, or sell them to others.

Treatment of syphilis (by any of the standard regimens in Tables 17.6 and 17.7) should be monitored by serology. A minimum fourfold fall in titers (quantitative VDRL or RPR test) or conversion to seronegativity may be expected 6 to 12 months after successful treatment. Titers increasing fourfold and more or remaining persistently high are indicative of reinfection or of asymptomatic neurosyphilis in latent syphilis. Retreatment should be given once CSF has been examined. Cross-reactions with yaws acquired in childhood have to be considered in the differential diagnosis of latent syphilis.

Table 17.1. Standard regimens for treatment of uncomplicated gonorrhea in areas where penicillinase-producing strains of *Neisseria gonorrhoeae* are rare

Antibiotic	Dosage
Procaine penicillin	4.8 MU i.m. + probenecid 1 g p.o. 30-60 min later SD
Ampicillin	3.5 g p.o. + probenecid 1 g p.o. SD
Amoxicillin	3 g p.o. + probenecid 1 g p.o. SD
Tetracycline	500 mg four times a day for 5 days or 2.5 g, followed by 500 mg four times a day for 4 days (Total dosage 10 g)
Erythromycin	500 mg four times a day for 7 days

SD, single dose.

Table 17.2. Standard regimens for treatment of uncomplicated gonorrhea in areas where penicillinase-producing strains of *Neisseria gonorrhoeae* are prevalent

Antibiotic	Dosage
Kanamycin	2 g i.m. SD
Spectinomycin	2 g i.m. SD
Thiamphenicol	2.5 g p.o. SD
Cephalosporins	
Cefoxitin	2 g i.m. or i.v. + probenicid 1 g p.o. SD
Cefuroxime	1.5 g i.m. or i.v. + probenicid 1 g p.o. SD
Ceftriaxone	0.25-0.5 g i.m. or i.v. SD
Cefotaxime[a]	0.5 g i.m. or i.v. SD
Quinolones	
Norfloxacin, pefloxacin	0.8 g p.o. SD
Ciprofloxacin	0.5-1.0 g p.o. SD
Ofloxacin	0.2-0.4 g p.o. SD
Co-trimetrole	4.8 g once a day for 2 days
Co-trimoxazole	1.92 g twice a day for 2 days, or 4.8 g once a day for 2 days

SD, single dose.
[a] Other second- and third-generation cephalosporins may also be prescribed.

Table 17.3. Treatment of chancroid

	Comment
Previous standard regimens	
Sulfadimidine 1 g four times daily for 7-14 days Tetracycline 500 mg four times daily for 7-14 days Streptomycin 1 g i.m. once daily for 7-14 days	Resistance to tetracycline and sulfonamides in 90% and 40%, respectively, of isolates of *Hemophilus ducreyi* was observed in Kenya [6]. Tetracycline resistance has also been noted in Swaziland, Singapore, and Vietnam [12, 27]. In Bangkok, all strains of *H. ducreyi* were recently found to be highly resistant to sulfonamides, and only 50% of them were susceptible to trimethoprim [28].
Alternative regimens (recently recommended)	
Single-dose oral co-trimetrole 3.84 g	Efficacy 96%, 1981, and 71%, 1985, in Kenya [21,39][a]
Single-dose i.m. ceftriaxone 250 mg	Efficacy 100% in Bangkok [28]
Co-trimoxazole 1.92 g twice daily for 2 days	Efficacy 92% in Swaziland [12]
Co-trimoxazole 0.96 g twice daily for 7 days	Efficacy 100% in Kenya [7], 66%-87% in Bangkok [28]
Co-trimetrole 0.92 g twice daily for 5 days	Efficacy 96% in Kenya [21]
Thiamphenicol 2.5 g once daily for 2 days	Efficacy 94% (with 2.5 g given the first day and 1.25 g the second day) [11]
Erythromycin 500 mg four times daily for 7-10 days	Efficacy 84%-95% [5, 20]
Single-dose oral ciprofloxacin 0.5-1 g	Efficacy 95%-100% in Kenya and South Africa [36, 39]
Ciprofloxacin 0.5 g twice daily for 3 days	Efficacy 100% in Kenya [39]

[a] Efficacy of single-dose oral co-trimoxazole 3.84 g was recently 55%-63% in Bangkok [28].

Table 17.4. Treatment of lymphogranuloma venereum

Tetracycline[a]	2 g daily for 2 weeks
Sulfonamide[b]	4 g daily for at least 3 weeks
Co-trimoxazole	480 mg twice daily for 2-3 weeks
Erythromycin	500 mg four times daily for 2-4 weeks

[a] This regimen is often recommended.
[b] Early stage only.

Table 17.5. Treatment of donovanosis (granuloma inguinale)

Co-trimoxazole	0.96 g twice daily for 2-3 weeks
Tetracycline	2 g daily for 2-3 weeks
Streptomycin	1 g i.m. twice daily for 10-20 days
Chloramphenicol	2 g daily for 10-14 days

No experience with modern antibiotics.

Table 17.6. Treatment of syphilis with penicillin. From [31]

	Benzathine penicillin		Procaine penicillin	
	Total dose (MU)	Total number of injections (once a week)	Total dose (MU)	Total number of injections (one a day)
Early syphilis (primary, secondary, early latent, i.e., of *not* more than 2 years duration)	2.4	1	6.0	10
Late latent syphilis and late benign syphilis	7.2	3	9.0	15
Cardiovascular syphilis and neurosyphilis	-	-	12.0	20
Congenital syphilis (up to 2 years of age)[a]	0.05/kg	1	0.5/kg	10

[a] These regimens apply to infants with normal CSF; infants with early congenital syphilis (up to 2 years of age) and abnormal CSF findings should be treated with benzylpenicillin, i.v. or i.m., 0.025 MU/kg twice daily for at least 10 days or with procaine penicillin, 0.05 MU/kg once daily for 10 days.

Table 17.7. Treatment of syphilis in patients allergic to penicillin. From [31]

	Tetracycline hydrochloride[a] 2 g daily p.o.		Erythromycin (*not* the estolate) 2 g daily p.o.	
	Total dose	Days	Total dose	Days
Syphilis of not more than 2 years duration	30 g	15	30 g	15
Syphilis of more than 2 years duration	60 g	30	60 g	30

[a] Not indicated in pregnancy and in children younger than 8 years.

Gonorrhea

Treatment

Standard treatment regimens for uncomplicated gonorrhea, where PPNG are rare, are given in Table 17.1. Single-dose treatment is generally preferable to a 5-day course because of problems with patient compliance. Tetracycline and erythromycin are usually used only for patients allergic to penicillin. Erythromycin may be given during pregnancy. The recommended dose of procaine penicillin (4.8 MU i.m.) cures incubating syphilis; tetracycline and erythromycin may mask it. Patients who fail to respond to any one of these regimens should be retreated with kanamycin, spectinomycin, cephalosporins, quinolones (cf. table 17.2).

Once the prevalence of PPNG increases to 10% or the failure rate of penicillins considerably exceeds 10%, the recommendation is to change to an alternative regimen (Table 17.2) for first-line treatment. Many strains of PPNG are also resistant to tetracycline and erythromycin! Agents used for treatment of PPNG are also active against sensitive strains.

Health budgets, costs and availability vary among different countries. Treatment policies, both in areas where PPNG are rare and where they are prevalent, should be devised at a local level. Such standard treatments should be based on monitoring of failure rates, costs and sensitivity testing.

See Table 17.2 for standard treatment regimens for uncomplicated gonorrhea in areas where PPNG are prevalent. The average failure rates for spectinomycin (0–5%) are less than those of kanamycin (5%–10%) and thiamphenical (approximately 10%). Treatment failure due to spectinomycin-resistant PPNG has been reported from Southeast Asia [1, 19, 35], where spectinomycin has been used extensively since the emergence of PPNG. Spectinomycin resistance has not, as yet, been reported in Africa. Single-dose treatment with parenteral cephalosporins is highly efficacious (cure rates over 95%) [3, 18, 29], but costs limit their use as standard treatment for gonorrhea in many developing countries [15].

Quinolones appear to be the drugs of choice in future. Limited data indicate very high cure rates with oral single-dose treatment [33, 34]. Side effects on the CNS (dizziness, drowsiness, headache) are apparently more frequent with rosoxacin, an older derivative. All single-dose regimens listed in table 17.2 for treatment of uncomplicated gonococcal urethritis (cephalosporins, quinolones etc.) fail to treat concurrent nongonococcal urethritis due to *Chlamydia trachomatis*. The efficacy of 2-day courses of high-dose trimethoprim-sulfa combinations appears to range between 90% and 95%; single-dose regimens are associated with failure rates of approximately 20% [24].

Acute *pelvic inflammatory disease* is often polymicrobial in origin, even if gonococci are detected in endocervical secretion. Regimens for treatment of gonococcal PID should, therefore, be used only in patients where gonococcal etiology is likely (Table 17.8). In treating outpatients in areas where PPNG are prevalent, use of tetracycline instead of co-trimoxazole is less advisable, since many strains of PPNG are tetracycline-resistant. Rifampicin is also active, but should be used with great reluctance. Quinolones will be prescribed more frequently in future. They are contraindicated in pregnancy and lactation. Acute PID of unknown etiology may be caused by gonococci, enterobacteria, streptococci, anaerobic bacteria, or chlamydiae, often in mixed infection. Treatment may be difficult; the usual regimens cover the spectrum of pathogens only incompletely (Table 17.9). Only regimens including tetracycline, or erythromycin (or quinolones) are effective against *Chlamydia trachomatis*, a major pathogen of acute PID in industrialized countries.

Oropharyngeal gonorrhea may be difficult to eradicate. Procaine penicillin (4.8 MU)+probenicid (1 g) p.o. has often been curative. Other single-dose regimens (ampicillin, spectinomycin) are not reliable. Treatment with multiple doses of tetracycline or co-trimoxazole, where PPNG are prevalent, is usually effective in eradicating the infection.

Tetracycline or, where PPNG are prevalent, co-trimoxazole for 7-10 days are standard regimens for treating *gonococcal epididymitis* or *prostatitis*.

Disseminated gonococcal infection, such as septicemia or arthritis, can initially be treated with high-dose benzylpenicillin for 7 days. In ambulatory patients, tetracycline for 7-10 days is an alternative. Spectinomycin, quinolones, second- or third-generation cephalosporins, or chloramphenicol for 5-7 days should be given in areas where PPNG are prevalent.

Table 17.8. Regimens for treatment of acute pelvic inflammatory disease of gonococcal origin

	Daily dosage
Outpatients (where PPNG are rare)	
Tetracycline	2 g p.o. for 5 days
Procaine penicillin	4.8 MU i.m.+probenecid 1 g p.o.
or Ampicillin	3.5 g+probenecid 1 g p.o.
or Amoxicillin	3 g+probenecid 1 g p.o. as initial dose
followed by	
Ampicillin *or* Amoxicillin	2-3 g p.o. for 10 days
Outpatients (where PPNG are prevalent)	
Spectinomycin	2 g i.m.
or Kanamycin[a]	2 g i.m. as initial dose
followed by	
Co-trimoxazole	1.92 g p.o. for 10 days
Hospitalized patients (where PPNG are rare)	
Procaine penicillin	3-5 MU i.m. for 7-10 days
Benzylpenicillin	4-8 MU (or more) for 7-10 days
Tetracycline	2 g for 7-10 days
Hospitalized patients (where PPNG are prevalent)[b, c]	
Spectinomycin	4 g i.m. in two doses for 5-7 days
Gentamicin	3 mg/kg i.m. in two doses for 4-7 days
Cefuroxime	2 g for 5-7 days
Cefotaxime	2 g for 5-7 days

[a] Or a second- or third-generation cephalosporin.

[b] Kanamycin, other aminoglycosides, other second- or third-generation cephalosporins, and quinolones are alternatives.

[c] In case of non-availability of other effective drugs, a therapeutic trial with chloramphenicol appears to be justified.

Table 17.9. Regimens for treatment of acute pelvic inflammatory disease of unknown origin

	Daily dosage	
Outpatients (where PPNG are rare)		WHO list
Tetracycline	1–2 g p.o. for 10 days	WHO list
Outpatients (where PPNG are prevalent)		WHO list
Chloramphenicol	2–3 g p.o. for 10 days	WHO list
Spectinomycin	2 g i.m. single-dose + tetracycline 2 g p.o. for 10 days	WHO list
Hospitalized patients		WHO list
Benzylpenicillin	4–6 MU i.v. or i.m. + chloramphenicol 2–3 g in three to four doses	WHO list
Tetracycline	1–2 g p.o. + metronidazole 1–1.5 g p.o., both for 10 days	WHO list
Hospitalized patients (suspicion of PPNG)		
Cefoxitin	3–6 g i.m. or i.v.	
Cefotaxime[a]	1.5–3–6 g i.m. or i.v.	
Ofloxacin	400 mg p.o. in two doses	
Ciprofloxacin	0.5–1 g p.o. in two doses	

[a] Other second- or third-generation cephalosporins are alternatives. Combination with metronidazole to cover anaerobes *(Bacteroides fragilis)* may be advantageous.

One drop 1% silver nitrate into each conjunctival sac is the standard prophylaxis against *gonococcal ophthalmia neonatorum*. Tetracycline-containing eye drops provide additional prophylaxis against inclusion conjunctivitis, but carry the risk of breakthrough infections by tetracycline-resistant PPNG. Gentamicin 0.3% eye ointment may also be used for prophylaxis.

Topical antibiotics do not provide sufficient treatment of established ophthalmia neonatorum. Procaine penicillin (0.2 MU i.m. daily for 3–7 days) should be prescribed. Where PPNG are prevalent, cefuroxime or cefotaxime (60 mg/kg i.m. daily in two doses) should be prescribed. Erythromycin for 5 days is associated with a high failure rate. Single-dose therapy is preferable under limited conditions. WHO [32] recommends a single i.m. dose of either 100 mg/kg cefotaxime or 25 mg/kg kanamycin + tetracycline 1% or erythromycin 0.5% eye ointment for 10 days.

Frequent saline irrigations of the conjunctival sac are advisable when antibiotic eye ointment is not available.

A controlled clinical trial in proven gonococcal ophthalmia neonatorum found no treatment failures among 68 neonates treated with 75 mg or 150 mg kanamycin as a single dose i.m. + gentamicin 1% eye ointment for 3 days [8]. There were six treatment failures and relapses among the 34 neonates treated with single-dose kanamycin and saline eye washes instead of gentamicin eye ointment for 3 days. Topical treatment was administered by the mothers at home. The percentage of PPNG among the isolates was in the range of 20%–26% in the different treatment groups [8]. A second clinical trial showed that treatment of culture-proved gonococcal ophthalmia neonatorum with a single i.m. dose of 125 mg ceftriaxone was highly effective and obviated the need for topical treatment [9a]. Fifty-five of the 61 newborns who were treated with ceftriaxone returned for follow-up, and all were cured. Three of the 50 newborns who were treated with a single i.m. dose of 75 mg kanamycin and topical antimicrobial therapy for 7 days and returned for follow-up had persistent or recurrent gonococcal conjunctivitis [9a].

Gonococcal vulvovaginitis in children should be treated with procaine penicillin (0.1 MU/kg i.m.)+probenecid (25 mg/kg p.o.). Where PPNG are prevalent, give spectinomycin (40 mg/kg i.m.), cefuroxime (100 mg/kg i.m.), or other second- or third-generation cephalosporins. These single-dose regimens are preferable to either erythromycin or (for children 8 years and older) tetracycline for 5 days.

Other Sexually Transmitted Diseases

Chlamydia trachomatis is the commonest cause of *nongonococcal urethritis* (NGU). Tetracycline for 7 days or erythromycin are the recommended treatments. Certain quinolones may prove to be an alternative in the future. When NGU persists despite adequate treatment, reinfection, candidiasis, tuberculosis, or trichomoniasis should be considered. Treatment regimens for *chancroid* are given in Table 17.3. Co-trimoxazole, 0.96 g twice daily for 7 days, is the treatment often used. Modern quinolones are very active against *Hemophilus ducreyi,* and first experiences with single-dose regimens are promising [36, 38, 39]. Erythromycin can be given during pregnancy. Progression of buboes to fluctuance and suppuration may occur despite effective antimicrobial therapy. Treatment is by aspiration to prevent rupture. Also in treating *lymphogranuloma venereum* (Table 17.4), fluctuant buboes should be aspirated. Dressings should be applied to open lesions to prevent superinfection and autoinoculation. Surgery is often necessitated by late complications. Antibiotic cover preceding the operation is indicated because secondary bacterial infection is common.

Duration and treatment of *granuloma inguinale* (donovanosis) varies according to microbiological or clinical response (Table 17.5). Strains relatively resistant to tetracycline and streptomycin have occurred in Papua New Guinea.

The recommended treatment regimens for *syphilis* vary according to the stage of the disease (Table 17.6). In patients allergic to penicillin, tetracycline or erythromycin are usually recommended (Table 17.7). Congenital syphilis of 2 or more years' duration should be treated with penicillin in a dosage that does not exceed that used for late acquired syphilis (Table 17.6). Single-dose benzathine penicillin is not recommended except for asymptomatic infants with normal CSF. Syphilis in pregnancy is treated with a schedule that is appropriate for the stage of the disease, as in nonpregnant patients. Pregnant patients allergic to penicillin should be treated with erythromycin according to the stage of the disease (Table 17.7). Babies born to mothers treated with erythromycin should receive penicillin as if for congenital syphilis. Bacteriolysis may lead to fever after the first dose (Herxheimer's reaction). For syphilis patients allergic to penicillin, an European alternative consists of treatment with cefazolin (2 g i.v. twice daily) for 2 weeks, which is more reliable, particularly in neurosyphilis (rare cross-allergy must be excluded).

Episodes of *genital herpes* can be aborted by topical or oral acyclovir, but repeated treatment is necessary for each recurrence. Symptomatic therapy (lesions should be kept dry and clean) is sufficient in most instances. As in scabies and pediculosis, antibiotics are only indicated in case of bacterial superinfection.

For management of *trichomoniasis* and *candidiasis,* see the section on treatment of vaginal discharge in the following chapter.

References

1. Anonymous (1983) Spectinomycin-resistant penicillinase-producing *Neisseria gonorrhoeae*. Morbid Mortal Weekly Rep 32: 51-52
2. Arya OP, Osoba AO, Bennett FJ (1980) Tropical venerology. Churchill and Livingstone, Edinburgh
3. Bello CSS, Fakunle YM (1983) Single-dose treatment of acute gonorrhoea with cefuroxime: experience from Zaria, Nigeria. East Afr Med J 60: 104-107
4. Catterall RD (1983) STD - third African regional conference. Br J Vener Dis 59: 337-339
5. Duncan MO, Bilgeri YR, Fehler HG, Ballard RC (1983) Treatment of chancroid with erythromycin. Br J Vener Dis 59: 265-268
6. Fast MV, Nsanze H, D'Costa LJ, Karasira P, Maclean IW, Piot P, Albritton WL, Ronald AR (1983) Antimicrobial therapy of chancroid: an evaluation of five treatment regimens correlated with in vitro sensitivity. Sex Transm Dis 10: 1-6
7. Fast MV, Nsanze H, Plummer FA, D'Costa LJ, Maclean IW, Ronald AR (1983) Treatment of chancroid. A comparison of sulphamethoxazole and trimethoprim-sulphamethoxazole. Br J Vener Dis 59: 320-324
8. Fransen L, D'Costa L, Ronald AR, Nsanze H, Brunham RC, Piot P (1984) Single-dose kanamycin therapy of gonococcal ophthalmia neonatorum. Lancet II: 1234-1237
9. Hira SK, Ratnam AV, Sehgal D, Bhat GJ, Chintu C, Lulenga RC (1982) Congenital syphilis in Lusaka. I. Incidence in a general nursery ward. East Afr Med J 59: 241-245

9a. Laga M, Naamara W, Brunham RC, D'Costa LJ, Nsanze H, Piot P, Kunimoto D, Ndinya-Achola JO, Slaney L, Ronald AR, Plummer FA (1986) Single-dose therapy of gonococcal ophthalmia neonatorum with ceftriaxone. N Engl J Med 315: 1382-1385

10. Lal S, Gary BR (1980) Further evidence of the efficacy of co-trimoxazole in granuloma venereum. Br J Vener Dis 56: 412-413
11. Latif AS (1982) Thiamphenicol in the treatment of chancroid in men. Br J Vener Dis 58: 54-55
12. Meheus A, Ursi JP, Van Dyck E, Ballard R (1981) Treatment of chancroid with single-dose doxycycline compared with a two-day course of co-trimoxazole. Ann Soc Belg Med Trop 61: 119-124
13. Meheus AZ (1984) Practical approaches in developing countries. In: Holmes KK, Mardh PA, Sparling F, Wiesner PJ (eds). Sexually transmitted diseases. McGraw-Hill, New York, pp 998-1008
14. Meheus A, Widy-Wirsky R, D'Costa J, Van Dyck E, Delgadillo R, Piot P (1984) Treatment of gonorrhoea in males in the Central African Republic with spectinomycin and procaine penicillin. Bull WHO 62: 89-94
15. Nsanze H (1983) Practical therapeutics: sexually transmitted diseases. East Afr Med J 60: 346-349
16. Osoba AO (1981) Sexually transmitted diseases in tropical Africa - a review of the present situation. Br J Vener Dis 57: 89-94
17. Panikabutra K, Ariyarit C, Chitwarakorn A, Warnnissorn T, Saensanoh C (1982) Sensitivity to penicillin, thiamphenicol, kanamycin, cefoxitin and spectinomycin of penicillinase-producing *Neisseria gonorrhoeae* (PPNG) in Bangkok. Relation to the results of treatment. J Med Assoc Thai 65: 316-323
18. Panikabutra K, Ariyarit C, Chitwarakorn A, Saensanoh C, Wongba C (1985) Randomized comparative study of ceftriaxone and spectinomycin in gonorrhoea. Genitourin Med 61: 106-108

19. Piziak MV, Woodbury C, Berliner D, Takafuji E, Kirkpatrick J, Opal S, Tramont E (1984) Resistance trends of *Neisseria gonorrhoeae* in the Republik of Korea. Antimicrob Agents Chemother 25: 7-9
20. Plummer FA, D'Costa LJ, Nsanze H, Maclean IW, Karasira P, Piot P, Fast MV, Ronald AR (1983) Antimicrobial therapy of chancroid: effectiveness of erythromycin. J Infect Dis 148: 726-731
21. Plummer FA, Nsanze H, D'Costa LJ, Karasira P, Maclean IW, Ellison RH, Ronald AR (1983) Single-dose therapy of chancroid with trimethoprim-sulfametrole. N Engl J Med 309: 67-71
22. Pradinaud R, Grosshans E, Basset A, Bertin Ch (1981) Etude de 24 cas de donovanose en Guyane Francaise. Bull Soc Pathol Exot Filiales 74: 30-36
23. Rajan VS, Pang R, Tan NJ, Sng EH (1979) Kanamycin in the treatment of penicillinase-producing gonococcal infections. Asian J Infect Dis 3: 37-39
24. Ratnam AV, Patel MI, Hira SK, Mulenga RC, Ellison RH (1982) Sulfametrol/trimethoprim (Lidaprim) in the treatment of gonorrhoea in men. East Afr Med J 59: 737-740
25. Ratnam AV, Patel MI, Mulenga RC, Hira SK (1982) Penicillinaseproducing gonococcal strains in Zambia. Observations on treatment failures. Br J Vener Dis 58: 29-31
26. Singh KG, Singh G, Kaur P, Pandey SS, Pal D (1984) Beta-lactamase producing *Neisseria gonorrhoeae*. Indian J Med Res 79: 337-339
27. Tan T, Rajan VS, Koe SL, Tan NJ, Tan BH, Goh AJ (1977) Chancroid: a study of 500 cases. Asian J Infect Dis 1: 27-28
28. Taylor DN, Pitarangsi C, Echeverria P, Panikabutra K, Suvongse C (1985) Comparative study of ceftriaxone and trimethoprim-sulfamethoxazole for the treatment of chancroid in Thailand. J Infect Dis 152: 1002-1006
29. Tupasi TE, Crisologo LB, Torres CA, Calubiran OV, De Jesus I (1983) Cefuroxime, thiamphenicol, spectinomycin, and penicillin G in uncomplicated infections due to penicillinase-producing strains of *Neisseria gonorrhoeae*. Br J Vener Dis 59: 172-175
30. WHO (1981) Nongonococcal urethritis and other selected sexually transmitted diseases of public health importance. Tech Rep Ser No. 660. WHO, Geneva
31. WHO (1982) Treponemal infections. Tech Rep Ser No. 674. WHO, Geneva
32. WHO (1985) Working group on the prevention and treatment of ophthalmia neonatorum at the primary level. WHO, Geneva (to be published) In: Fransen L, D'Costa L, Ronald AR, Nsanze H, Brunham RC, Piot P (1984) Single-dose kanamycin therapy of gonococcal ophthalmia neonatorum. Lancet II: 1234-1237
33. Ariyarit C, Panikabutra K, Chitwarakorn A, Wongba C, Buatiang A (1986) Efficacy of ofloxacin in uncomplicated gonorrhea. Infection 14 (Suppl 4): S311-S313
34. Bogaerts J, Tello WM, Verbist L, Piot P, Vandepitte J (1987) Norfloxacin versus thiamphenicol for treatment of uncomplicated gonorrhea in Rwanda. Antimicrob Agents Chemother 31: 434-437
35. Boslego JW, Tramont EC, Takafuji ET, Diniega BM, Mitchell BS, Small JW, Khan WN, Stein DC (1987) Effect of spectinomycin use on the prevalence of spectinomycinresistant and of penicillinase-producing Neisseria gonorrhoeae. New Engl J Med 317: 272-278
36. Duncan MO, Ballard RC, Dangor Y, da L Exposto F, Fehler HG (1986) A comparison of two single dose oral regimens of ciprofloxacin in the treatment of chancroid. Abstract, 2nd World Congress on Sex Trans Dis, Paris
37. Mabey DCW, Wall RA, Bello CSS (1987) Aetiology of genital ulceration in the Gambia. Genitourin Med 63: 312-315
38. McDonald K, Cameron DW, D'Costa LJ, Ndinya-Achola JO, Plummer FA, Ronald AR (1987) Evaluation of fleroxacin (Ro 23-6240) as single dose oral therapy of culture-proven chancroid in Nairobi, Kenya. Abstracts of ICAAC, New York; Abstract No. 821
39. Naamara W, Plummer FA, Greenblatt RM, D'Costa LJ, Ndinya-Achola JO, Ronald AR (1987) Treatment of chancroid with ciprofloxacin. A prospective, randomized clinical trial. Am J Med 82 (Suppl 4A): 317-320

18 Obstetric and Gynecological Infections

Topical chemotherapy is usually adequate for superficial infections of the vagina or vulva. Pelvic inflammatory disease (PID), puerperal sepsis, febrile abortion, and bartholinitis, however, require systemic chemotherapy in high dosage. The causative organisms are usually not known at the start of treatment, and mixed infections are common. Therefore, treatment with combinations covering the spectrum of potential organisms as far as possible is indicated. During pregnancy, the choice of antibiotics is limited due to the risk of toxicity to the fetus. Benzylpenicillin, phenoxymethylpenicillin, ampicillin, cefazolin, cefoxitin, cefotaxime, and erythromycin are generally safe throughout pregnancy.

Bartholinitis

Bartholinitis is usually caused by gonococci or staphylococci (Gram's stain of the aspirate). Mixed anaerobic infections are relatively rare. Early cases may respond to systemic antimicrobial therapy for 5–7 days; other require drainage in addition to chemotherapy. Co-trimoxazole is suggested for initial, empirical therapy. Oral cloxacillin, or, alternatively, chloramphenicol or erythromycin should be used for treatment of staphylococcal infections and when no organisms are found. Gonococcal infection is treated with benzylpenicillin or tetracycline. Penicillinase-producing Neisseria gonorrhoeae (PPNG) may be treated with second- or third-generation cephalosporins, co-trimoxazole, or quinolones.

Vulvovaginitis in Children

Vulvovaginitis in children is often gonococcal in origin, being transmitted either by intermediary objects or sexually [7]. Close contacts and relatives should be examined for gonorrhea and treated, when indicated.

The possibility of infection caused by pneumococci, staphylococci, *Hemophilus influenzae,* streptococci, gram-negative rods, *Candida, Trichomonas, Enterobius* or foreign bodies has to be considered.

Systemic chemotherapy should be given according to the pathogen (smear, slide of discharge). For treatment of gonococcal vulvovaginitis, see previous chapter. Education on personal hygiene is advisable.

Vaginitis, Vulvitis in Adults

Vulvitis often coexists with vaginitis. Recent antibiotic chemotherapy, diabetes mellitus, carcinoma, oral contraceptives, tampons or pessaries, and lack of estrogens in women beyond the menopause are possible predisposing factors.

Microscopic examination of vaginal discharge (methylene blue stain, wet mount) is usually efficacious in diagnosing trichomoniasis, candidiasis, and nonspecific vaginitis (clue cells). If neither pathogens, polymorphonuclear neutrophilic leukocytes (PMNs), nor clue cells are found, gonococcal cervicitis, in particular, but also syphilis and tuberculosis, should be excluded. If microscopy cannot be performed, treatment has to be guided by the appearance of vaginal discharge (although there is some overlap). A white, creamy discharge with pruritus is typical for candidiasis. Profuse, purulent discharge is usually seen in trichomoniasis and, with a fishy odor, in nonspecific vaginitis due to *Gardnerella vaginalis*. Treatment with multiple doses of metronidazole covers both of these conditions, which often coexist. If there is no response, gonococcal cervicitis (recent urethral discharge in a contact) should be considered.

An alternative approach has been proposed by Meheus [3, 4], who suggests initial treatment for gonorrhea in order to prevent gonococcal PID and its sequelae (Fig. 18.1). An overall prevalence of gonorrhea of more than 20% in women with vaginal discharge, as reported from eastern and southern Africa, forms the basis for this approach.

Low abdominal pain in addition to vaginal discharge suggests PID. Vaginal discharge due to infection with *N. gonorrhoeae* was significantly associated with abdominal pain in a study of women presenting with vaginal discharge at an STD clinic in Nairobi, Kenya [5].

Types of Vaginal Disease

Trichomoniasis can usually be treated adequately with a single dose of metronidazole, tinidazole, ornidazole, or nimorazole (2 g in adults, 1 g in children older than 12 years). Concurrent treatment of contacts (urethritis or asymptomatic infection) is indicated. The conventional treatment doses of metronidazole are 200 mg three times daily for adults, 100 mg three times daily for girls 6-10 years old, 50 mg twice daily for girls 2-5 years old, for 5-7 days. The treatment should not be repeated before 4-6 weeks.

The systemic use of these agents should be avoided in the first trimenon of pregnancy and during lactation. Intravaginal metronidazole or povidone-iodine are the alternatives.

Candidiasis should be treated with nystatin, one tablet intravaginally twice daily for 7-14 days, or with clotrimazole, miconazole or other azole derivatives intravaginally, preferably at night, for 3-5 days. Candida balanitis and urethritis in contacts should be treated to prevent reinfection. Relapses are common, and predisposing factors should be taken into account.

Nonspecific vaginitis due to *Gardnerella vaginalis*, often in mixed infection with anaerobic bacteria, responds to topical use of metronidazole or tetracycline. Treat-

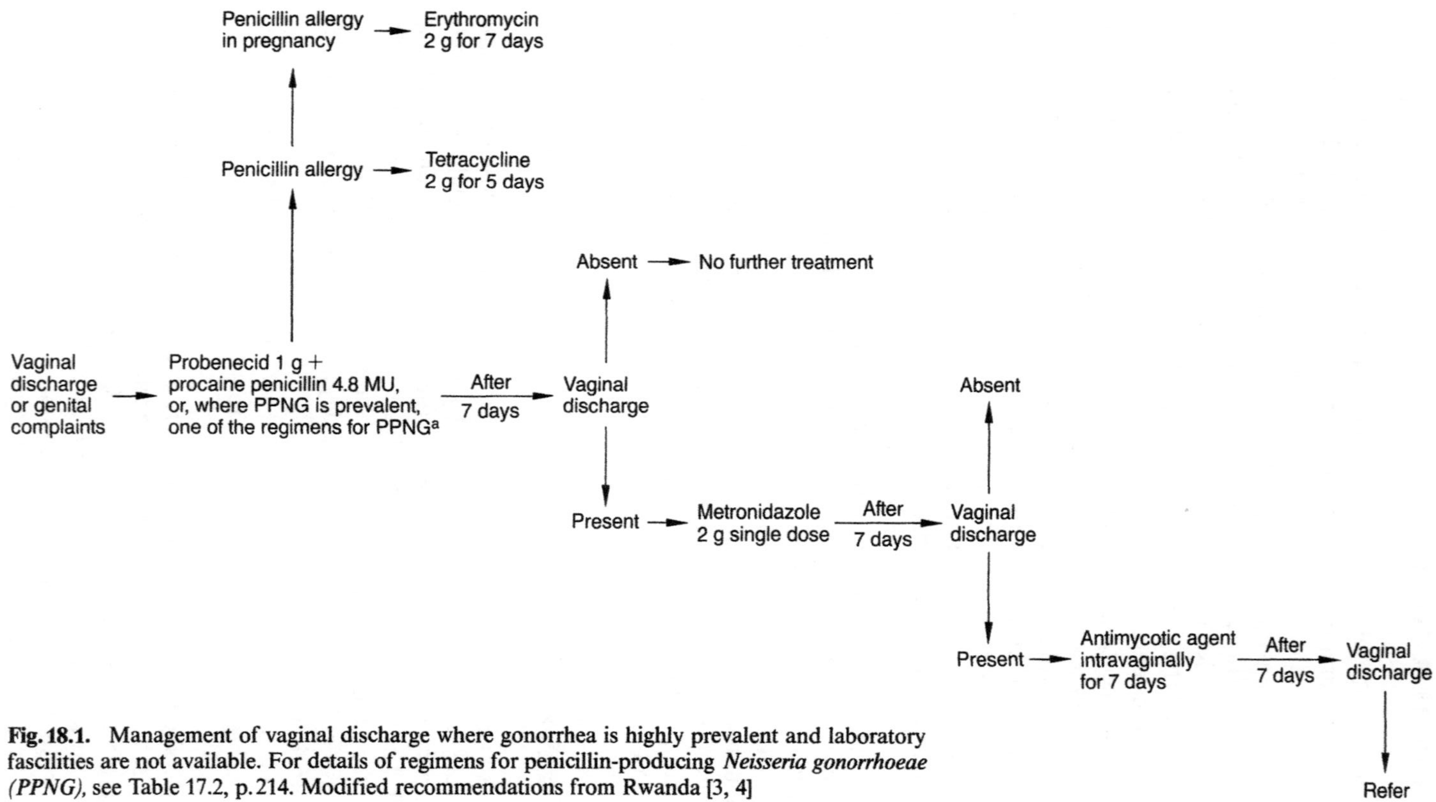

Fig. 18.1. Management of vaginal discharge where gonorrhea is highly prevalent and laboratory fascilities are not available. For details of regimens for penicillin-producing *Neisseria gonorrhoeae (PPNG)*, see Table 17.2, p. 214. Modified recommendations from Rwanda [3, 4]

ment with oral metronidazole for 5 to 7 days is suggested if topical treatment is not effective.

Herpetic vulvovaginitis due to herpesvirus type 2, may be aborted by acyclovir cream or oral acyclovir in more severe cases (expensive). Repeated treatment is required for recurrent episodes. Symptomatic therapy with prevention of secondary bacterial infection (lesions should be kept dry and clean) usually suffices.

Pelvic Inflammatory Disease

PID is a common cause of gynecological admissions in hospitals in developing countries [2, 6]. Comparison of the limited data available suggests that acute (gonococcal) salpingitis, chronic salpingitis, and puerperal sepsis are the main causes of PID in Africa, whereas in India these are chronic tuberculous salpingitis, septic abortion, and puerperal sepsis. The incidence of sequelae (infertility, ectopic pregnancy, chronic pain, etc.) appears to vary widely between different countries [6].

Acute PID (salpingitis, pelvioperitonitis) is often polymicrobial in origin. Enterobacteria, streptococci, gonococci, and anaerobic bacteria including *Bacteroides fragilis* are the pathogens commonly encountered. In countries with a high incidence of gonorrhea, *N. gonorrhoeae* is isolated from more than 40% of patients with PID [9, 10]. *Chlamydia trachomatis* has only recently been recognized as a major pathogen of PID in industrialized countries; its importance in developing countries remains to be evaluated. Filariasis and *Schistosoma* infection may involve the female genital tract and render it more susceptible to bacterial infection. Bacteriology of cervical secretion, pus, blood culture, or material obtained at laparoscopy or operation may be difficult to interpret, since mixed infections are common and the organism isolated need not be the primary pathogen. Treatment is, therefore, empirical in most instances (cf. tables 17.8 and 17.9).

Treatment of Acute PID

Tetracycline in ambulatory patients and parenteral penicillin + chloramphenicol in hospitalized patients are the regimens often used for treatment of acute PID under conditions of limited resources.

In severe infections, treatment with cefoxitin (2 g two to three times daily), a cephalosporin with activity against *Bacteroides fragilis,* is now widely recommended in Europe and the United States. Clindamycin (1.2 g a day) + gentamicin (240 mg a day), is an alternative. This combination may be preferred in postabortal infections, a percentage of which are caused by penicillin-resistant staphylococci and resistant enterobacteria. Both regimens are active against penicillinase-producing *N. gonorrhoeae* (PPNG), but not against *Chlamydia trachomatis* (tetracycline for at least 10–14 days). Ampicillin + gentamicin, tetracycline + gentamicin, metronidazole + gentamicin, and chloramphenicol + gentamicin are somewhat less expensive alternatives. Since gentamicin covers enterobacteria and PPNG, an agent active against anaerobic bacteria is added.

Benzylpenicillin (1-2 MU every 6 h) or procaine penicillin (2 MU i.m. per day) + chloramphenicol (500 mg every 6 h) is often used for initial therapy of PID in developing countries. If there is no response after 48 h, suppurative complications requiring surgical intervention (pelvic abscess, ovarian abscess, or pyosalpinx) or infection with resistant enterobacteria should be considered. In nonresponse, gentamicin or a second- or third-generation cephalosporin should be used. Co-trimoxazole may be sufficient in less severe nonresponding cases. Intrauterine contraceptive devices should be removed in patients with PID. Quinolones like ciprofloxacin cover enterobacteria, PPNG and chlamydiae, but are only weakly active against streptococci and *Bacteroides* spp. (cf. tables 17.8 and 17.9).

Treatment of Chronic PID

Frequently tuberculous in origin, chronic PID is treated with long-term triple-drug combinations. Histopathological examination of material obtained by endometrial curettage, culdoscopy, laparoscopy, or at operation is more often diagnostic than is culture of menstrual blood.

Other Obstetric and Gynecological Infections

Septic Abortion

Mixed infection with enterobacteria, streptococci, enterococci, and anaerobic bacteria, including *Bacteroides fragilis,* are common. Staphylococci and gonococci are less frequently found. Infection with clostridia, gas gangrene, and tetanus may also occur in septic abortion.

Swabs of the cervix and the placental tissue as well as blood cultures should be taken for bacteriological investigation. Mild cases may be treated with benzylpenicillin in high dosage (8-15 MU per day), which covers streptococci and clostridia.

In more severe cases with signs of peritonitis, therapy should cover enterobacteria as well, particularly if septic shock supervenes.

Under conditions of limited resources, chloramphenicol or gentamicin should be added. These combinations should be supplemented with metronidazole for treatment of *Bacteroides fragilis* infection.

Broad-spectrum combinations like cefotaxime + metronidazole, mezlocillin + gentamicin, or clindamycin + gentamicin, are suggested in Europe and the United States.

Vigorous treatment of shock and coagulation disorders (DIC) is also important. Retained placental tissue must be removed.

Puerperal Sepsis

When fever occurs during the puerperium, swabs of the lochia and blood cultures should be taken. Chemotherapy should be instituted with benzylpenicillin in high dosage, as soon as signs of puerperal sepsis develop. This therapy covers the dangerous infections with group A-streptococci and clostridia. The spectrum of pathogens encountered in puerperal sepsis is similar to that of septic abortion. In case of non-response to benzylpenicillin, combination therapy as for septic abortion is necessary.

Treatment with benzylpenicillin (6 MU i.v. in six doses a day) for 3 days, followed by procaine penicillin (1.2 MU daily i.m.) for an additional 7 days + chloramphenicol (500 mg orally or i.v. every 6 h) for 10 days has proven significantly more effective in limiting the duration of fever than benzylpenicillin in the same dosage combined with streptomycin (1 g daily i.m. for 10 days) in a study in Addis Ababa, Ethiopia [1]. The organisms most commonly isolated were enterobacteria, gonococci, and beta-hemolytic streptococci. The isolation rate of anaerobic bacteria was low due to technical difficulties [1, 8]. In a study of 114 women with puerperal sepsis from Papua New Guinea, anaerobes (*Bacteroides* spp. and peptostreptococci) were isolated in 70% of patients, either as pure growth (58%) or in mixed infection (12%). Group A streptococci, ranking second, were isolated in 19% of cases [12].

Mastitis

Mastitis in lactation is mostly caused by *Staphylococcus aureus*. Other pathogens, e.g., beta-hemolytic streptococci, are rare [11].

Antimicrobial therapy should, therefore, be directed at staphylococci (isolation from milk or pus of an abscess) and treatment be started with penicillinase-resistant penicillin such as cloxacillin (2–3 g a day). In case of nonavailability, chloramphenicol (2–3 g) or erythromycin (2 g) may be used. Clindamycin is a valuable alternative in patients who are allergic to penicillin or have infections due to (cloxacillin)-resistant staphylococci. Penicillin-sensitive staphylococci are best treated with phenoxymethylpenicillin or parenteral benzylpenicillin. Parenteral high-dose treatment started early in the course of the illness may prevent abscess formation. Once an abscess has formed, incision with drainage is required in addition to chemotherapy. Symptomatic therapy includes suppression of lactation and cold compresses. The infant should not be fed from an infected breast, since bacteria contained in the milk may cause disease in the child.

Mixed anaerobic infection with peptostreptococci and *Bacteroides* species is typical for the much less common mastitis of a nonlactating breast and should be treated with penicillin, tetracycline, or penicillin + metronidazole. Clindamycin or cefoxitin are more expensive alternatives.

Pyrexia During Labor

Where fever with signs of chorioamnionitis occurs during prolonged labor or after premature rupture of the membranes, there is a considerable risk that the child will develop septicemia after birth.

The range of pathogens is similar to that of septic abortion. Chemotherapy is, therefore, indicated. Only penicillins and cephalosporins are both nontoxic to the fetus and achieve high concentrations in the amniotic fluid. Parenteral ampicillin should be used when cephalosporins are not available.

After birth, blood cultures should be taken and the newborn should be monitored carefully for signs of infection during the first few days. Gentamicin + ampicillin or another combination appropriate for treatment of neonatal septicemia may be given if there is any indication of infection. Alternatively, combination therapy for suspected septicemia is instituted after blood cultures have been taken. Treatment is discontinued after several days when the child remains well and blood cultures are sterile.

Persistent infection in the mother should be treated as for puerperal sepsis.

Treatment with Antibiotics in Pregnancy

Antibacterial agents should be given during pregnancy only when their safety for the patient and for the developing embryo or fetus has been established. Tetracyclines and rifampicin may be hepatotoxic in pregnancy, particularly at high dosage and in underlying renal disease. Potentially nephrotoxic antibiotics (aminoglycosides) should be used only with great care in pregnant women with preexisting renal disease or eclampsia.

The use of antibiotics in pregnancy is to a large extent limited by the potential risks to the developing child. Only the minority of antibiotics can safely be used throughout pregnancy. Antibiotics can be divided into different groups according to their degree of safety (Table 18.1).

Only classical penicillins (benzylpenicillin, phenoxymethylpenicillin, ampicillin, amoxicillin), classical cephalosporins (cefazolin, cefalexin, cefuroxime), and erythromycin (except the estolate) are generally regarded as safe for pregnant women. If possible these agents should be used for treatment. Sulfonamides are also generally regarded as safe, with one exception; sulfonamides can displace bilirubin from its binding sites at the serum proteins, thereby increasing the serum concentration of free unbound bilirubin. Because of this risk of hyperbilirubinemia in the newborn, which may lead to bilirubin encephalopathy, sulfonamides should not be given during the last week of pregnancy.

The safety in pregnancy of the newer agents has not been proven in many cases. There are only weak arguments against the use of new cephalosporins or acylaminopenicillins (azlo-, mezlo-, piperacillin). The risk of certain derivatives with reactive side chains (cefamandole, cefmenoxime, ceftriaxon) may be higher.

A number of antibacterial agents are potentially teratogenic or cytotoxic in different in vitro tests. The mainly teratogenic risk is highest in the first trimester of pregnancy. In every patient the potential harm has to be weighed against the damage caused by an acute severe infection. The risk at which the pregnancy is put by such an infection should not be underestimated. The teratogenic and toxic risks of trimethoprim and of chloramphenicol are apparently low. Despite widespread use of these agents there have been no reports of teratogenicity.

Substances not mentioned in Table 18.1 should generally be considered as unsuitable for use during pregnancy. Tetracyclines may impair skeletal growth and stain teeth yellow. They should be given only during the second and third trimes-

Table 18.1. Use of important antimicrobials in pregnancy

Period	Generally safe	Safety not demonstrated	Potentially teratogenic or cytotoxic	Toxic
First trimester	Benzylpenicillin, phenoxymethyl-penicillin, ampicillin, amoxicillin, cefazolin, cefuroxime, cefoxitin, cefotaxime, erythromycin, sulfonamides	New penicillins, new cephalosporins (risk obviously very small), other new substances, clindamycin	Trimethoprim, tetroxoprim, pyrimethamine, rifampicin, ethionamide, chloramphenicol, flucytosine, nitrofurantoin, metronidazole and other nitroimidazoles, griseofulvin	Tetracyclines, aminoglycosides, nalidixic acid (and other quinolones), amphotericin B, flucytosine
Second and third trimester	As first trimester	Safe	Risk decreases with duration of pregnancy	As first trimester
Last week before delivery	As first trimester (except for sulfonamides)	Safe	Risk decreases with duration of pregnancy	As first trimester plus sulfonamides, co-trimoxazole, nitrofurantoin

ter when other antibiotics are either unavailable or inactive (for instance in rickettsial disease). Aminoglycosides during pregnancy may damage the child's inner ear and should thus be substituted by less toxic alternatives whenever possible. Antibacterial agents of the quinolone group (nalidixic acid, norfloxacin, ofloxacin, ciprofloxacin, etc.) may damage the articular cartilage of growing animals and should be avoided, if at all possible, during all stages of pregnancy. Nitrofurantoin is cytotoxic, teratogenic and may lead to hemolytic anemia in the newborn as a result of immaturity of enzymatic systems. It should not be used in pregnancy.

Suggestions for Treatment

Respiratory and urinary infections are best treated with ampicillin or amoxicillin. If treatment is considered necessary it should not be withheld for fear of teratogenicity or toxicity, since infections themselves can also harm the embryo and fetus. Skin and wound infections should be treated with penicillin or cloxacillin. Erythromycin is an alternative in respiratory and skin infections.

Chorioamnionitis is difficult to treat. Usually there are mixed infections with streptococci, gram-negative enteric bacilli, and anaerobic bacteria, but *Listeria monocytogenes,* chlamydiae, and clostridia may also be involved. Antibiotics reach the amniotic fluid via fetal renal excretion after having crossed the placenta. With penicillins and cephalosporins, high levels of antibiotic are achieved in the amniotic fluid, in contrary to many other agents (erythromycin, aminoglycosides, tetracycline). Chloramphenicol crosses the placenta best but should be avoided if possible because of its hematotoxicity. Intra-amniotic infections require parenteral treatment at high dosage. In developed countries treatment is usually carried out with broad-spectrum cephalosporins and modern penicillins. Ampicillin, chloramphenicol, and possibly high-dose benzylpenicillin are relevant alternatives for the tropics. Treatment of severe life-threatening infections in pregnant women with potentially harmful agents should be preferred to non-treatment in case of non-availability of safe antibiotics.

References

1. Duncan ME, Perine PL, Krause DW, Awoke S, Zaidí AA (1980) Pelvic inflammatory disease and puerperal sepsis in Ethiopia. II. Treatment. Am J Obstet Gynecol 138: 1059-1063
2. McGoldrick IA (1982) Pelvic inflammatory disease. Papua New Guinea Med J 25: 37-43
3. Meheus AZ (1984) Practical approaches in developing nations. In: Holmes KK, Mardh PA, Sparling PF, Wiesner PJ (eds) Sexually transmitted diseases. McGraw Hill, New York, p 998-1008
4. Meheus A, Habimana A (1982) Maladies transmises par voie sexuelle. In: Meheus A, Butera S, Eylenbosch W, Gatera G, Kivits M, Musafili I (eds) Sante et maladies au Rwanda. Universitaire Instelling Antwerpen p 481-496
5. Mirza NB, Nsanze H, D'Costa LJ, Piot P (1983) Microbiology of vaginal discharge in Nairobi, Kenya. Br J Vener Dis 59: 186-188
6. Muir DG, Belsey MA (1980) Pelvic inflammatory disease and its consequences in the developing world. Am J Obstet Gynecol 138: 913-928
7. Osoba AO, Alausa KO (1974) Vulvovaginitis in Nigerian children. Niger J Paediatr 1: 26-32
8. Perine PL, Duncan ME, Krause DW, Awoke S (1980) Pelvic inflammatory disease and puerperal sepsis in Ethiopia. I. Etiology. Am J Obstet Gynecol 138: 969-973
9. Ratnam AV, Din SN, Chatterjee TK (1980) Gonococcal infection in women with pelvic inflammatory disease in Lusaka, Zambia. Am J Obstet Gynecol 138: 965-968
10. Rooke PW (1982) Pelvic inflammatory disease in Port Moresby General Hospital. Papua New Guinea Med J 25: 29-32
11. Sarkar R, Agarwal DS, Mukherjee P (1979) Bacteriological study of breast abscess with special reference to Staph. aureus. Indian J Med Res 70: 384-390
12. Vacca A, Henderson A (1980) Puerperal sepsis in Port Moresby, Papua New Guinea, Papua New Guinea Med J 23: 120-125

19 Genitourinary Infections

Urinary Tract Infections

Incidence and Etiology

Information on the frequency and clinical importance of urinary tract infections (UTI) in developing countries is scanty and incomplete [16]. Apparently, UTI are frequent in the tropics.

Several surveys have been undertaken to investigate the prevalence of significant bacteriuria in certain, well-defined groups, such as schoolchildren, pregnant women, or people living in areas where schistosomiasis is endemic (Tables 19.1-19.3). The overall prevalence of significant bacteriuria in these surveys does not appear to be much higher than the USA and Europa. However, several studies indicate a considerably higher incidence in malnourished children (Table 19.4).

In general, UTI are apparently not considered a major health problem in developing countries, at least as long as other infections and infectious diseases are not under control, but severe UTI are not rare: UTI (mostly acute pyelonephritis) accounted for 0.5%-3% of medical admissions to hospital in four African studies [7, 11, 18, 24].

In a survey of bacterial infections in a Nigerian teaching hospital, 9.1% were UTI [3]. Of a total of 36 cases of UTI, 22 were observed in the department of gynecology. Overall, UTI were observed in 1.9% of the 1931 patients admitted to hospi-

Table 19.1. Prevalence of significant bacteriuria in schoolchildren

Study location	[Reference]	No. of children screened		Prevalence of bacteriuria (%)
India				
Bombay	[5]	190 girls	aged 5 to 10 years	2.1[a]
		810 boys		0.8
Chandigarh	[31]	1020 girls	aged 4 to 15 years	0.3[a]
		480 boys		0.9
Nigeria				
Kaduna	[1]	355 girls	aged 6 to 16 years	1.7[a]
		245 boys		0.0
Benin	[2]	678 girls	aged 10 to 19 years	1.2

[a] Significant bacteriuria (10^5 or more organisms/ml urine) was "confirmed" by at least one consecutive sample.

Table 19.2. Prevalence of significant bacteriuria in pregnancy

Study location	[Reference]	No. of women screened	Prevalence of asymptomatic bacteriuria (%)	Prevalence of urinary infection (symptomatic bacteriuria) (%)
Ibadan, Nigeria	[23]	2508	9.7	–
Nairobi, Kenya	[19]	1700	5.5	2.5
Calcutta, India	[15]	1000	8.0	2.2

Table 19.3. Prevalence of significant bacteriuria in urinary schistosomiasis: results of three community surveys

Country	[Reference]	Subjects		No. of subjects	Prevalence of bacteriuria (%)[a]
Gambia	[32]	Males only			
		<25 years:	schistosomiasis endemic	152	6.6
			schistosomiasis nonendemic	153	0.0
		>25 years:	schistosomiasis endemic	124	0.8
			schistosomiasis nonendemic	143	0.7
Egypt	[17]	School boys,	aged 5 to 15 years		
			egg excreters	260	6.5
			nonexcreters	130	2.3
Nigeria	[26]	Boys aged 5+	egg excreters	553	0.0
			nonexcreters	439	0.0
		Girls aged 5+	egg excreters	185	0.5
			nonexcreters	626	0.2
		Males aged 21+	egg excreters	79	0.0
			nonexcreters	557	0.0
		Females aged 21+	egg excreters	76	1.3
			nonexcreters	582	1.4

[a] Significant bacteriuria was determined by two consecutive samples yielding 10^5 or more organisms/ml urine.

Table 19.4. Incidence of significant bacteriuria in children with severe protein-energy malnutrition

Country	[Reference]	No. of children	Incidence of significant bacteriuria (%)
Uganda	[25]	75	10.7
Jamaica	[10]	95	9.5
Thailand	[20]	35	34.3
Bangladesh	[12]	91	29.6
Nigeria	[22]	65	23.1
South Africa	[8]	16	31.3

tal during the period of the survey [3]. UTI were diagnosed in 3.5% of 7643 patients hospitalized on the medical wards from 1970 to 1974 in a teaching hospital in Zaire [29]. A distinction between community acquired and nosocomially acquired infections was not made in either of the studies.

Enterobacteria, particularly *E. coli,* are the most important pathogens of UTI. Gram-positive bacteria, *Staphylococcus* spp. and enterococci, and *Pseudomonas aeruginosa* are isolated with lesser frequency (Table 19.5). *E. coli* is the predominant pathogen in initial episodes of UTI and cystitis in females (Table 19.6 and 19.7). *Proteus, Klebsiella, Enterobacter,* and *Pseudomonas aeruginosa* are frequently isolated in infections occurring in underlying urological diseases or where there is a history of previous infections, prior instrumentation of the urinary tract, catheterization, or recent urological surgery (Tables 19.6 and 19.7). Local factors and nosocomial infections with resistant hospital strains also play a role.

Table 19.5 see p. 234

Table 19.6. Percentage of bacterial isolates in first vs chronic recurrent urinary tract infections (UTI). India [28]

	First UTI	Chronic recurrent UTI
No. of patients	61	39
Bacteria isolated (%)		
Escherichia coli	85.2	48.7
Klebsiella spp.	8.2	17.9
Proteus spp.	4.9	15.4
Pseudomonas aeruginosa	–	7.7
Staphylococcus aureus	1.6	7.7
Enterococci	–	2.5

Table 19.7. Percentage of bacteria isolated in three types of urinary tract infection (UTI) at the Hospital Rosales, San Salvador, 1973. From [9]

Bacteria isolated (%)	Acute UTI	Chronic pyelonephritis, no obstruction	UTI with obstruction
Escherichia coli	71	60.3	28.3
Klebsiella spp.	1.4	0.9	2.6
Enterobacter spp.	7.3	10.0	12.0
Proteus spp.	4.3	7.5	20.1
Citrobacter spp.	5.8	9.2	16.0
Pseudomonas spp.	4.4	8.4	21.0
Providencia spp.	1.4	0.9	–
Hafnia spp.	–	1.9	–
Streptococci	–	0.9	–
Staphylococci	4.4	–	–

Table 19.5. Bacterial isolates (%) in urinary tract infections: results of six studies

Country [Reference]	India [27]	Nigeria [22]	Upper Volta [4]	Nigeria [13]	San Salvador [9]	Ethiopia [14]			
No. of urine samples	690	64	724	60	1000	241	278	801	413
Source	No data	Pediatric outpatients	Outpatients	Pregnant women, inpatients	Inpatients	Outpatients, male/female		Inpatients, male/female	
Bacteria isolated									
Escherichia coli	68.7	45.3	42.3	80.0	43.8	36.1	53.6	21.8	37.8
Klebsiella spp.	13.0	37.5	22.6	16.7	2.6	19.5	13.7	22.3	21.3
Enterobacter spp.	-	-	7.6	-	17.9	6.6	2.2	10.2	13.1
Proteus spp.	1.0	-	4.8	-	10.4	16.2	7.6	23.7	11.3
Citrobacter spp.	-	4.7	0.6	-	2.6	1.7	0.4	7.6	3.1
Pseudomonas spp.	5.1	-	3.6	-	11.7	4.6	4.7	9.0	3.9
Providencia spp.	-	-	0.4	-	0.9	-	-	-	-
Betahemolytic	-	-		1.7	2.6				
streptococci			1.9						
Enterococci	2.5	-		1.7	-				
Staphylococcus aureus	1.6	3.1	11.3[a]	-	0.8	15.3	18.0	5.2	9.5
Alcaligenes faecalis	4.2	1.6	1.7	-	0.1				
Acinetobacter spp.	-	3.1	1.0	-	-				
Others	2.6	1.6	1.5	-	0.6				
Mixed infections	-	3.1	-	-	-	-	-	-	-

[a] *Staphylococcus* spp.

Staphylococcus saprophyticus, a coagulase-negative species, was recognized as an important pathogen of UTI in young women in the USA and Europe during the 1970s. Its isolation in urinary infections has not yet been reported from developing countries.

Diagnosis

Diagnosis of UTI is based on typical symptoms - dysuria, urgency, frequency of micturition - and laboratory findings of significant bacteriuria and/or pyuria. Fever and loin pain indicate an infection of the kidney (pyelonephritis), but asymptomatic involvement of the kidney also occurs. Several laboratory tests (concentrating test, detection of antibody-coated bacteria in the urine, etc.) have been proposed to distinguish between upper and lower UTI, but none of them has yet been established in clinical routine.

Clinical symptoms of UTI are usually not characteristic in infants and young children (Table 19.8).

Significant bacteriuria of 10^5 or more bacteria/ml urine is almost always found in acute UTI due to gram-negative bacteria. Counts of less than 10^5/ml are commonly due to contamination and a second specimen should be examined. Semi-quantitative culture by dip slide is usually preferred in office practice to quantitative cultures on agar plates. Antibiograms should be performed in cases of persistent or relapsing infections and in cases of obstructive UTI.

Where microbiological cultures are not possible, microscopy of uncentrifuged, freshly voided, well-mixed urine is helpful. A drop of urine is stained (Gram's stain or methylene blue) and examined microscopically. A diagnosis of significant bacteriuria ($\geqslant 10^5$ bacteria/ml) can be made when at least two bacteria per high-power field (magnitude 1000) are seen [30]. Findings consistent with those of quantitative cultures can be expected in at least 90% of cases [30].

Examination of uncentrifuged, unstained urine for pyuria (defined as 20-25 or more leukocytes/mm^3) in a counting chamber is a reliable method, but cannot be

Table 19.8. Symptomatic urinary tract infections (UTI): presenting complaint and frequency of diagnosed UTI in pedriatric outpatients in Nigeria. Modified from [22]

Presenting complaint	Number visiting department in 6 months	Number providing urine sample (% of total)	Diagnosed UTI	Frequency of UTI (%)
PUO	1800	87 (5)	9	10
Mild diarrhea	400	20 (5)	2	10
Severe diarrhea	53	53 (100)	14	27
Dysuria	42	42 (100)	18	43
Failure to thrive (under 6 months of age)	9	9 (100)	1	12
Miscellaneous[a]	61	61 (100)	2	3

PUO, pyrexia of unknown origin.

[a] Various nonspecific symptoms including abdominal pain and enuresis.

Table 19.9. Urinary tract infections (UTI): selection of antibiotics from the WHO list

	First choice	Second choice	Third choice	Effective but not recommended	Poor efficacy	Ineffective	Not on WHO list
Acute cystitis (women only)	Co-trimoxazole, ampicillin (single dose)	Sulfisoxazole (single dose)	Gentamicin (single dose)	Chloramphenicol	Tetracycline, nitrofurantoin	Penicillin, erythromycin, rifampicin	Cefotaxime, quinolones (single dose)
Acute pyelonephritis (men and women)	Co-trimoxazole (14–20 days)	Ampicillin	Gentamicin	Chloramphenicol	Tetracycline, sulfonamides, nitrofurantoin	Penicillin, erythromycin, rifampicin	Cephalosporins, quinolones, amoxicillin + clavulanic acid
Chronic pyelonephritis	Co-trimoxazole	Ampicillin	Sulfonamides	Chloramphenicol, gentamicin	Nitrofurantoin, tetracycline	Penicillin, erythromycin, rifampicin	Oral cephalosporins, quinolones
UTI in pregnancy	Ampicillin	Sulfonamide	Co-trimoxazole	Many agents	Tetracycline	Penicillin, erythromycin, rifampicin	Cephalosporins
UTI in course of urological disease	Gentamicin, amikacin	Co-trimoxazole	Chloramphenicol		Ampicillin, tetracycline, nitrofurantoin	Sulfonamides	Cefotaxime, piperacillin, mezlocillin
Renal abscess	Cloxacillin	Erythromycin	Chloramphenicol	Rifampicin	Penicillin, tetracycline	Nitrofurantoin	Cephalosporins
Low-dose prophylaxis of recurrent infections	Co-trimoxazole	Nitrofurantoin	Sulfonamides		Ampicillin, doxycyclin		Oral cephalosporins, pipemidic acid

used in clinical conditions associated with increased urinary white cell loss, e.g., schistosomiasis or tuberculosis [21].

Clean midstream urine and prompt examination (within 1 h if specimen not kept in a refrigerator) are prerequisites for correct diagnosis of UTI.

Principles of Chemotherapy

1. Treatment of symptomatic UTI and of asymptomatic bacteriuria in children and pregnancy is generally recommended.
2. Duration of treatment: Studies carried out in North America, Europa, and Australia have shown the efficacy of single-dose therapy of *acute cystitis in women* [6]. A wide range of agents are effective (Table 19.10). If there is no rapid relief of symptoms, further diagnostic procedures and a conventional course of antibiotic therapy are indicated.
   Acute pyelonephritis should be treated for at least 14 days. Longer duration of treatment is suggested for UTI occurring in males.
3. Choice of agent (Table 19.9): Short-acting sulfonamides - sulfafurazole, sulfadimidine, sulfadiazine - are still often recommended in developing countries for treatment of UTI, since they are cheap and stable. However, single-dose treatment with co-trimoxazole or ampicillin may often be more cost-effective than a conventional course of sulfonamides, for reasons of resistance. Due to increasing resistance and slow onset of action, limited use of sulfonamides (treatment of acute cystitis only) is suggested. Acute pyelonephritis and UTI complicated by underlying urological disease should be treated with more active bacterial agents, such as co-trimoxazole, ampicillin, or gentamicin (see Table 19.9).
4. Dosage of oral and parenteral antibiotics in UTI are given in Tables 19.10 and 19.11. Parenteral antibiotics are preferable when there is the risk of urosepsis.
5. Treatment of complicated UTI, recurrent UTI, and cases not responding to initial antibiotic therapy should be guided by culture with antibiograms (Table 19.12). Quinolones will play an important role in these indications in future.
6. Low-dose prophylaxis of recurrent UTI may be needed in girls. Low-dose therapy is also carried out in females with recurrent episodes of cystitis as prophylaxis of ascending infections (Table 19.13).

Table 19.10. Dosage of oral antibiotics in urinary tract infections

Drug	Average daily dose on continuous therapy		Dose interval (h)	Reduce dose in renal failure	Dose for single-dose treatment (g)
	Children (mg/kg)	Adults (g)			
Sulfadimidine	[a]	4	6	Yes	-
Sulfisoxazole (≙sulfafurazole)	[a]	4	6	Yes	2
Sulfadiazine	20	1	12	Yes	Unsuitable
Ampicillin	80	3	6-8	No	3
Amoxicillin	50	1.5	6-8	No	1.5-3.0
Bacampicillin	60	2.4	6-8	No	1.6-3.2
Cefalexin	100	4.0	6-8	No	-
Co-trimoxazole	40	1.92	12	Slightly	1.92
Doxycycline	2	0.1	24	No	Unsuitable
Tetracycline	20	1.0	8-12	Yes	Unsuitable
Nalidixic acid	Contrain-dicated	4.0	6	Slightly	Unsuitable
Nitrofurantoin	5	0.15-0.3	8	Yes	Unsuitable
Pipemidic acid	Contrain-dicated	0.8	12	?	Unsuitable
Norfloxacin, pefloxacin	Contrain-dicated	0.4-0.8	8	Yes	0.4
Ofloxacin	Contrain-dicated	0.2-0.4	12	Yes	0.1-0.2
Ciprofloxacin	Contrain-dicated	0.5-1.0	12	Yes	0.25

[a] Average daily dose on continuous therapy in children: 0-1 year: 1.0-2.0 g; 1-6 years: 2.0-3.0 g; 6-12 years: 3.0-4.0 g.

Table 19.11. Dosage of parenteral antibiotics in urinary tract infections

Drug	Average daily dose on continuous therapy		Dose interval (h)	Reduce dose in renal failure	Dose for single-dose treatment (g)
	Children (mg/kg)	Adults (g)			
Cephalothin	100	6	6-8	Slightly	Unsuitable
Cefazolin	60	3-6	6-8	Slightly	Not known
Cefuroxime	60	1.5-4.5	6-8	Slightly	1.5-3.0
Cefotaxime	60	3.0-6.0	6-8	Slightly	1.0-2.0
Azlocillin, mezlocillin, piperacillin	100	6.0-10.0	6-8	Slightly	2.0
Kanamycin, amikacin	15	0.5-1.0	12	Yes	0.5
Gentamicin, tobramycin	2-3	0.16-0.24	8-12	Yes	0.12

Table 19.12. Antibiotic treatment of urinary tract infections, based on culture and antibiogram

Causative organism	Oral antibiotics	Oral antibiotics, second choice	Parenteral antibiotics	Parenteral antibiotics, second choice	Urinary antiseptic
Escherichia coli	Ampicillins, co-trimoxazole, quinolones[a]	Tetracycline, chloramphenicol, amoxicillin + clavulanic acid, sulfonamide	Cefazolin, mezlocillin	Cefotaxime, gentamicin	Nitrofurantoin, nalidixic acid[a], pipemidic acid[a]
Klebsiella spp.	Co-trimoxazole, quinolones[a]	Amoxicillin + clavulanic acid, oral cephalosporins, chloramphenicol	Gentamicin, cefotaxime	Mezlocillin, amikacin	Nitrofurantoin, nalidixic acid[a], pipemidic acid[a]
Enterobacter spp.	Co-trimoxazole, quinolones[a]	Chloramphenicol	Cefotaxime, gentamicin	Ceftazidime, amikacin	Nitrofurantoin, nalidixic acid[a], pipemidic acid[a]
Serratia marcescens	Quinolones[a], co-trimoxazole	Chloramphenicol	Ceftazidime, gentamicin	Amikacin, mezlocillin	Nitrofurantoin, nalidixic acid[a], pipemidic acid[a]
Proteus mirabilis	Ampicillin, co-trimoxazole, quinolones[a]	Amoxicillin + clavulanic acid	Cefazolin, gentamicin	Amikacin, mezlocillin	Nalidixic acid[a], pipemidic acid[a]
Proteus vulgaris, morganii, rettgeri	Quinolones[a]	Co-trimoxazole, amoxicillin + clavulanic acid	Cefotaxime, piperacillin, gentamicin	Amikacin, mezlocillin	Nalidixic acid[a], pipemidic acid[a]
Pseudomonas aeruginosa	Quinolones[a]	Carfecillin, carindacillin	Azlocillin, piperacillin, tobramycin, gentamicin	Ceftazidime, cefsulodin, amikacin	Pipemidic acid[a]
Enterococci	Ampicillin	Erythromycin, tetracyclines, quinolones[a]	Mezlocillin	–	Nitrofurantoin
Staphylococci	Cloxacillin	Erythromycin, oral cephalosporins, quinolones[a]	Benzylpenicillin, cefazolin	Vancomycin	Nitrofurantoin

[a] Adults only: norfloxacin, ofloxacin, pefloxacin, ciprofloxacin, nalidixic acid, pipemidic acid.

Table 19.13. Agents suitable for low-dose prophylaxis of recurrent urinary tract infections

Agents	Single daily dose	
	Adults	Children
Nitrofurantoin[a]	50-100 mg	2-3 mg/kg
Co-trimoxazole	240-480 mg	5-10 mg/kg
Cefalexin	250 mg	5 mg/kg
Short-acting sulfonamide (e.g., sulfafurazole)	0.5-1 g	

[a] Long-term use of nitrofurantoin is not advisable because of severe and dangerous side effects (pulmonary fibrosis, neuropathy).

Urethritis

The general management of urethral discharge is discussed on p. 211 of the chapter "Sexually Transmitted Diseases" (see Fig. 17.1 of that chapter).

Gonococcal Urethritis. See Tables 17.1 and 17.2 of chapter "Sexually Transmitted Diseases".

Nongonococcal Urethritis Due to Chlamydia Trachomatis. See on p. 219 of the chapter "Sexually Transmitted Diseases".

Candida Urethritis. Irrigation with nystatin is recommended. Cutaneous candidiasis and *Candida* balanitis can be treated with an antifungal cream (clotrimazole, miconazole). Systemic antifungal therapy, namely oral ketoconazole or fluconazole (investigational) may be successful in resistant cases or when endogenous reinfections occur.

Trichomonal Urethritis. Single-dose treatment with 2 g metronidazole, ornidazole or tinidazole, as in trichomonal vaginitis, is recommended. Contacts should also be treated.

Urethritis Due to Other Bacterial Causes. Urethritis caused by *E. coli, Proteus,* enterococci, and staphylococci requires effective antibiotic treatment for 2-3 weeks to avoid complications such as periurethral abscesses, urethral stricture, ascending infection, and epididymitis. Predisposing factors are meatal stenosis, foreign bodies, diverticula, tumor, and periurethral abscess.

Prostatitis, Epididymitis

Both prostatitis and epididymitis may complicate untreated gonococcal urethritis. *Chlamydia trachomatis* has also been recognized as a cause of both diseases in sexually active men in industrialized countries. Other bacterial causes comprise enterobacteria, mainly *E. coli, Pseudomonas aeruginosa*, enterococci, anaerobic bacteria, *Ureaplasma urealyticum*, and *Mycobacterium tuberculosis*.

Comparative quantitative culture of early-stream urine, midstream urine, expressed prostatic secretions, and the first portion of urine after prostatic massage is the procedure usually recommended for etiological diagnosis of chronic prostatitis. Culture of prostatic secretions is considered sufficient in acute prostatitis. If there is concomitant urethritis, examination of urethral smear for gram-negative diplococci is useful.

Acute Prostatitis. Treatment should consist of oral co-trimoxazole, 1.92 g in two divided doses for 4 weeks, if there is initial clinical response. This regimen is also efficacious in gonococcal prostatitis. Parenteral treatment with gentamicin, 3 mg/kg in two divided doses, plus tetracycline for 1 week, followed by a suitable oral antibiotic, usually co-trimoxazole, for another 3 weeks, is an alternative. Quinolones are also active in prostatitis. Penicillins and cephalosporins are often disappointing.

Chronic Bacterial Prostatitis. Co-trimoxazole, 1.92 g in two divided doses, for 3 months is recommended. An extended course of antibiotic therapy is suggested in order to avoid relapses.

Acute Epididymitis

In Association with Urethritis. Tetracycline, 500 mg every 6 h, or doxycycline, 100 mg twice daily for 3 weeks, covers sensitive strains of *Neisseria gonorrhoeae* and *Chlamydia trachomatis*. Where penicillinase-producing strains of *N. gonorrhoeae* are prevalent, resistance to tetracycline is also commonly found. Oral co-trimoxazole, 1.92 g in two divided doses for 3 weeks, or, more expensive, a 10-day course of a second- or third-generation cephalosporin, is then suggested.

In Association with UTI and/or Prostatitis. The recommended treatment is co-trimoxazole, 1.92 g in two divided doses for 4 weeks.

Chronic Epididymitis. Tuberculosis of the genitourinary tract must be excluded.

References

1. Abdurrahman MB, Chakrabarty DP, Ochoga SA (1978) Bacteriuria and other urinary abnormalities among school children in Kaduna. Niger Paediatr 5: 21-24
2. Abu-Bakare A, Oyaide S (1982) Asymptomatic bacteriuria in Nigerian schoolgirls. J Trop Med Hyg 82: 95-97
3. Alausa OK, Montefiore D (1978) Bacterial infections, sensitivity patterns, and chemotherapy among hospital patients in the tropics. Scand J Infect Dis 10: 295-302
4. Albert J-P, Le Mao, Menard M, Etienne J, Ricosse J-H (1974) Réflexions pratiques sur les urocultures effectuées au Centre Muraz (Bobo-Dioulasso) de 1966 à 1972. Méd Trop 34: 59-72
5. Badami PV, Deodhar LP (1976) Asymptomatic bacteriuria in school children. J Postgrad Med (Bombay) 22: 130-134

6. Bailey RR (ed) (1983) Single dose therapy of urinary tract infection. ADIS Health Science, Sydney
7. Barr RD (1972) A two-year prospective analysis of emergency admissions to an adult medical unit at the Kenyatta national hospital, nairobi. East Afr Med J 49: 772-782
8. Berkowitz FE (1983) Infections in children with severe protein-energy malnutrition. Ann Trop Pediatr 3: 79-83
9. Bloch M, Soundy J, deTaura YM (1977) Las bacterias responsables de las infecciones urinarias en el Hospital Rosales. Rev Inst Invest Med El Salvador 6: 289-297
10. Brooke, OG, Kerr DS (1973) The importance of routine urine culture in malnourished children. J Trop Pediatr 19: 348-349
11. Brown KGE (1975) Analysis of admissions to the adult medical wards at Queen Elizabeth Central Hospital, Blantyre, Malawi. East Afr Med J 52: 509-528
12. Brown KH, Gilman RH, Gaffar A, Alamgir ShM, Strife JL, Kapikian AZ, Sack RR (1981) Infections associated with severe protein-calorie malnutrition in hospitalized infants and children. Nutr Res 1: 33-46
13. Diejomaoh FME (1981) Acute urinary tract infection in pregnancy in a Nigerian community. East Afr Med J 58: 199-207
14. Gedebou M (1983) Urinary tract infections: aetiologic patterns and resistance. Ethiop Med J 21: 3-13
15. Gosh AK, Dey NC, Guha AC (1975) Significance of bacteriuria in pregnancy. J Indian Med Assoc 64: 63-68
16. Hendrickse RG (1980) Epidemiology and prevention of kidney disease in Africa. Trans R Soc Trop Med Hyg 74: 8-16
17. Laughlin LW, Farid Z, Mansour N, Edman DC, Higashi GI (1978) Bacteriuria in urinary schistosomiasis in Egypt. Am J Trop Med Hyg 27: 916-918
18. Lester FT, Tsega E (1976) The pattern of adult medical admission in Addis Ababa, Ethiopia. East Afr Med J 53: 620-634
19. Mati JKG (1974) Pregnancy bacteriuria in Nairobi, Kenia. East Afr Med J 51: 837-843
20. Morehead CD, Morehead M, Allen DM, Olson RE (1974) Bacterial infections in malnourished children. J Trop Pediatr Environ Child Health 20: 141-147
21. Morton RE, Lawande R (1982) The diagnosis of urinary tract infection: comparison of urine culture from suprapubic aspiration and midstream collection in a children's out-patient department in Nigeria. Ann Trop Pediatr 2: 109-112
22. Morton RE, Lawande R (1982) II. Frequency and clinical features of urinary tract infection in paediatric out-patients in Nigeria. Ann Trop Pediatr 2: 113-117
23. Ojo OA, Akinkugbe OO (1976) The significance of asymptomatic bacteriuria in pregnancy in Ibadan. W Afr J Surg 1: 23-28
24. Patel KM, Lwanga SK (1971) A study of medical admissions to Mulago Hospital, Kampala. East Afr Med J 48: 76-84
25. Phillips I, Wharton B (1968) Acute bacterial infection in kwashiorkor and marasmus. Br Med J 1: 407-409
26. Pugh RNH, Gilles HM (1979) Malumfashi endemic diseases research project, X; *Schistosoma haematobium* and bacteriuria in the Malumfashi area. Ann Trop Med Parasitol 73: 349-354
27. Rayan AR, Thara Bai K, Unnisa A, Lalitha K (1978) Drug sensitivity of urinary pathogenic bacteria. J Indian Med Assoc 70: 245-247
28. Singh SN, Singh S, Singh KN (1979) Nalidixic acid against urinary tract pathogens. J Indian Med Assoc 73: 151-154
29. Tshiani K, Nyomba B (1978) Infections urinaires dans la pratique médicale. Méd Afr Noire 25: 77-81
30. Washington JA II (1985) Bacteria, fungi, and parasites. In: Mandell GL, Douglas RG, Bennett JE (eds) Principles and practice of infectious diseases, 2nd ed
31. Whig J, Khuran SS, Khurana S, Sharma BK (1980) Significant bacteriuria in school children in different socioeconomic strata. Indian Pediatr 17: 507-510
32. Wilkins HA (1977) *Schistosoma haematobium* in a Gambian community. III. The prevalence of bacteriuria and of hypertension. Ann Trop Med Parasitol 71: 179-186

20 Infections of the Ear, Nose, and Throat

External Otitis

External otitis is a condition frequently seen at ENT clinics in tropical and subtropical areas. It can be classified into an infective group due to bacteria, fungi, or viruses (herpes) and a group of various dermatological disorders, e.g., neurodermatitis, psoriasis, or seborrheic dermatitis, which in turn may become superinfected.

Staphylococcus epidermidis and diphtheroids are the normal commensals of the external auditory canal. *Staphylococcus aureus,* gram-negative bacteria, *Proteus* spp., *Pseudomonas aeruginosa, Klebsiella* spp., and fungi - *Aspergillus* spp. and *Candida* spp. - predominate in external otitis. Treatment with antibiotic ointment or drops, neomycin, polymyxin B, bacitracin, or pimaricin in case of fungal infection, is usually recommended after cleaning of the external auditory canal (provided the tympanic membrane has remained intact). Daily application of 3% acetic acid eardrops appears to be an effective alternative [10].

Malignant External Otitis

Malignant external otitis is a typical but rare condition occurring in longstanding diabetes mellitus. The main pathogen is *Pseudomonas aeruginosa.* Left untreated, malignant external otitis is usually fatal due to the development of chronic osteomyelitis and involvement of cranial nerves, medulla or the brain. Long-term parenteral treatment with azlocillin and gentamicin in high dosage is necessary. Quinolones (ciprofloxacin 1.5 g/day) orally for several months are also effective.

Otitis Media

Otitis media in its various forms is common in many third world countries, although data on its incidence are scanty.

The complication of severe measles by suppurative otitis media developing within 1-2 weeks after the onset of the rash is well described [14]. Otitis media is particularly common in children, but it also affects a considerable number of adults. The suppurative forms, acute or chronic, are by far more common than the serous ones [12]. Twenty-eight percent of children under 5 years of age with a temperature of 37.8 °C or over who attended a clinic in Ilesha, West Nigeria, were found to have otitis media, i.e., one or two red tympanic membranes with loss of light reflex [6]. A survey of antibiotic use in primary care clinics carried out in Harare, Zimbabwe, showed that 8.7% of total antibiotic use was for otitis media [18].

However, late attendance when spontaneous perforation has occurred is common [16]. At ENT clinics, chronic otitis media with otorrhea is a condition frequently seen. From a total of 80–100 new patients with otitis media seen daily at the otorhinolaryngology department of a major medical center in Kinshasa, Zaire, 98% had otorrhea [12].

Acute Otitis Media

Antibiotic treatment of acute otitis media is directed at sterilization of the infected middle ear effusion. Pneumococci and *Hemophilus influenzae* are the main pathogens. Group A streptococci are less frequently isolated.

International Standard Therapy. Systemic treatment with oral antibiotics for 10 days is advised. Amoxicillin is usually considered the drug of choice. Amoxicillin is preferred to ampicillin, since higher concentrations in middle ear fluids are achieved.

Other drugs with proven efficacy include erythromycin, co-trimoxazole, tetracycline (in adults), cefaclor (the only oral cephalosporin with adequate activity against *H. influenzae*), and phenoxymethylpenicillin (in older children, when infection with *H. influenzae* is less likely). The fixed combination of erythromycin and sulfisoxazole is frequently used in the United States, because the concentration in middle ear fluid achieved by erythromycin is occassionally not adequate for infections with *H. influenzae*. This gap is covered by the sulfonamide, which is less active against pneumococci.

Limited Resources. In conditions of limited resources, parenteral benzylpenicillin may serve as an alternative. Thus, in Papua New Guinea, pediatric otitis media is treated by daily i.m. injections of procaine penicillin for 5 days. Data on the efficacy of this regimen have, however, not been published. Oral short-acting sulfonamides may be used if the agents mentioned above are either unavailable or too expensive. Chloramphenicol should not be used for treatment of acute otitis media unless complications like brain abscess or sinus thrombosis have supervened.

Chronic Suppurative Otitis Media

Chronic suppurative otitis media with discharge does not usually respond to systemic antibiotics.

Pseudomonas aeruginosa, Proteus mirabilis, Proteus vulgaris, Klebsiella spp. and *Staphylococcus aureus* are frequently isolated from the discharge [4, 12]. Mixed infection with anaerobic bacteria is also encountered. Treatment consists of regular and thorough cleaning of the ear and local administration of mild antiseptics in the form of powder by insufflation, or, more practically, as drops by instillation.

Application of topical aluminium acetate (Burow's solution) on wads of cotton placed against the tympanic membrane and changed daily proved to be an effective method of mass control of the disease without antibiotics [12].

Prolonged topical administration of aminoglycosides such as neomycin, kanamycin, or gentamicin carries the risk of cochlear damage.

When there is *bone involvement* (associated with marginal perforation of the tympanic membrane, in contrast to the central perforation in mucosal infections), surgical intervention is required to prevent intracranial complications like brain abscess, subdural empyema, or otogenic meningitis.

The differential diagnosis of a chronic discharging ear should also include tuberculosis. A tuberculous etiology should be considered particularly in children and young adults presenting with chronic discharging ear accompanied by cervical lymphadenitis. Lymph node biopsy or biopsy of granulation tissue obtained at surgery is useful in establishing the diagnosis. Tuberculous otitis media usually responds well to routine chemotherapy of tuberculosis.

Mastoiditis

Antibiotic treatment can occasionally abort *mastoiditis* in its early stage. Surgical intervention by antrotomy or mastoidectomy is, however, required in many instances in addition to chemotherapy. Parenteral benzylpenicillin in high dosage is appropriate in pneumococcal infections. Chloramphenicol, which covers both *H. influenzae* and pneumococci, may be used if parenteral ampicillin or a modern cephalosporin active against *H. influenzae,* such as cefuroxime or cefotaxime, is not available.

Gram's stain of aspirated effusion allows directed treatment. Tympanocentesis, aspiration of pus, is further indicated in patients with acute otitis media who are critically ill and do not respond to empirical treatment, and when intracranial complications have supervened.

Recommendations for the simple treatment of ear infections by primary health care workers have been made [9]. "Treatment kits" contain spirit drops (0.1% acriflavine in 50% alcohol), fungicidal drops (1% clotrimazole), ephedrine drops, a rubber bulb syringe, water purification tablets, and instruction sheets [9]. A flowchart to assist primary health workers in the treatment of ear disease has also been designed (Fig. 20.1).

Sinusitis

Episodes of *acute sinusitis* are best treated with amoxicillin, ampicillin, co-trimoxazole, erythromycin, or cefaclor. As in acute otitis media, pneumococci and *H. influenzae* are the main pathogens. Nasal decongestants are often prescribed as an adjunct to chemotherapy. *Chronic sinusitis* often responds to appropriate antimicrobial chemotherapy and repeated sinus lavage [16]. In cases of chronic sinusitis not responding to empirical chemotherapy sinus puncture and aspiration in order to determine the causative agent may be considered. Gram-negative enteric bacilli or anaerobic bacteria, but also *Aspergillus* spp., are frequently isolated in chronic cases. When medical therapy fails, surgical intervention may be necessary.

	Patient's complaint	Pain	Discharge	Itching	Deafness	Questions ?	Test	Probable disease	First treatment without seeing eardrum	Treatment after otoscopy: Appearance of eardrum	Treatment after otoscopy: Action to be taken	Further treatment: Medical officer	Further treatment: E.N.T. clinic
1	Pain	+	– Drop of blood	–	(+)	After injury?	*	Traumatic perforation	Keep ear dry and clean	Some blood	Keep ear dry and clean until healed	Failure to heal in 6 months	Very few
2	Pain; Discharge	+	–/+ Blood then pus	–	(+)	After a cold?	Pressure above & behind ear: May hurt	Acute otitis media	Antibiotic (systemic) and ephedrine nose drops	Drum swollen red then pus	Continue treatment if necessary	Drain surgically	Some for surgery
							Pressure above & behind ear: Does hurt & swelling	Acute Mastoiditis					
3	Pain; Discharge	+	+/–	+	–	With itching now or earlier?	Pain on moving ear	Acute otitis externa	Clean ear - syringe if wet antibiotic (systemic) antifungal (local)	Meatal walls swollen	Continue treatment especially cleaning	*	Very few
4	Pain; Discharge	+	+	–	(+)	Pain (dizziness) foul discharge	Pressure does hurt	Acute on chronic otitis media (attic)	Antibiotic (systemic)	Pus-foul smelling	Antibiotic (S)	All soon	Probable early surgery
5	Discharge	–	+	–	(+)	Foul discharge	*	Chronic otitis media (attic)	Syringe gently spirit drops		Continue treatment	All in due course	Possible later surgery
6	Discharge	–	+	–	(+)	Variable discharge possibly foul	*	Chronic otitis media (intermediate)	Syringe gently spirit drops		Continue treatment until dry	Those that discharge after 6 weeks	Some for assessment
7	Discharge	–	+	–	(+)	Mucoid discharge	*	Chronic otitis media (mucoid)	Syringe gently spirit drops children: ephedrine nose drops	or	Continue treatment & repeat treatment until dry	Those that discharge after 12 weeks	Very few
8	Discharge; Itching	–	+	+	–	Itching and discharge	Possible pain on moving ear	Otitis externa (wet)	Syringe gently spirit drops ± fungicide	Wall swollen white/ yellow debris	Continue treatment cleaning important	Those that have symptoms after 4 months	Very few
9	Itching	–	–	+	–	Itching		Otitis externa (dry)	Keep dry, clean antibiotic (locally) ± fungicide	Perhaps black fungus threads	Continue treatment cleaning important		
10	Deaf	–	–	–	+	Sudden hearing loss: one or both ears	*	Wax	Oil for 4 days syringe firmly	Black or brown mass	Repeat if necessary	Failure to clear	Very few

Fig. 20.1. Treatment of ear disease: a chart to assist primary health workers and health workers without otological training. From [8]

Stomatitis

Stomatitis is a common condition in many developing countries, particularly in children, and may be due to viral, bacterial, or fungal infection.

Viral Stomatitis

Antibiotics are not indicated in viral stomatitis (herpetic gingivostomatitis, herpangina, and hand, foot, and mouth disease). Anesthetic lozenges can be useful if the "sore mouth" interferes with adequate fluid and food intake, especially in herpetic gingivostomatitis in young children. Use of local disinfectants (1% gentian violet) several times a day may be helpful in preventing bacterial superinfection.

Candida Stomatitis

Candida stomatitis (thrush) is frequently seen in young babies or after a prolonged course of antibiotics. Prematurity, malnutrition, and treatment with steroids are other predisposing factors. The "sore mouth" in severe measles after the initial 3 days may also be due to Candida infection [15]. Persistent thrush in adults with no known predisposing factors may be one of the earliest symptoms of AIDS, especially when associated with dysphagia, indicative of *Candida* esophagitis.

Treatment

Oral antifungal agents such as nystatin or amphotericin B suspension, 1-2 ml three to four times a day, comprise the recommended therapy. Antibiotics may be required to treat secondary bacterial infection. In adult patients with AIDS, ketoconazole has been found useful in the treatment of recurrent *Candida* stomatitis and esophagitis.

Necrotizing Ulcerative Gingivitis

Necrotizing ulcerating gingivitis is a common disease of children in the tropics and subtropics. It is characterized by ulceration, necrosis of one or more interdental papillae, and easy bleeding on being touched. Grayish necrotic sloughs on the gums and the buccal mucosa may be seen in severe cases. It is the general opinion that necrotizing ulcerative gingivitis results from infection with *Borrelia vincenti* and *Fusobacterium fusiforme,* anaerobic bacteria that are normally present in the oral cavity. Malnutrition, hypovitaminosis, low socioeconomic status, and poor oral hygiene with food debris lodging in the interdental spaces have all been incriminated in the etiology of this condition.

Treatment

Treatment of severe cases with phenoxymethylpenicillin (125 mg or 250 mg four times taily) plus metronidazole (100 mg or 200 mg twice daily) for 5 days (dosage depending on body weight) was effective in 25 of 28 children with chronic periodontal disease [11]. Two children with early cancrum oris, mucosal involvement of the cheek adjacent to the affected gums with exposure of the underlying bone, also responded to this regimen, though the bone remained exposed for several weeks after clearance of the infection. The same authors state that two doses of long-acting injectable penicillin were effective when compliance with oral medication was poor [11].

Noma (Cancrum Oris)

It is widely accepted that necrotizing ulcerative gingivitis precedes the development of cancrum oris (noma), the infective process spreading rapidly to involve the adjacent mucosa of the cheek and the underlying bone. Noma is still common in many countries of the tropics and subtropics; this is probably due to the prevalence of malnutrition and decreased resistance to infection. Occurrence of cancrum oris after attacks of malaria, smallpox, and, particularly, measles has been described [7, 15, 19].

Treatment

Combined treatment with metronidazole and penicillin can be expected to be effective in small ulcers early in the course of the disease. In larger ulcers (over 2 cm in diameter), high-dose metronidazole plus i.m. penicillin did not prevent the formation of contractures with gross disfigurement of the face in more than half the cases [3].

Osteomyelitis of the Jaw

Osteomyelitis of the upper and lower jaw is usually due to *Staphylococcus aureus,* streptococci, or anaerobic bacteria, often in mixed infection. In a series from Nigeria, necrotizing ulcerative gingivitis or cancrum oris preceded the majority of such cases of osteomyelitis seen in children [1]. Extension of an infection from the middle ear was also not uncommon in pediatric patients. Periodontal disease, trauma, and odontogenic infection (i.e., pulpal disease) were the probable predisposing factors, in that order of frequency, in a Nigerian series of 124 adult cases [2].

Treatment

Benzylpenicillin is the first-line drug. Clindamycin covers *Staphylococcus aureus*, streptococci, and anaerobic bacteria. Surgical procedures, drainage, sequestrectomy, or even partial excision of the jaw are required in many instances [2].

Cervical Lymphadenitis

Acute suppurative lymphadenitis is treated with phenoxymethylpenicillin (streptococcal infection) or oral erythromycin or cloxacillin (staphylococcal or streptococcal infection). When abscess formation has occurred, surgical drainage is indicated. An initial focus of infection (larynx, nasopharynx, and tonsils if the superior cervical nodes are affected, lower jaw and teeth if the submental and auricular nodes) should be eradicated whenever possible.

In developing countries, persistent or nonresponsive cervical lymphadenopathy is frequently due to tuberculosis, particularly in children [5, 13, 17]. Hodgkin's disease, other lymphomas, metastases, and reactive hyperplasia are other common causes of persistent cervical lymphadenopathy. Toxoplasmosis accounted for 6% of 152 cases of cervical lymphadenopathy in a recent study from Nigeria [17]. Cervical lymphadenopathy is common in patients with HIV infection (lymphadenopathy syndrome).

In the absence of other diagnostic clues, excisional lymph node biopsy is required to establish the cause of persistent lymphadenopathy.

References

1. Adekeye EO, James JD, Shamia RI (1973) Osteomyelitis of maxilla and mandible in children. In: Adekeye EO, Shamia RJ (1976) Pyogenic osteomyelitis of the mandible in adults: a review of 124 cases and a report of 4 cases. Niger Med J 6: 477-485
2. Adekeye EO, Shamia RI (1976) Pyogenic osteomyelitis of the mandible in adults: a review of 124 cases and a report of 24 cases. Niger Med J 6: 477-485
3. Ana JR (1975) The effect of metronidazole on the healing of cancrum oris lesions in Nigerian children. Niger Med J 1975 5: 52-56
4. Coker AO, Ijaduola GTA, Odugbemi TO (1983) Bacterial isolates from chronic discharging ears in nigerian children. East Afr Med J 60: 462-466
5. Ejeckam GC, Nwabueze ED (1984) Pathology of superficial lymph node in children and young adults in Enugu, Nigeria. J Trop Pediatr 30: 307-309
6. Elton P, Cornell J (1979) Study of otitis media and malaria among pyrexic attenders of an under-fives clinic. J Trop Med Hyg 82: 110-112
7. Emslie RD (1963) Cancrum oris. Dent Practnr Dent Rec 13: 481-495
8. Holborow C (1986) Treatment of otitis media and ear infections. Trop Doct 16: 32-33
9. Holborow C, McPherson B (1986) Simple treatment for the infected ear. Trop Doct 16: 31-32
10. Kuijlen K, Manni JJ, Kanyawana JZ, Mhalu FS (1983) External otitis in Dar Es Salaam, Tanzania. East Afr Med J 60: 150-155
11. Lamb HW, Prentice AM (1983) A prospective survey of gingivitis in Keneba, a rural West African community. Ann Trop Pediatr 3: 137-142
12. Mahoney JL (1980) Mass management of otitis media in Zaire. Laryngoscope 90: 1200-1208
13. Mbise RL (1984) Peripheral lymphadenopathy in children in Dar es Salaam, Tanzania. A study from biopsy material. Ann Trop Pediatr 4: 83-85

14. Morley D (1969) Severe measles in the tropics, I. Br Med J 1: 297-300
15. Morley D (1973) Paediatric priorities in the developing world. Butterworth, London, p 215
16. Ntabomvura JP, Vande Weghe JP (1982) Morbidité en ORL. In: Meheus A, Butera S, Eylenbosch W, Gatera G, Kiritis M, Musafili I (eds) Santé et maladies au Rwanda. Universitaire Instelling, Antwerpen, pp 601-612
17. Oluwole SF, Odesanmi WO, Kalidasa AM (1985) Peripheral lymphadenopathy in Nigeria. Acta Trop 42: 87-96
18. Stein CM, Todd WTA, Parirenyatwa D, Chakonda J, Dizwani AGM (1984) A survey of antibiotic use in Harare primary care clinics. J Antimicrob Chemother 14: 149-156
19. Tempest MN (1966) Cancrum oris. Br J Surg 53: 949-969

21 Eye Infections

Ulcerating Blepharitis

Ulcerating blepharitis is caused by *Staphylococcus aureus* or streptococci. Topical treatment with antibiotic eye ointment (e.g., tetracycline or gentamicin) requires prior removal of crusts by application of wet compresses dipped in saline or in clean water. Cloxacillin or benzylpenicillin are suggested for systemic treatment of severe cases.

Hordeolum

External hordeolum (stye; abscess of an external gland of the eyelid) and the rarer internal hordeolum (abscess of a Meibomian gland) are usually due to *Staphylococcus aureus*. Local treatment consists in application of dry heat or warm compresses to fasten abscess formation. If spontaneous perforation fails to occur, incision is needed. Local antibiotic treatment is not usually efficacious. Systemic treatment with cloxacillin may abort early cases.

Abscess and Cellulitis of Lid

Abscess and cellulitis of the lid are, in most instances, staphylococcal or streptococcal in origin. They are either consequent upon trauma or arise from continuous spread from sinusitis or osteomyelitis. Systemic antibiotic treatment with cloxacillin, erythromycin or, in more severe cases, a first- or second generation cephalosporin, is recommended because of the risk of complications like orbital cellulitis and septic venous thrombosis. Incision of a lid abscess may also be required.

Orbital Cellulitis

Orbital cellulitis is usually secondary to ulcerating blepharitis, dacryocystitis, sinusitis, or osteomyelitis of the upper jaw. In septicemia, hematogenous spread to the orbit is also possible. The range of pathogens comprises *Streptococcus pneumoniae, Hemophilus influenzae, Staphylococcus aureus,* streptococci, and anaerobic bacteria. The rare cases of fungal infection are accounted for by sinuorbital aspergillosis and mucormycosis.

When possible, blood cultures should be taken and pus obtained by drainage or aspiration of the primary focus for examination by Gram's stain and culture.

Ideally, antibiotic therapy should be directed against the causative agent. Surgical intervention may be needed for drainage of a primary focus, for relief of an orbital abscess, and for treatment of intracranial complications such as epidural abscess, subdural empyema, brain abscess, meningitis, or septic sinus cavernosus thrombosis.

Initial, empirical therapy should be guided by the likely pathogens according to the primary focus, i.e., cloxacillin (staphylococci), ampicillin *(H. influenzae),* chloramphenicol (pneumococci, *H. influenzae,* streptococci, anaerobic bacteria). Combination therapy with cloxacillin plus chloramphenicol is suggested for severely ill patients.

Fungal infections require treatment with amphotericin B.

Acute Dacryocystitis

Acute dacryocystitis is often associated with an obstruction of the nasolacrimal duct and arises when secondary infection with *Staphylococcus aureus, Streptococcus pneumoniae,* or other streptococci supervenes. Local heat and treatment with systemic antibiotics (cloxacillin, erythromycin, first- and second-generation cephalosporins) suffices in early cases. When a lacrimal abscess has formed, incision is frequently required (risk of spontaneous perforation through the skin with fistula formation).

Acute Bacterial Conjunctivitis

Acute bacterial conjunctivitis, a very common condition in the tropics, particularly among children [1], frequently occurs in epidemics. Poverty, lack of clean water, a dry, dusty climate, and an abundance of flies are contributory factors. The practice of native healers to apply various nonsterile or irritant substances or mixtures to the eyes of their patients also favors the development of acute bacterial conjunctivitis. Although spontaneous healing often occurs, permanent sequelae such as impaired vision due to corneal defects (leukomas) may complicate more severe infections that have been left untreated. *Streptococcus pneumoniae, Haemophilus* spp., *Staphylococcus aureus, Staphylococcus epidermidis, Neisseria gonorrhoeae,* beta-hemolytic streptococci and chlamydiae are the most common pathogens of bacterial conjunctivitis [2, 3, 10]. In young adults, *N. gonorrhoeae* may be the commonest cause of purulent conjunctivitis [19]. *E. coli,* other gram-negative enteric bacilli, and *Pseudomonas aeruginosa* are comparatively rare causes of bacterial conjunctivitis. Ideally, the purulent discharge should be examined by Gram's stain and, at least in cases not responding to treatment, by culture and sensitivity testing.

Many cases respond to topical antibiotic treatment alone. Antibiotic eye ointment is preferable to antibiotic drops because of its more prolonged action. Tetracycline eye ointment (1%) administered four times daily is often used under limited conditions. Chloramphenicol eye ointment (1%), which is cheap and stable in tropical climates, is active against the main pathogens of bacterial conjunctivitis, including gonococci and chlamydiae. There is no relevant risk of hematotoxicity in case of topical administration. Chloramphenicol and its water soluble analogue azidamphenicol administered two to four times daily are therefore suggested as topical treatment of choice for external eye infections. Ophthalmological preparations of sulfonamides, e.g., sulfacetamide eye drops (10%), are also cheap, but their lower antibacterial activity and more limited antibacterial spectrum renders them less suitable alternatives. Gentamicin eye ointment (0.3%) and eye ointments containing mixtures of neomycin, bacitracin, or polymyxin B are more expensive alternatives, but may be needed in case of resistant staphylococci, gram-negative enteric bacilli, or *Pseudomonas aeruginosa*. Penicillin eye drops should not be used because of the high risk of sensitization and poor stability. Further local measures include the application of hot, wet compresses.

If there is no definite improvement within 48 h, additional systemic antibiotic therapy should be instituted guided by Gram's stain and culture.

If *gonococcal infection* is suspected, systemic treatment with benzylpenicillin (5-20 MU daily in adults) should be started immediately together with local treatment (risk of blindness). When infection due to penicillinase-producing *N. gonorrhoeae* (PPNG) is likely, treatment with cefuroxime or cefotaxime (2-6 g daily in adults) is usually recommended. Under conditions of limited resources, where these or other second- or third-generation cephalosporins are not available, kanamycin or gentamicin should be used.

Ophthalmia Neonatorum

Neisseria gonorrhoeae appears to be the most common pathogen of conjunctivitis in newborns in Africa [8, 16]. There is a rising incidence of cases due to PPNG [8, 12]. *Staphylococcus aureus,* which is more common than *N. gonorrhoeae* as cause of ophthalmia neonatorum in developed countries, was isolated with variable frequency from African infants, i.e., in 37% and 3% respectively in two studies from Nigeria and Kenya [8, 15]. *Chlamydia trachomatis* was isolated from 13%, 19%, and 35% of infants with ophthalmia neonatorum in Nairobi, Kenya; Bangui, Central African Republic; and Gambia, respectively [8, 13, 14]. Mixed infection with *N. gonorrhoeae* and *C. trachomatis* was found in 4% of the Nigerian infants [8]. Other pathogens - pneumococci, beta-hemolytic streptococci, gram-negative enteric bacilli, and *Pseudomonas aeruginosa* - are less commonly encountered.

Prophylaxis

Prophylaxis of gonococcal ophthalmia neonatorum with 1% silver nitrate drops (Credé) is recommended [9]. Exsiccation of the drops, however, must be prevented. Prophylaxis with tetracycline or erythromycin eye ointment which, in contrast to silver nitrate, are also effective as prophylaxis of chlamydial ophthalmia neonatorum carries the risk of failure in infections caused by PPNG. "Chemical" conjunctivitis associated with silver nitrate prophylaxis occurs in many infants. It starts within hours and subsides within 48 h after use. Ocular inflammation commencing more than 24 h after application of 1% silver nitrate is therefore usually of a different etiology. Otherwise, incubation periods are unreliable in the differential diagnosis of ophthalmia neonatorum.

Treatment

Therapy of ophthalmia neonatorum should be preceded by a Gram's stain of the discharge which differentiates between the main bacterial agents with the exception of *C. trachomatis*. *Topical treatment* of established ophthalmia with gentamicin eye ointment (0.3%) is often suggested in developed countries, but is usually too expensive under conditions of limited resources, where tetracycline or chloramphenicol eye ointment (1%) or erythromycin eye ointment (0.5%) is used [7]. Although staphylococcal ophthalmia commonly responds to topical treatment alone, the frequent resistance of *Staphylococcus aureus* to tetracycline, and, to a lesser extent, to erythromycin should be borne in mind.

Systemic treatment should be instituted without delay in all cases of *gonococcal ophthalmia neonatorum* (risk of blindness) and in severe cases of other etiology. For details of treatment of gonococcal ophthalmia, see the chapter on sexually transmitted diseases.

Treatment of chlamydial ophthalmia neonatorum: Chlamydial eye infection (neonatal inclusion conjunctivitis) is characterized by a mucopurulent discharge that frequently starts between day 5 and day 13 post partum but may occur earlier or later. Topical administration of either sulfacetamide drops (10%) or tetracycline eye ointment (1%) four times daily for 14 days, though effective for treating the conjunctivitis, does not suffice to eliminate the organism. The conjunctivitis may be part of a generalized infection. Chlamydial pneumonia develops at 3–13 weeks in some of these infants. For this reason, additional systemic treatment with erythromycin (40 mg/kg/day for 2 weeks) is recommended.

Inclusion conjunctivitis in adults should be treated with systemic tetracyclines, sulfonamides, or erythromycin for 3 weeks. In the presence of a mucopurulent discharge, topical antibiotics should be added. Infected sexual partners should also be treated.

Trachoma

Trachoma, the leading cause of blindness worldwide, is estimated to affect 500 million people. It is endemic in the Indian subcontinent, the Middle East, and North and Subsaharan Africa.

Trachoma is a chronic follicular keratoconjunctivitis due to infection with *Chlamydia trachomatis* serotypes A, B, Ba, and C. Single infections usually take an acute or subacute, but self-limiting course. The development of trachoma depends on repeated reinfections with *C. trachomatis*. The chronically inflamed conjunctiva is also more susceptible to bacterial superinfections which aggravate the course of the disease. Seasonal epidemics of bacterial conjunctivitis are well known from the endemic trachoma areas of North Africa. If the cycle of repeated episodes of reinfection and bacterial superinfection is not broken, conjunctival (tarsal) scarring develops, leading to trichiasis and entropion. Corneal ulceration, opacities, and blindness are the consequences of the severe form of the disease.

The primary infection is usually acquired by the age of 2 years in endemic areas; subsequent blindness occurs in middle age. Direct (eye to eye) and indirect (hands, clothes) transmission is facilitated by low-standard, crowded living conditions and poor personal and environmental hygiene, with lack of clean, easily accessible water. Flies attracted by ocular secretions also transmit the infection.

According to WHO [4], a *clinical diagnosis* of trachoma can be made if any two of the following four signs are present in an individual patient:

1. Follicles on the upper tarsal conjunctiva
2. Limbal follicles or their sequelae, Herbert's pits
3. Typical conjunctival scarring
4. Vascular pannus most marked at the superior limbus

Chemotherapy

Chemotherapy of trachoma relies on sulfonamides, tetracyclines, erythromycin, and related macrolides. Rifampicin, though active against *C. trachomatis,* rapidly induces emergence of resistant strains.

Topical chemotherapy can be carried out with ophthalmological preparations of tetracycline, chloramphenicol or erythromycin [5]. Sulfonamides are apparently less efficacious when applied topically.

"Intermittent" trachoma chemotherapy as used in mass treatment programs consists in administration of 1% tetracycline eye ointment twice daily for 5 consecutive days, or, alternatively, once daily for 10 days, every month for 6 months, to be repeated as necessary [4].

Systemic chemotherapy is not recommended for mass treatment campaigns because of a high percentage of side effects and expense [4]. Topical tetracycline continues to be used for mass treatment. On the other hand, systemic chemotherapy appears to be advantageous, since eradication of the organism is not usually achieved by topical chemotherapy.

Systemic treatment with oral sulfonamides requires full therapeutic dosage [e.g., a daily dosage of sulfisoxazole (sulfafurazole) of 6 g in adults and 150 mg/kg in children, in four divided doses]. A rate of side effects of 5% is to be expected with 3 weeks' duration of treatment. Tetracyclines (2 g daily in adults divided into four doses) are not recommended for treatment in pregnancy or for children under the age of 7 years. Oral erythromycin (2 g daily in adults and 40 mg/kg/day in children in four divided doses) provides an alternative. Three weeks' duration of treatment is required with any of these drugs for eradication of the organism. Clinical improvement will take 2 weeks or more after the institution of treatment to become apparent.

Systemic chemotherapy combined with topical antibiotics in case of coexisting bacterial conjunctivitis is effective when treating individual patients. In endemic areas, however, patients will rapidly become reinfected. Mass treatment campaigns aim at breaking transmission and reducing the infectious pool of the community in order to decrease the number of reinfections and to avoid blinding trachoma. Apart from antibiotics, prevention also requires education of the community to improve standards of personal and environmental hygiene [11].

The complications of chronic trachoma, trichiasis and entropion, should be corrected to prevent corneal ulceration and subsequent loss of vision.

Keratitis

Bacterial Keratitis

Bacterial keratitis is usually caused by pneumococci, and to a lesser degree by staphylococci, *Pseudomonas aeruginosa,* streptococci, or gram-negative enteric bacilli [3 a]. Mixed infections are not uncommon. With the exception of gonococci and *Corynebacterium diphtheriae,* bacteria do not invade intact corneal epithelium. Infection commonly follows corneal injuries, e.g., those due to foreign bodies or trauma. Microscopy, culture, and sensitivity testing are important for improvement of therapy and also to exclude keratomycosis, but treatment must be started without delay.

Treatment

Local treatment takes the form of ophthalmological preparations of gentamicin (staphylococci + gram-negative enteric bacilli + *Pseudomonas aeruginosa*), tetracycline (pneumococci + streptococci), chloramphenicol (pneumococci + streptococci + sensitive staphylococci + gram-negative enteric bacilli), or combinations of polymyxin B + neomycin (staphylococci + *Pseudomonas aeruginosa* + gram-negative enteric bacilli) or bacitracin + neomycin (staphylococci + streptococci + pneumococci + gram-negative enteric bacilli + *Pseudomonas aeruginosa*), given as available and as appropriate according to the pathogen.

Deep corneal ulcers with hypopyon often require subconjunctival injections (to be administered by an ophthalmologist). Benzylpenicillin 0.3-0.5 MU, gentamicin 20 mg, ampicillin 100 mg, or azlocillin 50 mg in 0.3-0.5 ml solution is administered once daily for 3 days.

Systemic treatment in high dosage with benzylpenicillin (pneumococci, streptococci, and sensitive staphylococci), cloxacillin (penicillin-resistant staphylococci), or chloramphenicol (sensitive staphylococci, gram-negative enteric bacilli, pneumococci, streptococci) is chosen according to the pathogen. *Pseudomonas aeruginosa* requires treatment with a modern acylaminopenicillin (e.g., azlocillin) in combination with an aminoglycoside. Further therapeutic measures include instillation of a mydriatic agent.

Children with corneal ulceration should receive vitamin A in addition. A first dose of 200000 IU orally or of 100000 IU aqueous vitamin A parenterally should be given immediately. This should be followed by a second dose on the next day and by a third dose a week later [7].

Herpetic Keratitis

A presumptive diagnosis of herpetic keratitis can be based on diminished corneal sensation and a typical dendritic ulcer as outlined by fluorescein. Dendritic ulcers can be treated with (1) chemical cauterization, (2) cryotherapy, (3) administration of topical antiviral agents such as idoxuridine (IDU), or (4) systemically with acyclovir. In addition, a mydriatic agent should be given until the ulcer has healed. Steroids are contraindicated.

Fungal Keratitis

Fungal keratitis due to *Candida* spp. or *Aspergillus* spp. may supervene on primary bacterial or viral keratitis, follow trauma, or complicate topical treatment with corticosteroids.

Topical chemotherapy with the ophthalmological preparation of an antifungal agent such as pimaricin should be combined with systemic amphotericin B in severe cases. Oral therapy with ketoconazole is possible in *Candida* infection.

Phlyctenular Keratitis

Phlyctenular keratitis is an allergic reaction occurring in tuberculosis particularly in malnourished children. Apart from treatment for tuberculosis, topical antibacterial therapy is only needed in case of bacterial superinfection.

Measles Keratitis

Superficial punctate keratitis due to measles takes a mild, self-limiting course in well-nourished children in developed countries. However, "1% to 4% of African children with acute measles will develop corneal ulceration," which is often bilateral with resulting blindness [7]. Measles affecting malnourished young children appears to precipitate xerophthalmia with corneal ulceration, but secondary herpes simplex infection may also play a role. Prophylactic administration of a single oral dose of 200000 IU vitamin A to all children with severe measles or protein-energy malnutrition has, therefore, been recommended [7].

Bacterial Endophthalmitis

Bacterial endophthalmitis carries a poor prognosis. It usually occurs as complication of a penetrating trauma to the eye or postoperatively. Staphylococci are the main pathogens, but *Pseudomonas aeruginosa,* streptococci, and anaerobic bacteria are also occasionally isolated. Hematogenous endophthalmitis is a rare complication in septicemia. Diagnostic aspiration of the vitreous body for culture and sensitivity testing is recommended to guide chemotherapy. Unfavorable results of chemotherapy are due to poor penetration of antibiotics in the vitreous body, with the exception of chloramphenicol. Intravitreal injections of gentamicin or tobramycin 0.2 mg, benzylpenicillin 600 units, or certain cephalosporins 0.25 mg may be given by an experienced ophthalmologist. Systemic antibiotic treatment in high dosage is always required in addition to administration of antibiotics by topical, subconjunctival, and intravitreal routes.

Combinations that have been widely used for systemic chemotherapy of endophthalmitis are benzylpenicillin or chloramphenicol + gentamicin or tobramycin. Experience with combinations including modern cephalosporins, such as cefotaxime plus an aminoglycoside, is still limited.

In many developing countries, endophthalmitis is still one of the major complications after cataract extraction. Although prophylaxis of postoperative endophthalmitis with antibiotics is practised worldwide, there are no controlled studies on the efficacy of different regimens. Incidence rates of postoperative endophthalmitis were 3.6% and 0.53%, respectively for prophylaxis with oral chloramphenicol or with parenteral penicillin plus topical chloramphenicol preoperatively followed by topical chloramphenicol at each change of dressings postoperatively [17, 18]. The incidence fell from 3.6% to 0.37% in the Indian study with the use of intracameral gentamicin at operation [18].

References

1. Abiose A (1985) Paediatric ophthalmic problems in Nigeria. J Trop Pediatr 31: 30-35
2. Amoni SS (1979) Acute purulent conjunctivitis in Nigerian children in Zaria. J Pediatr Ophthalmol Strabis 16: 308-312
3. Ayanru JO (1978) Affections of conjunctiva and cornea in Bendel state of Nigeria. Trop Geogr Med 30: 69-74
3a. Carmichael TR, Wolpert M, Koornhof HJ (1985) Corneal ulceration at an urban African hospital. Br J Ophthalmol 69: 920-926
4. Dawson CR, Jones BR, Tarizzo ML (1981) Guide to trachoma control. WHO, Geneva
5. Dawson CR, Daghfous T, Whitcher J, Messadi M, Hoshiwara T, Triki F, Chadgrah F, Briones O, Yoneda C, Schachter J (1981) Intermittent trachoma chemotherapy: a controlled trial of topical tetracycline or erythromycin. Bull WHO 59: 91-97
6. Dawson CR, Daghfous T, Hoshiwara I, Ramdhane K, Kamoun M, Yoneda C, Schachter J (1982) Trachoma therapy with topical tetracycline and oral erythromycin: a comparative trial. Bull WHO 60: 347-355
7. Foster A, Sommer A (1986) Childhood blindness from corneal ulceration in Africa: causes, prevention, and treatment. Bull WHO 64: 619-623
8. Fransen L, Nsanze H, Klauss V, Van der Stuyft P, D'Costa L, Brunham RC, Piot P (1986) Ophthalmia neonatorum in Nairobi, Kenya: The roles of *Neisseria gonorrhoeae* and *Chlamydia trachomatis*. J Infect Dis 153: 862-869
9. Galega FP, Heymann DL, Nasah BT (1984) Gonococcal ophthalmia neonatorum: the case for prophylaxis in tropical Africa. Bull WHO 62: 95-98
10. Hassan AM, Mahfouz RA (1973) Some bacteriological studies of acute conjunctivitis in Mansoura. Bull Ophthalmol Soc Egypt 66: 7-15
11. Jones BR (1975) The prevention of blindness. Trans Ophthalmol Soc UK 95: 12-33
12. Latif AS, Nathoo KJ (1983) Ophthalmia neonatorum due to penicillinase-producing *Neisseria gonorrhoeae* - two case reports. Centr Afr J Med 29: 103-104
13. Mabey DCW, Whittle HC (1982) Genital and neonatal chlamydial infection in a trachoma endemic area. Lancet II: 300-301
14. Meheus A, Delgadillo R, Widy-Wirski R, Piot P (1982) Chlamydial ophthalmia neonatorum in Central Africa (letter). Lancet II: 882
15. Obi JO, Wemambu SNC (1977) Ophthalmia neonatorum in a Nigerian community. Niger Med J 7: 418-421
16. Onyango-Ogony PJ, Hetland-Eriksen J (1975) Purulent ophthalmia neonatorum at Mulago hospital. East Afr Med J 52: 640-643
17. Osuntokun O, Olurin O (1973) Cataract and cataract extraction in Nigerians. An evaluation of 567 extractions. Br J Ophthalmol 57: 27-33
18. Peyman GA, Sathar ML, May DR (1977) Intraocular gentamicin as intraoperative prophylaxis in South India eye camps. Br J Ophthalmol 61: 260-262
19. Sangawe JLF, Mhalu FS, Mtanda AT (1984) The growing problem of purulent conjunctivitis in adults and its management. East Afr Med J 61: 802-805

22 Salmonella Infections

Salmonellae are aerobic, nonsporulating gram-negative bacilli. The recognized species compose more than 1700 serotypes according to antigenic analysis by the Kaufmann-White Schema. For clinical and therapeutic reasons enteric fever (typhoid and paratyphoid fever) is distinguished from nontyphoidal *Salmonella* infections.

Typhoid Fever

Typhoid fever is an acute septicemic illness caused by *Salmonella typhi*. Paratyphoid fever is a clinically similar, but milder illness due to *S. paratyphi* A, B, or C. Other *Salmonella* serotypes can rarely cause the clinical picture of paratyphoid fever. In the following, "typhoid fever" is used to mean both typhoid and paratyphoid fever.

Typhoid fever constitutes a considerable public health problem in many developing countries. Whereas the annual incidence of typhoid fever has fallen to about 0.25–3.7 cases per 100000 of the population in Western Europe, Japan, and the USA, its present annual incidence in the developing world (excluding China) has been estimated at 540 cases per 100000 population [7]. In endemic areas, typhoid fever is largely a disease of children and young adults [7, 14]. Man is the only reservoir of *S. typhi* and the main reservoir of *S. paratyphi* A, B, and C. Transmission occurs by the fecal-oral route via contaminated food, drink, and water or, less often, by direct contact from person to person. The main reason for its endemicity in the developing world is the lack of sanitation, i.e., the water used for drinking or irrigation is frequently contaminated by raw sewage [7]. Infected food handlers also contribute to transmission in the community. Poor domestic hygiene and inadequate follow-up of chronic carriers are less important.

The diagnosis of typhoid, when suspected clinically, is confirmed by culture of the pathogen. Culture of blood is most useful during the first two weeks of illness, culture of urine or stool (rectal swabs) from the 2nd week onward. Culture of bone marrow aspirate is also possible. The Widal reaction can serve as an useful adjunct to diagnosis. Culture of bile, apparently as sensitive as blood culture during the acute illness, has also been used to detect chronic carriage [7]. Sensitivity testing is recommended, since R-plasmid-mediated resistance of *S. typhi* to chloramphenicol has been reported from Mexico, India, and Southeast Asia [1, 19, 21].

These plasmids usually confer resistance to chloramphenicol, sulfonamides, tetracycline, streptomycin, and, in some cases, ampicillin. Although resistance to sulfonamides is conferred, combined resistance of *S. typhi* to sulfonamide and trimethoprim has remained rare.

However, disease due to resistant strains has remained rather the exception, despite the large Mexican epidemic and the outbreak in Peru. In general, *S. typhi* is still susceptible to the commonly used antibiotics.

Treatment

Principles of treatment consist in adequate antibiotic therapy, good nursing care and bed rest, maintenance of fluid and electrolyte balance and nutrition, the use of high-dose glucocorticosteroids in severely ill, toxic patients, and the diagnosis and treatment of complications.

During the past decade many comparative therapeutic trials have been conducted. The results have occasionally been conflicting, but the following generalizations may be made.

Chloramphenicol remains the drug of choice for typhoid fever in developing countries since it is effective, stable, inexpensive, and widely available [4, 13]. Disadvantages of chloramphenicol comprise its lack of influence on the relapse rate, its high convalescent excretion rate and, the major drawback, the occurrence of chloramphenicol-resistant strains. The risk of chloramphenicol-induced aplastic anemia appears small when weighed against the mortality of untreated typhoid. Compared to chloramphenicol, amoxicillin and co-trimoxazole display lower relapse and convalescent excretion rates, similar failure rates, and take about the same time or somewhat longer to produce defervescence [12]. Ampicillin is inferior to amoxicillin, but also to co-trimoxazole and chloramphenicol.

Time to defervescence varies from country to country, but is usually between 4 and 6 days on average with chloramphenicol [12]. Treatment with co-trimoxazole or chloramphenicol should be monitored for hematotoxicity by regular blood counts.

With any of the agents mentioned, duration of therapy is usually recommended for 7-10 days after defervescence. For dosage see Table 21.1. Relapses can be treated with a second course of the same antibiotic, since they are generally not due to resistant organisms.

The third-generation cephalosporins and modern quinolones - ciprofloxacin, pefloxacin, and ofloxacin - show excellent in vitro activity against *S. typhi*, but because of poor correlations between in vitro activity and clinical efficacy with other agents in the past, proper comparative clinical trials should be conducted before general recommendations are made. Preliminary results of open studies with ceftriaxone, cefotaxime and ciprofloxacin are encouraging [20, 26].

There are convincing arguments for the use of glucocorticosteroids in severe cases of typhoid fever. Treatment with high-dose dexamethasone, 3 mg/kg initially followed by 1 mg/kg every 6 h for eight doses altogether, was shown to reduce mortality in severe chloramphenicol-treated typhoid fever significantly, from 55.6% to 10%, when compared to placebo in a double-blind, randomized clinical trial in Jakarta [15]. The rates of complications were similar in the two groups. The authors recommend dexamethasone for all patients with suspected typhoid fever who are delirious, obtunded, stuporous, comatose, or in shock. According to these

Table 21.1. Antibiotic treatment of typhoid fever

Antibiotic	Route of administration	Total daily dose (mg/kg)	Usual daily dose in adults (g)	No. of doses per day
Chloramphenicol	p.o./i.v.	50[a]	3 (−4)[a]	3–4
Co-trimoxazole	p.o./i.v.	5–8[b]	(1.92–)2.88[c]	2
Amoxicillin	p.o.	75–100	4	3–4
Ampicillin	p.o./i.m./i.v.	100	(4–)6	3–4

[a] Many authorities recommend halving the daily dose 2 days after defervescence.
[b] This refers to the trimethoprim constituent.
[c] This is equal to (4–)6 tablets co-trimoxazole each containing 80 mg trimethoprim plus 400 mg sulfamethoxazole.

criteria, dexamethasone was not given to 83% of patients with typhoid seen in the study period. The case-fatality rate in that group was 1.4% [15].

Intestinal hemorrhage, when massive, is managed conservatively by blood transfusion.

The treatment of perforations which still carry high mortality, remains controversial with regard to the choice between prompt surgical intervention, with wedge excision or resection, and mainly conservative management [5, 18]. In any case, treatment of peritonitis is essential. The addition of metronidazole or clindamycin plus an aminoglycoside to the antibiotic regimen appears warranted once perforation has occurred.

Other complications - myocarditis, renal failure, pneumonia, etc. - are treated according to standard medical practice. Metastatic infections are treated with chloramphenicol in high dosage, 100 mg/kg/day, for prolonged periods of time, i.e., usually 3 weeks in the case of meningitis and 4–6 weeks in osteomyelitis or arthritis.

Chronic carriage, defined as continued excretion of *S. typhi* in stool or urine for more than 1 year, develops in 1%–3% of patients. The incidence of biliary tract disease tends to be higher in this collective than in the general population. *Clonorchis sinensis* and *Schistosoma haematobium* infections also predispose to chronic carriage.

Chloramphenicol is not effective in the treatment of chronic carriers. Treatment regimens of either ampicillin or amoxicillin, 100 mg/kg/day, plus probenecid, 30 mg/kg/day, or of co-trimoxazole, two tablets (equal to 0.96 g) twice daily for periods up to 3 months, have been suggested for cases of chronic carriage. In most instances of cholelithiasis or nonfunctioning gallbladder, cholecystectomy has to be performed first for one of these regimens to be effective. Similarly, predisposing *Schistosoma haematobium* infection should be treated first. Quinolones (ciprofloxacin) are promising for the treatment of chronic carriers.

Prevention

Measures of control of typhoid consist in the improvement of sanitation, mainly the provision of a reliable water supply and adequate sewage disposal, and vaccination [14]. Field trials using a new oral vaccine, the Ty 21a strain, in different formulations and immunization schedules have so far shown a lower level of protection than the 95% vaccine efficacy initially observed with this strain in Egypt [7].

Nontyphoidal Salmonella Infections

Nontyphoidal *Salmonella* infections are highly epidemic in developing countries. In contrast to *S. typhi,* chronic carriage is rare in man. Nevertheless, high enteric carriage rates have been reported in several surveys from tropical countries (see Table 14.1 in chapter "Infections of the Gastrointestinal Tract"). *Salmonella* septicemia and meningitis are both more common in developing countries than in Europe or United States (see Table 9.1 of chapter "Septicemia" and Table 12.1 of chapter "Infections of the Central Nervous System"). The pattern of predominant serotypes varies from country to country and due to the enzootic background of the infection. *Salmonella typhimurium* is probably the commonest serotype worldwide. *S. cholerae-suis* frequently leads to septicemia. Asymptomatic intestinal infection with nontyphoidal salmonellae is common. Clinically, nontyphoidal *Salmonella* infections may present as acute enteritis, paratyphoid fever, septicemia, and focal infection subsequent upon an episode of bacteremia.

Several diseases are associated with an increased risk of invasive salmonellosis, i.e., AIDS, malignancy, particularly lymphoma and leukemia, sickle cell disease and other hemolytic anemias, and schistosomiasis. Infants are also more likely to develop systemic disease from intestinal infections. In a recent report from South Africa, *S. enteritidis* accounted for 24 of 78 episodes of bacteremia in children with severe protein-energy malnutrition compared to 30 of 217 episodes in children with normal nutritional status or nutritional growth retardation [2]. Of the 51 isolates serotyped, 92% were nontyphoidal salmonellae. Nontyphoidal salmonellae have a wide range of animals and birds as reservoir, and transmission via contaminated food, such as meat and meat products, poultry, and unpasteurized milk, has long been recognized as "*Salmonella* food poisoning." Direct person-to-person spread by the fecal-oral route is now assumed to be of greater importance in crowded, unsanitary conditions of living. Nosocomial salmonellosis by continuous spread is also well recognized, and outbreaks have been reported from hospitals worldwide. Neonatal care units, nurseries, and pediatric departments are at particular risk. Furthermore, hospitals may actually act as reservoirs for multiresistant *S. typhimurium* causing infection in the community [23].

Treatment

Whereas antibiotics are generally not recommended for treatment of *Salmonella* enteritis, chloramphenicol, ampicillin, co-trimoxazole, and gentamicin are still the drugs of choice for treatment of systemic salmonellosis. The emergence of multiresistant strains of *S. typhimurium* and other nontyphoidal serotypes since the early 1970s is therefore of great concern, particularly to developing countries. Large outbreaks of infections with a high incidence of septicemia and considerable mortality have been caused by these strains in most parts of the world [6, 8, 9, 11, 17, 22, 24, 27].

Although some of these strains are still susceptible to co-trimoxazole, R-plasmid-mediated resistance of *S. typhimurium* to co-trimoxazole in addition to chloramphenicol and ampicillin was reported from India a few years ago.

Strains of *S. typhimurium* resistant to co-trimoxazole are now widespread in India [16, 22, 25] and have also been reported from Rwanda, Zaire, South Africa, and Sao Paulo, Brazil [10, 23, 24]. Experience with the expensive second- and third-generation cephalosporins is still limited to case reports and to small number of patients, respectively [3]. Cefuroxime, latamoxef, ceftriaxone, ceftazidime, and cefotaxime have been used for treatment of *Salmonella* meningitis (though the serotypes were not always stated) [3]. Ceftazidime, cefoperazone, and cefotaxime have apparently been used with success in nontyphoidal *Salmonella* septicemia [3]. The modern quinolones are promising oral agents for the treatment of *Salmonella* septicemia and local abscesses in adults.

References

1. Anderson ES (1975) The problem and implications of chloramphenicol resistance in the typhoid bacillus. J Hyg 74: 289-299
2. Berkowitz FE (1984) Bacteremia in hospitalized black South African children. Am J Dis Child 138: 551-556
3. Bryan JP, Rocha H, Scheld WM (1986) Problems in salmonellosis: rationale for clinical trials with newer beta-lactam agents and quinolones. Rev Infect Dis 8: 189-207
4. Butler T, Rumans L, Arnold K (1982) Response of typhoid fever caused by chloramphenicol-susceptible and chloramphenicol-resistant strains of Salmonella typhi to treatment with trimethoprim-sulfamethoxazole. Rev Infect Dis 4: 551-561
5. Cook GC (1985) Management of typhoid. Trop Doct 15: 154-159
6. Dube SD, Bhagwat AG (1983) Non-typhoidal *Salmonella* infection in Zambian infants. Trans R Soc Trop Med Hyg 77: 336-337
7. Edelman RE, Levine MM (1986) Summary of an international workshop on typhoid fever. Rev Infect Dis 8: 329-350
8. Filloy L, Borjas GE (1978) Epidemia por *Salmonella poona* (del grupo G) resistente a altas concentraciones de ampicilina, cloranfenicol y otros agentes antimicrobianos. Bol Med Hosp Infant Mex 35: 355-359
9. Ghafour SY, Khuffash FA, Majeed HA (1982) An epidemic of group B salmonellosis in the newborn nursery of a regional hospital in Kuwait. J Kuwait Med Assoc 16: 195-202
10. Goossens H, Vanhoof R, De Mol P, Grados O, Ghysels G, Butzler JP (1984) In-vitro susceptibility to antimicrobial agents. J Antimicrob Chemother 13: 559-565
11. Hadfield TL, Monson MH, Wachsmuth IK (1985) An outbreak of antibiotic-resistant *Salmonella enteritidis* in Liberia, West Africa. J Infect Dis 151: 790-795
12. Herzog C (1980) New trends in the chemotherapy of typhoid fever. Acta Trop 37: 275-280
13. Herzog C, Geddes AM (1982) Chloramphenicol in the treatment of enteric fever. Trans R Soc Trop Med Hyg 76: 848-849
14. Hornick RB (1985) Selective primary health care: Strategies for control of disease in the developing world. XX Typhoid fever. Rev Infect Dis 7: 536-546

15. Hoffman SL, Punjabi NH, Kumala S, Moechtar MA, Sri Pandam Pulungsih, Rifajati Rivai A, Rockhill RC, Woodward TE, Loedin AA (1984) Reduction of mortality in chloramphenicol-treated severe typhoid fever by high-dose dexamethasone. N Engl J Med 310: 82-88
16. Koshi G (1981) Alarming increase in multi-drug resistant *Salmonella typhimurium* in southern India. Indian J Med Res 74: 635-641
17. Lepage P, Bogaerts J, Nsengumuremyi F, Hitimani DG, Van Goethem C, Vandepitte J, Butzler JP (1984) Severe multiresistant *Salmonella typhimurium* systemic infections in Central Africa - Clinical features and treatment in a paediatric department. J Antimicrob Chemother 14 [Suppl B]: 153-159
18. Mulligan TO (1972) The treatment of typhoid perforation of the ileum. J R Coll Surg 17: 364-368
19. Olarte J, Galindo E (1973) *Salmonella typhi* resistant to chloramphenicol, ampicillin, and other antimicrobial agents: strains isolated during an extensive typhoid fever epidemic in Mexico. Antimicrob Agents Chemother 4: 597-601
20. Ramirez CA, Bran JL, Mejia CR, Garcia JF (1985) Open, prospective study of the clinical efficacy of ciprofloxacin. Antimicrob Agents Chemother 28: 128-132
21. Rangnekar VM, Banker DD, Jhala HI (1981) Plasmid-mediated multidrug resistance in *Salmonella typhi*. Lancet II: 364
22. Rangnekar VM, Banker DD, Jhala HI (1983) Antimicrobial resistance and incompatibility groups of R plasmids in *Salmonella typhimurium* isolated from human sources in Bombay from 1978 to 1980. Antimicrob Agents Chemother 23: 54-58
23. Riley LW, Ceballos BSO, Trabulsi LR, Fernandes de Toledo MR, Blake PA (1984) The significance of hospitals as reservoirs for endemic multiresistant *Salmonella typhimurium* causing infection in urban Brazilian children. J Infect Dis 150: 236-241
24. Robins-Browne RM, Rowe B, Ramsaroop R, Naran AD, Threlfall EJ, Ward EJ, Lloyd DA, Michel RE (1983) A hospital outbreak of multiresistant *Salmonella typhimurium* belonging to phage type 193. J Infect Dis 147: 210-216
25. Sharma PL, Sharma KB, Prakash K (1984) Incidence of conjugative plasmid amongst multidrug resistant *Salmonella typhimurium* strains isolated from different parts of India (1979-1982). Indian J Med Res 79: 473-478
26. Ti T-Y, Monteiro EH, Lam S, Lee H-S (1985) Ceftriaxone therapy in bacteremic typhoid fever. Antimicrob Agents Chemother 28: 540-543
27. WHO Scientific Working Group (1980) Enteric infections due to *Campylobacter, Yersinia, Salmonella* and *Shigella*. Bull WHO 58: 519-537

23 Rickettsial Diseases

Rickettsial diseases are often divided into the typhus group of diseases, the spotted fever group, scrub typhus, and Q fever (Table 23.1). Epidemic typhus due to *Rickettsia prowazekii* still causes thousands of cases per year in the cool highlands of Ethiopia. Infection due to *R. conorii* is common in all parts of Africa and a considerable risk to tourists in some areas ("fièvre boutonneuse"). Other rickettsial diseases may also constitute considerable health problems, particularly in developing countries, which to a large extent remain unrecognized because of limited diagnostic facilities [1, 13]. Thus, scrub typhus presenting as a nonspecific febrile illness is apparently a common cause of "pyrexia of unknown origin" in East Asia [2].

Serological diagnosis of rickettsial infections (Weil-Felix reaction, immunofluorescence assay, enzyme-linked immunosorbent assay, complement fixation) is of only limited value in the management of the individual patient. Antibodies are not usually detectable before the end of the 1st week of illness, by which time chemotherapy should have been instituted in order to prevent the serious complications and mortality of severe typhus. Major differential diagnoses, depending on the area, include typhoid fever, relapsing fever, leptospirosis, cerebral malaria, meningococcemia, and measles. Of the drugs suitable for treatment of typhoid fever, only chloramphenicol is effective in rickettsial infections. Treatment with co-trimoxazole, ampicillin, other beta-lactams, aminoglycosides, or sulfonamides does not "cover" rickettsial infections.

Treatment

The choice of agents is between tetracycline, doxycycline, and chloramphenicol. These agents are only rickettsiostatic, and relapses may occur after early discontinuation of treatment, particularly when chemotherapy has been started early in the course of the disease [9]. Quinolones may provide a treatment alternative in future [10].

The average daily dose of oral tetracycline is 2 g (children 25–50 mg/kg) in four divided doses. When doxycycline is used, 200 mg (4 mg/kg/day in children) are administered on the 1st day, followed by 100–200 mg (children 2–4 mg/kg) daily in one or two doses thereafter. The daily dose of chloramphenicol is 2(–3) g in three divided doses, either i.v. or orally. Children with *Rickettsia conorii* infection may be treated with erythromycin, but tetracycline was found to be superior in a controlled study [7]. Rifampicin also has excellent activity against rickettsiae.

Treatment is usually discontinued when the patient has been afebrile for 2–3 days [15]. Some recommend prolonged continuation of treatment after defervescence until 12–14 days have passed since the onset of illness, in order to pre-

Table 23.1. Synopsis of certain epidemiologic, clinical, and serologic features of important rickettsioses. From [11, 14]

Disease	Organism	Geographic area	Natural cycle		Rash		Weil-Felix				Duration of fever after chemotherapy (hours)
			Insect vector	Animal reservoir	Onset (day of disease)	Distribution	Eschar	OX-19	OX-2	OX-K	
Spotted fever group	*Rickettsia*										
RMSF	R. rickettsii	Western hemisphere	Tick	Wild rodents, dogs	3–5	Extremities to trunk	No	+[a]	+	0	<48–72
Boutonneuse	R. conorii	Africa, Mediterranean, India	Tick	Wild rodents, dogs	3–4	Trunk, extremities, face	Yes	+	+	0	-
Rickettsialpox	R. akari	USA, USSR, Korea, Africa	Mite	Mouse	1–3	Vesicular; trunk, extremities, face	Yes	0	0	0	-
Typhus group											
Epidemic typhus	R. prowazekii	Highland areas of South America, Africa, Asia, ? USA	Body louse	Man, flying squirrel	5–7	Trunk to extremities	No	+	±	0	48–72
Brill-Zinsser	R. prowazekii	Worldwide based upon immigration	None	Man (recurrence years after primary attack)	-	Trunk to extremities (may be absent)	No	±	0	0	48–72
Murine typhus	R. typhi	Worldwide in pockets	Flea	Small rodents	5–7	Trunk to extremities	No	+	±	0	48–72
Scrub typhus	R. tsutsugamushi	South Pacific, Asia, Australia	Mite	Wild rodents	4–6	Trunk to extremities	Yes	0	0	+/0[b]	24–36
Others	*Coxiella*										
Q fever	C. burnetti	Worldwide	? Ticks	Cattle, sheep, goats (inhalation of organism)	None	None	No	0	0	0	48 (occasionally slow)

RMSF: Rocky Mountain Spotted Fever.

[a] At least fourfold rise in titer, or a single convalescent titer $\geq 1/320$.

[b] Not likely to be reliably positive in second or subsequent reinfections with different serotypes.

vent relapse and recrudescence of fever. Relapses are frequent but respond to retreatment with the same drug.

Concurrent treatment with corticosteroids for several days is often recommended for patients severely ill with typhus.

Single-Dose Treatment of Louse-Borne Typhus. A single oral dose of 200 mg doxycycline is curative and is now considered by some the treatment of choice [4, 5, 14]. Single-dose treatment of scrub typhus with 200 mg doxycycline appears to be effective when given after day 7 of illness, but was associated with relapses requiring retreatment when given on day 3 of illness [3, 12].

Chemoprophylaxis. Only in scrub typhus has chemoprophylaxis been formally studied [8, 12]. A weekly dose of 200 mg oral doxycycline for 6 weeks after exposure was associated with an efficiency of 89% [12]. Chemoprophylaxis with a single dose of 100-200 mg doxycycline may be justified in an outbreak of epidemic typhus (in addition to louse control!).

Q Fever. Two weeks' treatment with doxycycline, 200 mg daily, tetracycline, or chloramphenicol is recommended for acute Q fever. Chronic Q fever with endocarditis presents a difficult therapeutic problem. Treatment with a combination of tetracycline (or doxycycline) and co-trimoxazole, lincomycin, or rifampicin for many months is suggested. Relapses may occur after discontinuation of antibacterial chemotherapy. Replacement of the affected cardiac valves is indicated for hemodynamic reasons. However, infection of the prosthetic valve is not uncommon.

References

1. Al-Awadi AR, Al-Kazemi N, Ezzat G (1982) Murine typhus in Kuwait in 1978. Bull WHO 60: 283-289
2. Brown GW, Robinson DM, Huxsoll DL, Ng TS, Lim KJ, Sannasey G (1976) Scrub typhus: a common cause of illness in indigenous populations. Trans R Soc Trop Med Hyg 70: 444-448
3. Brown GW, Saunders JP, Sadhu Singh, Huxsoll DL, Shirai A (1978) Single dose doxycycline therapy for scrub typhus. Trans R Soc Trop Med Hyg 72: 412-416
4. Huys J, Kayihigi J, Freyens P, Vanden Berghe G (1973) Single-dose treatment of epidemic typhus with doxcycline. Chemotherapy 18: 314-317
5. Krause DW, Perine PL, McDade JE, Awoke S (1975) Treatment of louse-borne typhus fever with chloramphenicol, tetracycline or doxycycline. East Afr Med J 52: 421-427
6. Laufer D, Lew PD, Oberhansli I, Cox JN, Longson M (1986) Chronic Q fever endocarditis with massive splenomegaly in childhood. J Pediatr 108: 535-539
7. Muñoz-Espin T, López-Parés P, Espejo-Arenas E, Font-Creus B, Martinez-Vila I, Travería-Casanova J, Segura-Porta F, Bella-Cueto F (1986) Erythromycin versus tetracycline for treatment of Mediterranean spotted fever. Arch Dis Child 61: 1027-1029
8. Olson JG, Bourgeois AL, Fanf RCY, Coolbaugh JC, Dennis DT (1980) Prevention of scrub typhus. Prophylactic administration of doxycycline in a randomized double blind trial. Am J Trop Med Hyg 29: 989-997
9. Olson JG, Bourgeois AL, Fang RCY, Dennis DT (1981) Risk of relapse associated with doxycycline therapy for scrub typhus. In: Burgdorfer W, Anacker RL (eds) Rickettsiae and rickettsial diseases. Academic, New York
10. Raoult D, Gallais H, De Micco P, Casanova P (1986) Ciprofloxacin therapy for Mediterranean spotted fever. Antimicrob Agents Chemother 30: 606-607

11. Saah AJ, Hornick RB (1985) Rickettsiosis. In: Mandell GL, Douglas G Jr, Bennett JE (eds) Principles and practice of infectious diseases 2nd edn. Wiley, New York, pp 1081-1095
12. Twartz JC, Shirai A, Selvaraju G, Saunders JP, Huxsoll DL, Groves MG (1982) Doxycycline prophylaxis for human scrub typhus. J Infect Dis 146: 811-818
13. WHO Working Group on Rickettsial Diseases (1982) Rickettsioses: a continuing disease problem. Bull WHO 60: 157-164
14. Wisseman CL Jr (1984) Rickettsial and *Bartonella* infections. In: Strickland GT (ed) Hunter's tropical medicine, 6th edn, Saunders, Philadelphia, pp 198-228
15. Ranjan S, Karanth S, Dickson C (1987) Scrub typhus along the Thai-Kampuchean border: new treatment regimen. Trop Doct 17: 104-107

24 Leptospirosis

Leptospirosis, a zoonotic disease of worldwide distribution, occurs in many animal species, including rodents, dogs, cattle, and swine. Infected animals shed leptospires in their urine. Depending on the ambient temperature, moisture, and pH, leptospires may survive up to 15 days on soil, e.g., in swampy areas of tropical jungle. The infection is transmitted to man either by direct contact with infected urine or indirectly through contaminated soil, surface water, mud, or vegetation. Leptospirosis is a common occupational risk in rice or sugarcane workers.

The genus *Leptospira* is divided into two complexes. The "*biflexa* complex" comprises the saprophytic leptospires, the "*interrogans* complex" the pathogenic strains. According to antigenic differences the *L. interrogens* complex is classified into 18 serogroups, each of which contains several serotypes (serovars) which share antigens common to the given serogroup. The virulence varies remarkably among different serotypes. Likewise, the spectrum of illness ranges from a mild, anicteric, flulike illness over aseptic meningitis to severe Weil's disease [1, 2, 3].

Leptospirosis may be diagnosed by culture of blood or cerebrospinal fluid during the first or "leptospiremic" phase or by culture of urine during the second or "immune" phase. Culture requires the use of special media. Serological diagnosis is possible from the 2nd week onward, by which time positive titers develop. Identification of serotypes is possible by serological means.

Treatment

Both penicillin and tetracycline have been used, though treatment with benzylpenicillin in high dosage, (4–)8–12 megaunits daily for 7 days in adults, is usually preferred. In general, chemotherapy must be instituted within the first few days of illness, preferably within the first 2 days, in order to have a positive effect, if at all.

Chemoprophylaxis with a weekly dose of 200 mg doxycycline was shown to have an efficacy of 95% in people exposed for a limited period of time [4].

References

1. Brown GW, Lee CK, Huxsoll DL, Ng TS, Lim KJ, Wan HS, Eeran JD, Sannasey G (1976) Leptospirosis in Malaysia: a common cause of short-term fever. Southeast Asian J Trop Med Public Health 7: 380-383
2. Damude DF, Jones CJ, White HS, Myers DM (1979) The problem of human leptospirosis in Barbados. Trans R Soc Trop Med Hyg 73: 169-177
3. De Geus A, Wolff JW, Timmer VEA (1977) Clinical leptospirosis in Kenya. I. A clinical study in Kwale District, Coast Province. East Afr Med J 54: 115-124
4. Takafuji ET, Kirkpatrick JW, Miller RN, Karwacki JJ, Kelley PW, Gray MR, McNeill KM, Timboe HL, Kane RE, Sanchez JL (1984) An efficacy trial of doxycycline chemoprophylaxis against leptospirosis. N Engl J Med 310: 497-500

25 Brucellosis

Brucellosis, a zoonosis of worldwide distribution, is of considerable public health importance in many developing countries [1, 2, 6, 7]. *Brucella melitensis* (goat, sheep) is the commonest species and occurs in the Mediterranean countries, East and West Africa, Central and South America and in parts of the Far East. *B. abortus* (cattle) is more common in temperate climates but occurs worldwide. Human infection with *B. suis* (pigs) is largely limited to North America.

Diagnosis of brucellosis is mainly based on standard agglutination tests, since recovery rates from blood cultures are low even under optimal conditions. Culture of *B. abortus* requires supplementation of the atmosphere with 10% CO_2.

Treatment

Brucellosis is difficult to treat because of frequent bone involvement, a high relapse rate, and the chronic course the disease often takes.

Treatment with oral tetracycline, 500 mg every 6 h, for 3-6 weeks plus intramuscular streptomycin, 1 g (children: 25 mg/kg) daily, for 14 (-21) days still appears to be the most effective regimen (Tables 25.1 and 25.2). In children, doxycycline, 4 mg/kg/day in two divided doses, is presumably preferable to tetracycline, 50 mg/kg/day in four divided doses, because of the smaller risk of dental staining. Mild disease may be treated with tetracycline alone (risk of inner ear damage from streptomycin), but a relapse rate higher than with combination therapy has then to be accepted. Relapses are not associated with increased resistance to tetracycline, and treatment with tetracycline for 4 weeks and streptomycin for 2 weeks at the dosage given above is suggested. The addition of rifampicin may be considered.

Co-trimoxazole is apparently less active; a prolonged treatment course is recommended. Rifampicin is active against *Brucella* spp.; it should, however, be used only in combination. Quinolones (ciprofloxacin, ofloxacin) are promising, but the data are still limited.

Localized brucellosis, sacroiliitis, spondylitis, prostatitis, orchitis, endocarditis, or pneumonitis, requires prolonged treatment for at least 6 weeks. If the aminoglycoside is given for much longer than 2 weeks, as is indicated in the case of osteomyelitis, gentamicin should be used instead of streptomycin for reasons of ototoxicity. Surgical measures, such as drainage of an abscess, are performed as necessary.

Rifampicin should be added to the treatment regimen of tetracycline plus aminoglycoside in patients with brucella endocarditis or meningoencephalitis for reasons of bactericidal activity and good penetration into cerebrospinal fluid. Prolonged treatment for at least 6 weeks is recommended [8].

Chronic brucellosis responds only poorly to chemotherapy, but a trial with tetracycline and streptomycin appears to be indicated.

Table 25.1. Data from two prospective, randomized trials of treatment of brucellosis: tetracycline-streptomycin versus doxycycline-rifampicin and tetracycline-streptomycin versus co-trimoxazole. From [3, 4]

	Tetracycline 500 mg p.o. every 6 h or doxycycline 100 mg p.o. every 12 h for 30 days plus streptomycin 1 g i.m. once daily for 21 days	Doxycycline 100 mg p.o. every 12 h for 30 days plus rifampicin 15 mg/kg orally once daily for 30 days	Tetracycline 500 mg p.o. every 6 h for 21 days plus streptomycin 1 g i.m. once daily for 14 days (WHO regimen)	Co-trimoxazole 1440 g p.o. every 12 h for 45 days
No. of patients	28	18	27	28
Sex (male, female)	22/6	14/4	20/7	25/3
Age (years; range/mean)	13-61/33	12-54/34	13-72/33	7-69/32
No. of patients with blood cultures positive for *B. melitensis* prior to treatment	24	17	27	28
Focal disease (no. of patients)	6[a]	4[a]	4	5
Follow-up (months; range/mean)	6-23/16	6-24/16	6-24/15	6-36/18
Defervescence (days; range/mean)	1-6/3.1	1-4/2.6	1-10/4.2	1-25/8.4
No. of patients with relapse (%)	2 (7.1)	7 (38.8)	4 (14.8)	13 (46.6)
No. of patients with positive blood cultures during relapse	2	7	4	12

[a] Patients with focal disease received prolonged treatment for 45 days with tetracycline or doxycycline and rifampicin. Streptomycin was given for 21 days. None of the 10 patients with focal disease relapsed.

Table 25.2. Treatment of brucellosis: data of a prospective, randomized study. From [5]

Treatment regimen	Cure	Initial failure	Relapse[a]
Rifampicin 900 mg p.o. once daily for 45 days + Doxycycline 200 mg p.o. once daily for 45 days	13	-	2
Doxycycline 200 mg p.o. once daily for 45 days + Streptomycin 1 g i.m. once daily for 21 days	14	1	-
Tetracycline 500 mg p.o. every 6 h for 21 days + Streptomycin 1 g i.m. once daily for 14 days (WHO regimen)	8	3	4

[a] Follow-up period was 12 months.

References

1. Alausa OK, Awoseyi A (1976) Brucellosis: the situation in Western Nigeria. Trop Geogr Med 28: 54-59
2. Alausa OK (1980) Brucellosis: Socio-economic problems and control in various countries. Trop Geogr Med 32: 5-11
3. Ariza J, Gudiol F, Pallarés R, Rufi G, Fernández-Viladrich P (1985) Comparative trial of rifampin-doxycycline versus tetracycline-streptomycin in the therapy of human brucellosis. Antimicrob Agents Chemother 28: 548-551
4. Ariza J, Gudiol F, Pallarés R, Rufi G, Fernández-Viladrich P (1985) Comparative trial of co-trimoxazole versus tetracycline-streptomycin in treating human brucellosis (letter). J Infect Dis 152: 1358-1359
5. Kosmidis J (1984) Comparative therapeutic trials in human brucellosis. 4th Mediterrean Congress on Chemotherapy, Rhodes, Greece, pp 669-670
6. Lubani M, Sharda D, Helin I (1986) Cardiac manifestations in brucellosis. Arch Dis Child 61: 569-572
7. Oomen LJA (1976) Human brucellosis in Kenya. Trop Geogr Med 28: 45-53
8. Al-Orainey IO, Laajam MA, Al-Aska AK, Rajapakse CN (1987) Brucella meningitis. J Infect 14: 141-145

26 Relapsing Fever

There are two forms of relapsing fever, louse-borne and tick-borne. Louse-borne relapsing fever is caused by only one species, *Borrelia recurrentis*, whereas the tick-borne variety is caused by several *Borrelia* species. *B. duttonii*, the most widely known agent of tick-borne relapsing fever, is found in Africa.

Several epidemics of louse-borne relapsing fever have occurred during this century, the last one in Europe during World War II. Epidemics tend to occur in conditions of famine or war which result in overcrowding and spread of the human body louse. Endemic foci still exist in Ethiopia and the South American Andes.

The tick-borne disease, whose distribution is largely determined by the biology of the vector (argasid tick of the genus *Ornithodoros*) is more widespread. It occurs in the Mediterranean basin and the Near East, in sub-Saharan Africa, South and Central America, and the southern parts of the United States.

The diagnosis of relapsing fever is established by the demonstration of spirochetemia. The borreliae, which are 5–25 μm in size, are found by microscopic examination of Giemsa-stained blood films or in wet preparations. Serological tests are difficult.

Treatment

Relapsing fever has been effectively treated with a conventional 5–10 days' course of tetracycline or chloramphenicol, the usual dose being 500 mg every 6 h.

Following earlier reports of successful treatment of relapsing fever with single-dose doxycycline, several single-dose regimens have been studied [2, 3, 4, 5]. A single 500-mg oral dose of either tetracycline, erythromycin, or chloramphenicol has been shown to be effective treatment for louse-borne relapsing fever [2, 5]. Patients unable to take oral medications were given a single i. v. dose of 250 mg tetracycline hydrochloride with equally good results [2, 5].

Antibiotic treatment of relapsing fever usually induces a Jarisch-Herxheimer reaction within 1 to 2 h after initiation. This reaction may be severe and even fatal. It starts with severe and distressing rigors associated with rises in temperature, blood pressure, and pulse and respiration rates. Subsequently, the temperature falls and sustained hypotension with vasodilatation develops. Supportive fluid therapy is frequently required to maintain an adequate blood pressure. Death usually occurs during the hypotensive phase. The reaction is most severe in louse-borne relapsing fever. Antipyretics and corticosteroids may ameliorate the severity of the reaction, but do not prevent it [2, 7].

Initial treatment with 0.6 megaunits procaine penicillin i. m. followed by single-dose oral tetracycline after defervescence may be preferable in patients with louse-borne disease. Benzylpenicillin, which takes longer to eliminate the borreliae from the blood, induced severe reactions less frequently than tetracycline. Since the use of penicillin has been associated with relapses after treatment, subsequent administration of tetracycline is advisable [1, 6]. Benzylpenicillin is not recommended for treatment of tick-borne relapsing fever, because a number of strains are apparently resistant.

References

1. Bryceson ADM, Parry EHO, Perine PL, Warrell DA, Vukotich D, Leithead CS (1970) Louse-borne relapsing fever. A clinical and laboratory study of 62 cases in Ethiopia and a reconsideration of the literature. Q J Med 39: 129-170
2. Butler T, Jones PK, Wallace CK (1978) *Borrelia recurrentis* infection: single-dose antibiotic regimens and management of the Jarish-Herxheimer reaction. J Infect Dis 137: 573-577
3. De Clercq AG, Meheus AZ, de Pierpont E, Nyirashema C (1975) Single-dose doxycycline treatment of tick-borne relapsing fever. East Afr Med J 52: 428-429
4. Perine PL, Awoke S, Krause DW, McDade JE (1974) Single-dose doxycycline treatment of louse-borne relapsing fever and epidemic typhus. Lancet II: 742-744
5. Perine PL, Teklu B (1983) Antibiotic treatment of louse-borne relapsing fever in Ethiopia: a report of 377 cases. Am J Trop Med Hyg 32: 1096-1100
6. Salih SY, Mustafa D (1977) Louse-borne relapsing fever: II. Combined penicillin and tetracycline therapy in 160 Sudanese patients. Trans R Soc Trop Med Hyg 71: 49-51
7. Teklu B, Habte-Michael A, Warrell DA, White NJ, Wright DJM (1983) Meptazinol diminishes the Jarisch-Herxheimer reaction of relapsing fever. Lancet I: 835-839

27 Anthrax

Anthrax is mainly a disease of cattle, sheep, horses, and goats. The limited evidence indicates that anthrax is endemic worldwide in the rural, livestock-farming parts of developing countries, and that epidemics continue to occur [1, 2, 5, 6, 8].

Cutaneous anthrax accounted for the majority of cases during an epidemic in Zimbabwe, whereas the rapidly fatal forms, inhalation and gastrointestinal anthrax, were much less common [1]. Anthrax meningitis consequent upon septicemia from cutaneous, pulmonary, or gastrointestinal anthrax is a rare and usually lethal complication [3].

Bacteriological diagnosis of anthrax is possible by detection of *Bacillus anthracis* in smear or culture of exudate, sputum, or cerebrospinal fluid. Bacteremia is frequently detected in cases of anthrax pneumonia and meningitis. However, the clinical diagnosis of cutaneous anthrax is readily made, and isolation of *B. anthracis* carries the risk of sporulation [7].

Treatment

Cutaneous anthrax is treated with 4-8 megaunits (MU) benzylpenicillin (children 0.05-0.1 MU/kg/day) daily in divided doses, which can be substituted by oral penicillin after a few days for a total of 7-10 days. Tetracyclines (2 g tetracycline daily in adults, 40 mg/kg/day in children above 7 years) are recommended as alternative for patients allergic to penicillin. *B. anthracis* is sensitive to many antibiotics (erythromycin, co-trimoxazole, chloramphenicol, all penicillins and cephalosporins). Single-dose treatment with a very high dose of 300-500 mg doxycycline was reported as effective [4]. Antibacterial chemotherapy does not prevent eschar formation, but is indicated for prevention of septicemia. High-dose treatment with 20 MU/day (children 0.5 MU/kg/day) benzylpenicillin is suggested for treatment of inhalation and gastrointestinal anthrax, but usually death ensues despite early treatment.

References

1. Davies JCA (1982) A major epidemic of anthrax in Zimbabwe. Cent Afr J Med 28: 291-298
2. Heyworth B, Ropp ME, Voos UG, Meinel HI, Darlow HM (1975) Anthrax in The Gambia; an epidemiological study. Br Med J 4: 79-82
3. Levy LM, Baker N, Meyer MP, Crosland P, Hampton J (1981) Anthrax meningitis in Zimbabwe. Cent Afr J Med 27: 101-104
4. Saggar SN, Joseph MM, Bell WJ (1974) Treatment of cutaneous anthrax with a single oral dose of doxycycline. East Afr Med J 51: 889-894
5. Sirisanthana T, Navachareon N, Tharavichitkul P, Sirisanthana V, Brown AE (1984) Outbreak of oral-pharyngeal anthrax: an unusual manifestation of human infection with *Bacillus anthracis*. Am J Trop Med Hyg 33: 144-150
6. Sirol J, Gendron Y, Condat M (1973) Le charbon humain en Afrique. Réflexions à partir de 22 nouveaux cas observés en Haute-Volta. Bull WHO 49: 143-148
7. Turner M (1980) Anthrax in humans in Zimbabwe. Centr Afr J Med 26: 160-161
8. Webber RH (1985) Anthrax in Lake Rukwa Valley, Tanzania: a persistent problem. J Trop Med Hyg 88: 327-331

28 Plague

Plague is still endemic in several countries of the Americas (e.g., the southern states of the USA, Brazil, Bolivia, Peru), Africa (e.g., Zaire, Madagascar, Kenya, Namibia, South Africa), and the Far East (Vietnam, Burma) [1].

Plague is primarily a zoonotic disease occurring in sylvatic and urban rodents, between which it is spread mainly by fleas. Rats - *Rattus rattus* and *R. norvegicus* - are the main urban reservoir of *Yersinia pestis*. The infection is usually transmitted to man incidentally, by the bite of an infected rat flea. Secondary transmission from person to person takes place via aerosols from individuals with pneumonic plague, whose respiratory secretions are highly infectious. Plague is one of the four quarantinable diseases. When caring for patients with plague, masks, goggles, gloves, and gowns should be worn. In the laboratory, precautions must be taken to avoid the formation of aerosols from infectious material.

Diagnosis of plague can be made bacteriologically by smear and culture of bubo aspirate, blood, sputum, and cerebrospinal fluid (CSF), or serologically.

Treatment

Rapid treatment is essential if the unfavorable prognosis is to be altered, especially in septicemic and pneumonic plague. Streptomycin and tetracycline are traditionally the drugs of choice [3, 4]. Streptomycin, 1 g twice daily (equal to 30 mg/kg/day), is given intramuscularly for 10 days or at least until 3 days after defervescence. Tetracycline is recommended in a large daily dose of 2-4(-6) g in four divided doses. The daily dosages recommended for children are also very high, namely 40-50 mg/kg/day streptomycin and 50 mg/kg/day tetracycline [2]. Intravenous administration is necessary in severely ill patients. Treatment with chloramphenicol, 50-70 mg/kg/day, is indicated in plague meningitis, where a drug with reliable CSF penetration is required. Clinical experience with modern antibiotics is lacking despite in vitro activity. *Chemoprophylaxis* with tetracyclines is advisable for contacts of patients with pneumonic plague, for household contacts of flea-borne plague, and for medical personnel.

References

1. Akiev AK (1982) Epidemiology and incidence of plague in the world 1958-1979. Bull WHO 60: 165-170
2. Burkle FM JR (1973) Plague as seen in South Vietnamese children. A chronicle of observations and treatment under adverse conditions. Clin Pediatr 12: 291-298
3. Butler T, Levin J, Linh NN, Chau DM, Adickman M, Arnold K (1976) *Yersinia pestis* infection in Vietnam. II. Quantitative blood cultures and detection of endotoxin in the cerebrospinal fluid of patients with meningitis. J Infect Dis 133: 493-499
4. WHO Expert Committee (1976) Plague, fourth report. Tech Rep Ser No 447. WHO, Geneva

29 *Listeria* Infections

Listeria monocytogenes occurs worldwide. It is widespread in nature and has been isolated from animals and asymptomatic humans, from soil, water, cheese, and sewage.

Listeria monocytogenes is an aerobic gram-positive rod which on Gram's stains from clinical specimens can appear coccoid. In the laboratory, the organism can be confused with enterococci, group B streptococci, and diphtheroids.

The organism is facultatively pathogenic and causes illness characteristically in pregnancy, in neonates, in patients with liver cirrhosis, and in immunocompromised subjects (usually the lymphocyte-monocyte system is defective).

Apart from the asymptomatic carrier state, infection with *L. monocytogenes* can present as nonspecific febrile illness in pregnancy, as granulomatosis infantiseptica following transplacental transmission, as sepsis of unknown origin and as meningoencephalitis both in neonates and in adults - the adults are typically immunocompromised - and as focal infection. It now appears to be the commonest cause of meningitis in patients with underlying malignant disease in the United States. Data on clinical illness due to *L. monocytogenes* from the tropics are scarce. In a prospective study from Zaria, Nigeria, the organism was isolated from 19 patients during a period of 1 year in which the patient turnover during this period was 750000 [1]. The clinical conditions comprised meningitis, meningoencephalitis, spontaneous peritonitis, septicemia, arthritis, and pelvic infection [1].

In neonates with granulomatosis infantiseptica, the diagnosis is made by Gram's stain and culture of the normally sterile meconium. Culture of blood, cerebrospinal fluid (CSF), and amniotic fluid is carried out in addition. In case of sepsis or meningoencephalitis, diagnosis is established by positive blood culture and CSF culture respectively.

Treatment

Controlled studies have never been conducted. The recommended drugs are either ampicillin or penicillin, with ampicillin usually preferred in Europe. Since in vitro synergism has been demonstrated between aminoglycosides and either benzylpenicillin or ampicillin, treatment with ampicillin and gentamicin is often suggested for *Listeria* infections occurring in neonates and for patients with meningoencephalitis, which carries a particularly poor prognosis. In case of meningoencephalitis some recommend both systemic and intrathecal administration of gentamicin. Chloramphenicol, erythromycin, or tetracyclines are now considered drugs of second choice. The currently available cephalosporins are not suitable for treatment of infections with *Listeria monocytogenes*. Prolonged therapy, 3 weeks or longer, is required, since relapses are frequent.

The dosage of ampicillin is 200–400 mg/kg/day in neonates and children, and 6–12 g daily in adults. The dosage of benzylpenicillin is 0.3 megaunits (MU)/kg/day (equal to 20 MU/day in adults). Gentamicin is given at conventional dosage, 5 mg/kg/day.

Reference

1. Onyemelukwe GC, Lawande RV, Egler LJ, Mohammed I (1983) *Listeria monocytogenes* in Northern Nigeria. J Infect 6: 141–145

30 Melioidosis

Melioidosis is a rare disease that is endemic in Southeast Asia, where the causative agent, *Pseudomonas pseudomallei,* is widely distributed in soil. Most cases have been reported from Vietnam, Malaysia, Burma, and Thailand, but the disease occurs worldwide between latitudes 20° north and 20° south. Subclinical infection is not rare, as indicated by serological surveys, and delayed reactivation may occur after years of latency.

Pseudomonas pseudomallei grows well on most ordinary media and can be isolated from blood, sputum, pus, and other body secretions. The characteristic bipolar staining by methylene blue, the resistance to colistin and gentamicin, and the wrinkled appearance of the colonies after several days are helpful for a presumptive diagnosis, apart from the clinical picture. Serology is of help in diagnosis if a fourfold rise in titers is demonstrated.

Treatment

In general, prolonged therapy for at least 4 weeks, according to the results of sensitivity testing, is required. Chronic suppurative disease, including pulmonary infections, requires treatment for several months for prevention of relapse. Combination treatment at very high dosage is recommended for acute septicemic melioidosis and for toxic patients with localized disease.

Pseudomonas pseudomallei is usually sensitive to sulfonamides, tetracyclines, chloramphenicol, co-trimoxazole, and kanamycin, though exceptions occur [2, 3, 6]. Resistance to penicillin, ampicillin, colistin, erythromycin, first and second-generation cephalosporins, gentamicin, tobramycin, and amikacin is typical [1, 2]. Modern third-generation cephalosporins (ceftazidime, ceftriaxone, cefotaxime) and an acylaminopenicillin, piperacillin, have been found to be active in vitro, but clinical experience is as yet limited to case reports [1, 5, 6]. For comparison, two modern quinolones, ciprofloxacin and ofloxacin, were only weakly active in vitro [1].

Monotherapy with tetracycline, 2–3 g daily, chloramphenicol, 2–3 g daily, or co-trimoxazole, 0.92–1.92 g daily, is suggested for chronic suppurative disease, including patients with pulmonary infection. Patients who are more severely ill should be treated with two of these agents initially.

Parenteral therapy with two or three agents at very high dosage appears to be necessary for acute septicemic melioidosis and for toxic patients who are extremely ill. Treatment with a combination of tetracycline, 4–6 g daily, chloramphenicol, 4–6 g daily, and co-trimoxazole, 3.84 g daily, is usually recommended. These high dosages are associated with appreciable toxicity and should be reduced as soon as clinical improvement allows. Alternatively, combination treatment with a third-generation cephalosporin (ceftazidime 6 g daily) and co-trimoxazole is suggested by some [4, 6].

References

1. Chau PY, Ng WS, Leung YK, Lolekha S (1986) In vitro susceptibility of strains of *Pseudomonas pseudomallei* isolated in Thailand and Hong Kong to some newer beta-lactam antibiotics and quinolone derivatives. J Infect Dis 153: 167-170
2. Everett ED, Nelson RA (1975) Pulmonary melioidosis. Observations in thirty-nine cases. Am Rev Respir Dis 112: 331-340
3. Guard RW, Khafagi FA, Bridgen MC, Ashdown LR (1984) Melioidosis in far North Queensland. A clinical and epidemiological review of twenty cases. Am J Trop Med Hyg 33: 467-473
4. Sanford JP (1985) *Pseudomonas* species (including melioidosis and glanders). In: Mandell GL, Douglas RG Jr, Bennett JE (eds) Principles and practice of infectious diseases, 2nd edn. Wiley, New York, pp 1250-1254
5. Shaefer CF, Trincher RC, Rissing JP (1983) Melioidosis: recrudescence with a strain resistant to multiple antimicrobials. Am Rev Respir Dis 128: 173-175
6. So SY, Chau PY, Leung YK, Lam WK, Yu DYC (1983) Successful treatment of melioidosis caused by a multiresistant strain in an immunocompromised host with third-generation cephalosporins. Am Rev Respir Dis 127: 650-654

31 Nonvenereal Treponematoses

The treatment schedule laid down by the WHO Scientific Group on Treponemal Infections in 1980 [5] applies to all of the three nonvenereal treponematoses, yaws, endemic syphilis, and pinta (Table 31.1). Doses of benzathine penicillin higher than those suggested by WHO were used in the antiyaws campaign in Ghana. Adults received 2.4 megaunits (MU), children less 15 years old were given 1.2 MU and those less than 6 years old received 0.3 MU [1].

The resurgence of yaws in parts of West and Central Africa has caused considerable concern [3, 4, 5]. The principles of mass treatment implemented during the campaigns in the 1950s and 1960s have been endorsed [5]. Treatment is given according to the prevalence of clinically active yaws. When the prevalence is 10% or more, all members of the community are treated. When the prevalence ranges between 5% and 10%, all patients and their contacts are treated as well as all children (juvenile mass treatment). Where the prevalence is less than 5%, only patients and their contacts are treated.

The problem of possibly more widespread allergy to penicillin in the 1980s than in the 1950s has received due attention [2, 4]. However, during the campaign in Ghana, where almost 700000 people were treated in the early 1980s, no anaphylactoid reactions were noted [1]. Nevertheless, in patients sensitive to penicillin a 5 days' course of oral tetracycline, 2 g daily in adults, is considered adequate according to available data [2].

Table 31.1. Schedule for treatment of the nonvenereal treponematoses according to WHO recommendations. From [4]

	Dose (megaunits of benzathine penicillin G)	
	Patients under 10 years	Patients 10 years and over
Early and late active cases Latent cases and contacts	0.6	1.2

References

1. Agadzi VK, Aboagye-Atta Y, Nelson JW, Perine PL, Hopkins DR (1983) Resurgence of yaws in Ghana. Lancet II: 389-390
2. Brown ST (1985) Therapy for nonvenereal treponematoses: review of the efficacy of penicillin and consideration of alternatives. Rev Infect Dis 7 [Suppl 2]: S318-S326
3. Garner PA, Talwat EN, Hill G, Reid MS, Garner MF (1986) Yaws reappears. Papua New Guinea Med J 29: 247-252
4. Widy-Wirski R (1985) Surveillance and control of resurgent yaws in the African region. Rev Infect Dis 7 [Suppl 2]: S227-S232
5. WHO Scientific Group (1982) Treponemal infections. Tech Rep Ser No 674. WHO, Geneva

32 Mycobacterial Infections

Tuberculosis

Tuberculosis was a major cause of morbidity and mortality in the industrialized countries of North America and Europe at the turn of the century, associated mainly with poverty and poor living conditions. Since then, incidence and prevalence of tuberculosis have decreased dramatically with improvement of living conditions. This is in contrast to the present situation in developing countries, where tuberculosis still causes millions of deaths in the prevailing conditions of malnutrition, crowded living conditions and poverty.

It has been estimated that, in developing countries, 4-5 million highly infectious cases of smear-positive tuberculosis occur each year, with an equal number of less infectious cases, including those positive by culture only and culture-negative cases, the latter being the most frequent form of pulmonary disease in children. Thus, each year about 10 million persons still develop tuberculosis and at least 3 million die from the disease. [16]

The annual risk of infection is estimated to be in the range of 2%-5%, a risk 20-50 times higher than that in industrialized countries [16]. Pulmonary tuberculosis is the commonest form of the disease, although the extrapulmonary forms are also much more common in developing countries. Most of the infections are caused by *Mycobacterium tuberculosis; M. bovis* is rare, the atypical mycobacteria even more so. (Differentiation of these mycobacteria is possible only by culture, not by microscopy).

Infection is usually acquired by inhalation of droplets containing *M. tuberculosis*. The risk of infection progressing to disease is in the range of 5%-10% within the first 5 years of infection. This risk is highest in children under the age of 5 years, and second highest in adolescents and young adults. Morbidity and mortality from tuberculosis in developing countries are highest in the young adult years. Malnutrition increases the severity of the disease.

In children, both poor nutrition and intercurrent infections increase the risk of tuberculous disease. Of the childhood diseases, measles in particular predisposes to tuberculous disease. The onset of symptoms was within 2 weeks of measles infection in 9% of all cases of pulmonary tuberculosis of childhood in one Nigerian study [1]. In another study of childhood tuberculosis from the same country, 57% of children had a history of measles in the preceding 3 months [30]. Tuberculosis is highly associated with AIDS and may be the first manifestation of HIV infection.

Ideally, the diagnosis of tuberculosis is confirmed by isolation of *M. tuberculosis* from sputum, gastric aspirates, cerebrospinal fluid (CSF), other body secretions, or from tissue. In developing countries, where sputum specimens constitute the bulk of samples sent to the laboratory, microscopical examination of Ziehl-Neelsen-stained smears is the most important tool of diagnosis and treatment. Approximately 80% of sputum-positive patients will be detected by microscopy. The usual recommendation is that three "first in the morning" sputum samples should be examined. Culture, which is 1000 times more sensitive, will detect the remaining 20% of sputum-positive patients who excrete only small numbers of organisms. Sensitivity testing is required for the detection of resistant strains. Histopathological examination of lymph nodes, peritoneum, or a liver biopsy is also of help in selected cases. Chest X-rays are another inportant adjunct to diagnosis.

Tuberculin skin tests are used to diagnose infection with *M. tuberculosis,* as done in epidemiological surveys. A positive skin test (induration of more than 10 mm 48–72 h after intradermal injection of 5 TU) is also useful in reaching a clinical diagnosis of tuberculosis in patients who have had no BCG vaccination previously. However, the skin test can be negative in tuberculosis for a variety of reasons, such as overwhelming tuberculosis, anergy associated with measles or whooping cough or, more recently, with AIDS, and also in protein-energy malnutrition.

Definitive diagnosis of childhood tuberculosis is difficult, since gastric aspirates are frequently negative. Tuberculin reactions may be negative and chest X-rays inconclusive. Therefore, a therapeutic trial may be started on the grounds of clinical suspicion (fever and cough, pneumonia that fails to resolve, a decline in the child's normal activities, failure to gain weight, painless enlargement of superficial lymph nodes, positive family history).

Chronic peritonitis with ascites or "idiopathic" pleurisy should also alert one to the diagnostic possibility of tuberculosis and a therapeutic trial is recommended.

Likewise, in tuberculous meningitis, treatment has often to be started on clinical grounds alone. A negative tuberculin test does not exclude the diagnosis. Acid-fast bacilli (AFB) are only seen in a minority of cases in smears of centrifuged CSF [17]. The characteristic changes of CSF – lymphocytic pleocytosis with an increased concentration of protein and low concentration of sugar – are also occasionally absent. Evidence of tuberculosis elsewhere in the body should be sought. One study showed that tuberculous meningitis followed tuberculosis infection elsewhere in 74% of patients [23].

Treatment

Isoniazid, rifampicin, pyrazinamide, streptomycin, and ethambutol are now considered the first-line drugs for primary treatment of tuberculosis in the USA and Europe. In many developing countries, however, only four antituberculous drugs are available: isoniazid, streptomycin, para-aminosalicylic acid, and thiacetazone [2].

Drugs like cycloserine, prothionamide or ethionamide, viomycin, kanamycin, and capreomycin are used only in the case of drug resistance. Dosage and side effects of the commonly used antituberculous agents are summarized in Tables 32.1 and 32.2

When the mode of action of antituberculous drugs is discussed, three different populations of organisms are usually distinguished:

(1) The large population of actively replicating mycobacteria found within a cavity, against which streptomycin, isoniazid, and rifampicin exert bactericidal activity.

Table 32.1. Dosage of antituberculous drugs

	Daily dosage[a]					Twice-weekly dosage in intermittent regimens	
	Children		Adults				
			Bodyweight				
			< 50 kg	> 50 kg			
Streptomycin[b] i.m.	¼–12 years: 0–¼ years:	20–30 mg/kg 10 mg/kg	0.75 g	1.0 g	(10–15 mg/kg)	1.0–1.5 g	(20–25 mg/kg)
Isoniazid[b] p.o., i.m.		10–20 mg/kg (max. 600 mg)	300 mg	300–450 mg	(5 mg/kg)	900 mg	(15 mg/kg)
Rifampicin[b] p.o.		10–20 mg/kg (max. 600 mg)	450 mg	600 mg	(10 mg/kg)	450–600 mg	(10 mg/kg)
Pyrazinamide[b] p.o.		20–35 mg/kg	1.5 g	2.0 g	(30–35 mg/kg)	3.0–3.5 g	(45–50 mg/kg)
Ethambutol[b] p.o.	For 2 months: Thereafter:	25 mg/kg 10–15 mg/kg	1.2 g[c]	1.5 g[c]	(15–25 mg/kg)[c]	3.4 g	(50 mg/kg)
Para-amino-salicylic acid p.o.		200–300 mg/kg (max. 12 g)	10 g	12 g	(150 mg/kg)	–	–
Thiacetazone[b] p.o.		2 mg/kg	150 mg	150 mg	–	–	–
Ethionamide/prothionamide p.o.	> 8 years: 4–8 years: < 4 years:	15 mg/kg 20 mg/kg 25 mg/kg	500 mg	750 mg	(10–15 mg/kg)		–

[a] With the exceptions of para-aminosalicylic acid and ethionamide/prothionamide, which are given in divided doses, the total daily dosage is given as one dose.
[b] WHO essential drug.
[c] The usual recommendation is 25 mg/kg/day for 2 months followed by 15 mg/kg/day.

Table 32.2. Side effects of antituberculous drugs

Drug	Side effects		Contraindications (c) and precautions
	Severity	Description	
Streptomycin	++	Vestibular and auditory VIIIth nerve toxicity: ataxia, vertigo, impairment of hearing	Impaired renal function, impaired hearing, pregnancy (c)
	+	Nephrotoxicity: in patients above the age of 60 years or with decreased renal function, the daily dose should be reduced to 0.5 g	
	+	Allergy: skin rashes and fever	
Isoniazid	+	Hepatotoxicity: a transient rise of SGOT that spontaneously resolves is observed frequently. When symptoms indicative of hepatitis occur, the drug should be withdrawn. Following a prolonged period of withdrawal, isoniazid is often tolerated again. The risk of serious hepatotoxicity is age-related, being more common in older people and rare in children. It is also more common in alcoholic liver disease	Preexisting liver damage, alcoholism, acute hepatitis (c), epilepsy
	+	Peripheral neuropathy due to competition with pyridoxine is more common with poor nutrition. Substition of pyridoxine, 30–50 mg daily, for improvement or prevention has no detrimental effect on the antituberculous activity of isoniazid	
	(+)	CNS toxicity: memory loss, personality disorders, psychosis	
	(+)	Allergy: skin rashes, fever, lupus-erythematosus-like syndrome	
	(+)	Hematotoxicity: pyridoxine-dependent anemia. Rarely, agranulocytosis	
	(+)	Pellagra in malnourished people	
Rifampicin	++	Hepatotoxicity: minor rises (5%–20%) of SGOT or of bilirubin are common, but usually resolve under continued therapy. A rise in SGOT beyond 100 U/L should result in discontinuation of the drug	Preexisting liver damage, first 5 months of pregnancy (c)
	(+)	Hypersensitivity reactions: fever, flushing, pruritus, neutropenia. With intermittent therapy, a "flulike syndrome", occasionally associated with thrombocytopenia and acute renal failure due to interstitial nephritis, may occur. It is rare at the now employed dosage of 600 mg twice weekly	
	(+)	Discoloration of urine and other body fluids	
	(+)	Hepatic enzyme induction: interference with many drugs (oral contraceptives, anticoagulants, etc.)	

Table 32.2 (continued)

Drug	Side effects		Contraindications (c) and precautions
	Severity	Description	
Pyrazinamide	+	Hepatotoxicity occurs in 1%-2% with the now recommended lower dosage of 20 to 35 mg/kg daily	Renal failure, preexisting liver damage, gout, pregnancy
	(+)	Hyperuricemia is common, clinical gout is rare	
	(+)	Allergy: rashes and fever	
	(+)	Arthralgia	
Ethambutol	+	Optic neuritis occurs in 0%-6% at the recommended dosage of 25 mg/kg daily for 2 months, followed by 15 mg/kg daily afterwards. Early symptoms of impaired color vision and reduction of visual acuity may progress to complete blindness when the drug is not withdrawn promptly	Alcoholism, eye disease (prior optic neuritis)
	(+)	Rarely: skin rashes, hyperuricemia and gastrointestinal intolerance	
Para-amino-salicylic acid	++	Gastrointestinal intolerance (nausea, vomiting, diarrhea) may be severe and often causes poor patient compliance	Renal failure, preexisting liver damage, peptic ulcer disease
	+	Allergy: skin rashes, fever, lupus-erythematosus-like syndrome	
	(+)	Hepatotoxicity	
Thiacetazone	+	Gastrointestinal intolerance	Preexisting Liver damage, allergy, pregnancy
	(+)	Bone marrow depression: anaemia, leucopenia, or, agranulocytosis. Serious reactions occur at 0.5%	
	(+)	Allergy: skin rashes, fever, Stevens-Johnson syndrome. The incidence of skin rashes is high in Asian countries. Thiacetazone is commonly used in Africa, Central and South America	
Ethionamide/ prothionamide	++	Gastrointestinal intolerance	
	+	Hepatotoxicity is reversible and is indicated by rising SGOT values	Preexisting liver damage, psychosis, epilepsy, diabetes mellitus, pregnancy (c)
	(+)	Allergy: skin rashes and fever	
	(+)	Neurologic and psychic abnormalities	

SGOT, serum glutamine oxaloacetic transaminase.

At a dosage of 15 mg/kg/day, ethambutol exerts a bacteriostatic effect on this population.
(2) Slowly multiplying organisms in closed cavities or abscesses. Only rifampicin and, to a lesser extent, isoniazid, act on these bacteria.
(3) The slowly dividing population of organisms located intracellularly in macrophages at an acid pH. This population is killed by pyrazinamide, isoniazid, and rifampicin. Reduction of duration of treatment to 6-9 months or less is only possible by combination of these bactericidal agents.

Para-aminosalicylic acid and thiacetazone are bacteriostatic agents with weak activity against extracellular organisms. These agents are combined with a bactericidal agent such as isoniazid to prevent the emergence of drug resistance. These combinations have to be given for a period of 12-18 months. The comparatively nontoxic combination of isoniazid and ethambutol also has to be taken for up to 18 months.

Effective antibacterial chemotherapy given uninterruptedly for an adequate treatment period is paramount for the cure of tuberculosis. Bed rest, hospitalization, or roborant measures like a special diet are not required when effective chemotherapy is given. Patients should continue to work when possible, according to their state of health. Hospitalization is indicated in severely ill patients, and is often instituted for the initial intensive phase of treatment. Daily administration of streptomycin is often difficult to organize in an ambulatory setting.

Effective chemotherapy of tuberculosis requires combination of at least two drugs to prevent the emergence of drug resistance. In many developing countries there are national standard regimens which have to be followed.

Several effective regimens are commented upon in Table 32.3.

The WHO-approved standard regimen of 12 months' treatment with isoniazid and thiacetazone supplemented initially by streptomycin has a cure rate of approximately 90%-95% under clinical trial conditions. Under programme conditions, however, with minimally supervised treatment, the success rate can easily drop to 70% or less. Thus, the 1974 Kenya survey showed an efficacy of 71% at 1 year of the aforementioned national standard regimens [9]. Indian data are somewhat more favorable. They indicate an effectiveness of 82% after 1 year and a relapse rate of 10% in the following year for the same regimen of isoniazid and thiacetazone [25].

Failure to complete antibacterial chemotherapy is the main reason for this substantial loss in efficacy. The proportion of patients receiving the 12 months' course of the national standard regimen was only 24%, 35%, and 58% respectively in surveys undertaken in Kenya, Tanzania, and Peru [7, 9, 14]. Data from the Kenya survey showed that 72% of patients received 6 months' treatment [9].

The failure of long-term continuation of self-administered chemotherapy stimulated research into intermittent treatment and short-course chemotherapy [28]. Many regimens were evaluated in clinical trials, the majority of them in developing countries. Intermittent chemotherapy, commonly twice weekly, has the advantage that all doses of treatment can be administered in a fully supervised fashion. In India, the reported efficacy of the WHO-recommended regimen of twice-weekly isoniazid and streptomycin for 12 months was 93% at 1 year [25]. The relapse rate was 7% in the 2nd year and the overall efficacy was estimated to be 86% un-

Table 32.3. Standard primary treatment of tuberculosis in developing countries and in the USA/Europe, and selected regimens of short course chemotherapy

Regimen		Comments
WHO recommendation		
2 STH/10 TH *or* 2 SPH/10 PH *or* 12 S_2H_2		Two regimens of 12 months' duration were recommended by WHO for use in programme conditions in developing countries [16]. The alternative to daily self-administration of TH is the twice-weekly, fully supervised administration of SH, preferably preceded by an initial daily treatment phase. T has largely been substituted for para-aminosalicylic acid, except in countries where use of T is associated with an unacceptably high rate of side effects [18]. When treatment is continued up to 18 months (16 TH instead of 10 TH), relapse rates are reduced to almost zero.
Standard treatment in USA/Europe		
9 RH *or* 1 RH/8 R_2H_2 *or* 2 ERH/7 RH *or* 2 SRH/7 RH		Duration of treatment can be reduced to 9 months if two bactericidal drugs, RH, are administered throughout. Following 1 month of daily therapy, the drugs may also be given twice weekly. In Europe, initial supplementation with E or, less often, S is common. These regimens are widely used in the USA/Europe as standard primary treatment of tuberculosis. They are not recommended for use in developing countries, where initial H resistance is more common [6].
Short-course chemotherapy		
2 HRZS *or* 2 HRZE	4 RH *or* 4 R_2-H_2	Combination of four bactericidal drugs for usually 2 months is followed by daily or twice weekly treatment with RH in short course regimens of tuberculosis. These regimens are highly effective and have the advantage of early sputum conversion. Hepatotoxicity is in the range of 1%–4% with these regimens, which contain three potentially hepatotoxic drugs. To reduce drug costs, treatment with HT may be substituted for RH whilst retaining efficacy against H-resistant disease [8].
or 2 HRZS/6 TH		

H, isoniazid; T, thiacetazone; S, streptomycin; R, rifampicin; E, ethambutol; Z, pyrazinamide. Figures preceding drugs indicate duration of treatment in months. Subscript figures stand for number of doses per week in intermittent treatment.

der programme conditions [25]. Programmes of fully supervised treatment can be operated in urban areas, but are not adequate for much less densely populated rural areas.

Short-course chemotherapy of tuberculosis became possible by combination of two or more bactericidal drugs (Table 32.3). In general, 6-month regimens include a first phase of intensive treatment with four bactericidal drugs for 2 months, usually rifampicin, isoniazid, pyrazinamide, and streptomycin, to be followed by a consolidation phase with daily or twice-weekly rifampicin and isoniazid for 4 months. It was shown that streptomycin can be replaced by ethambutol in the initial intensive treatment phase, when an oral regimen is preferred. In addition to being highly effective, these regimens have the added advantage of early sputum conversion. Furthermore, they are also effective in case of initial isoniazid resistance, which is more common in developing countries. An effective regimen of intermittent short-course chemotherapy is rifampicin + isoniazid + pyrazinamide + streptomycin (or ethambutol) given thrice weekly throughout for 6 months [13].

The associated drug costs of the short-course regimens are the main obstacle to their use in developing countries. A somewhat less expensive regimen consists in 2 months' intensive treatment with daily isoniazid + pyrazinamide + rifampicin + and streptomycin followed by 6 months' treatment with isoniazid + thiacetazone [8]. Being less expensive, and effective both in isoniazid-susceptible and isoniazid-resistant strains, this regimen is often considered the most suitable short-course regimen for treatment of tuberculosis in developing countries [2, 5, 6, 29]. The efficacy of this regimen was found to be 100% in a clinical trial where patients were hospitalized for the first 6 months [8]. When this regimen was used in a district of Tanzania with hospitalization for the first 2 months, a cure rate of 78% was achieved [21]. This compares favorably with the nation wide rate of 50%-55% achieved by standard treatment with isoniazid + thiacetazone, and an initial course of streptomycin [21].

But as long as drug costs are not lowered, especially those of rifampicin, treatment with isoniazid + thiacetazone supplemented by streptomycin in the initial phase will remain the standard treatment of tuberculosis in many developing countries. The dosage is as follows: streptomycin 1 g/day i.m. (0.75 g in individuals with body weight less than 50 kg) initially plus one tablet daily containing 300 mg isoniazid and 150 mg thiacetazone for 12 months. Care must be taken in the dosage prescribed to children. The dose of isoniazid contained in the combined formulation (300 mg isoniazid plus 150 mg thiacetazone) is too low in relation to thiacetazone for use in children. The following dosage scheme may be used in children:

- Under 12 kg: ¼ tablet + 50 mg isoniazid
- 12-29 kg: ½ tablet + 100 mg isoniazid
- Over 30 kg: 1 tablet (no extra isoniazid required)

The incidence of side effects caused by thiacetazone varies widely between different countries. For example, in several countries of Asia and West Africa thiacetazone is generally not well tolerated and para-aminosalicylic acid is used instead [18].

Outlook on Antituberculous Chemotherapy. There are promising developments in the field of antituberculous chemotherapy. Some quinolones (pefloxacin, ofloxacin, ciprofloxacin) are remarkably active against mycobacteria, and clinical trials concerning their efficacy in tuberculosis and leprosy are in progress. Investigational quinolones are even more promising. There is hope that this group of bactericidal agents will make a further reduction of treatment duration possible. It may well be that the final goal of 6-8 weeks' duration of antituberculous treatment will be achieved in future.

Although the various treatment regimens have been studied in pulmonary tuberculosis, antibacterial *chemotherapy of extrapulmonary tuberculosis* is largely the same as that for pulmonary tuberculosis. For example, controlled clinical trials have shown very good results of ambulant chemotherapy of spinal tuberculosis in developing countries [11]. Surgery improved prognosis only when a radical resection of the focus with anterior fusion was performed [11]. Drainage of pus in empyema or abscesses, and treatment of sequelae like hydrocephalus or constrictive pericarditis are other surgical procedures performed in the therapy of tuberculosis.

Tuberculous meningitis needs to be treated with agents that diffuse into the CSF, namely isoniazid, rifampicin, ethambutol, pyrazinamide, and prothionamide. Streptomycin is also usually added to the regimen. Where resources are limited, a combination of isoniazid, streptomycin, and ethambutol may be used [17]. The combination of isoniazid, streptomycin, and rifampicin is, however, superior. According to availability, pyrazinamide may be added in order to administer a combination of four bactericidal drugs initially. After 2-3 months treatment is continued with isoniazid + rifampicin, isoniazid + ethambutol, or isoniazid + thiacetazone, depending on resources.

Initial treatment at maximal dosage for 3-4 weeks (to be followed by conventional dosage) is suggested by many. The doses are as follows: isoniazid, 10 mg/kg/day in adults (max. 1 g) and 15-20 mg/kg/day in children (max. 0.5 g); rifampicin, 10 mg/kg/day (max. 0.75 g); streptomycin, 30 mg/kg/day (adults max. 1.5 g); pyrazinamide, 35 mg/kg/day. These doses are decreased to conventional dosage after 3-4 weeks. Ethambutol is given as usual: 25 mg/kg/day for 2 months, followed by 15 mg/kg/day.

Glucocorticosteroids are often recommended for initial treatment of severe cases of tuberculous meningitis and in case of tuberculous pericarditis and miliary tuberculosis, but their use is controversial. The dose of prednisone is 30-50 mg daily in adults, 0.5-1 mg/kg/day in children.

Antituberculous treatment in pregnancy can be performed with isoniazid, ethambutol, and para-aminosalicylic acid. Rifampicin, which is contraindicated in the first months of pregnancy, may be given in the second half of pregnancy. Streptomycin, prothionamide, and ethionamide are contraindicated throughout pregnancy. The safety of pyrazinamide and thioacetazone in pregnancy is not established.

Monitoring of treatment for efficacy and for occurrence of side effects is important. Ambulatory patients should be seen frequently during the first few months of treatment to ensure regular drug intake. Side effects are more likely to occur within the first 4-8 weeks of treatment. Patients should be informed to report symptoms suggestive of hepatitis (nausea, vomiting, anorexia, fever, and finally, jaundice) and be checked for signs of hypersensitivity. The value of regular determinations of liver function tests is controversially discussed in the USA and Europe (frequent, though insignificant transient rises) and usually not feasible in developing countries. Efficacy is best monitored by repeated bacteriological examination of sputum in pulmonary tuberculosis. This is preferably carried out by sputum culture, but usually only smear examinations will be possible in developing countries, be it for reasons of expense or lack of facilities. Repeated radiological examination is also usually too expensive. Progressive reduction in the number of acid fast bacilli seen on repeated smears is a good indicator of treatment efficacy. If sputum smears are still positive after 3-4 months of treatment, compliance with treatment should be checked. In Europe, noncompliance is often seen in patients with alcoholism. Treatment failure is certain when the sputum smear is still positive after 6 months of fully supervised treatment. Sensitivity testing should then be performed, if at all possible.

If drug resistance is suspected because of nonresponse to regular, uninterrupted treatment, a combination of agents of which at least two have not been used previously is chosen for treatment (see national guidelines). In case of proved resistance, most commonly to isoniazid, at least two other agents have to be used. In these instances, where retreatment with second-line drugs becomes necessary, supervised treatment, preferably at a referral center, is advisable. Failure to complete the prescribed treatment course is commonly the cause of relapse in developing countries. Retreatment with the initial regimens starting from the beginning is usually effective.

Chemoprophylaxis with isoniazid is recommended in the United States of America and in Europe in the following settings. The daily dosage of isoniazid is recommended at 6–10 mg/kg, the maximal dose being 300 mg:

1. Tuberculin-negative household contacts, in particular children under the age of 6 years, of patients with active disease are given isoniazid for 3 months. When they are still tuberculin-negative after 3 months, BCG vaccination is recommended. Otherwise, continuation for a further 3–9 months is recommended.
2. Recent "converters," that is individuals with documented conversion of the PPD reaction to positive within the last 2 years, are given isoniazid for 3–12 months.
3. Individuals with a positive tuberculin reaction or a history of past tuberculosis who are at special risk of reactivation should also be given isoniazid. This includes patients with leukemia, Hodgkin's disease, diabetes mellitus, AIDS and patients who receive glucocorticosteroids or other immunosuppressive therapy.

In developing countries, chemoprophylaxis with isoniazid should also be considered for children known to be infected with *M. tuberculosis* when they develop measles and/or suffer from protein-energy malnutrition.

Particular attention must be paid to newborns and infants of tuberculous mothers. While the mother is being treated, her child should receive chemoprophylaxis with isoniazid. When the mother is free of disease (sputum-negative), chemoprophylaxis can be discontinued and BCG vaccination be given.

The risk of isoniazid-induced hepatitis is less than 1% in persons aged less than 35 years, but rises to 3% or more in people beyond the age of 50 years.

Prevention and Control

Control of tuberculosis remains an aim yet to be achieved in most developing countries. For example, there is an estimate from India that of every 100 patients with active tuberculosis only eight are treated successfully [25]. This is so because on average only 30 of the 100 are diagnosed, only 10.5 of them receive the standard treatment for 12 months, and the estimated cure rate of the most widely employed regimen, isoniazid + thiacetazone, is 74% [25].

BCG vaccination of infants is strongly recommended in countries with a high prevalence of tuberculosis (WHO Expanded Programme on Immunization). Despite variable results regarding efficacy in different studies, BCG vaccination will prevent hematogenous spread and tuberculous meningitis in the majority of vaccinated children. BCG vaccination of infants has, however, little effect on the prevalence of tuberculosis, since the majority of children do not excrete tubercle bacilli.

Case finding, case holding, and effective treatment are the principle methods of control of tuberculosis [16]. Ideally, any individual complaining of cough of more than 3-4 weeks' duration should be offered to have his sputum examined for acid fast bacilli, and be treated adequately, when found sputum-positive. In this context, short-course chemotherapy is often considered to be superior to conventional treatment of tuberculosis as regards to efficacy and patient compliance, but extension of tuberculosis services seems at least equally important in terms of control, since:

> ... full involvement of the existing and operating primary health care centres in developing countries would immediately results in a tremendous increase in population coverage by tuberculosis services. Less than one-third of the peripheral health centres deliver tuberculosis services at present, which shows that integration of tuberculosis control into the general health services is far from satisfactory ... Thus it is possible to increase the number of highly infectious cases treated substantially; even if only with the standard treatment regimens that the majority of countries can afford. [24]

Leprosy

Leprosy still affects an estimated 15 million people in the world, the majority of whom live in developing countries. The social stigma that surrounds leprosy prevents early treatment. For fear of it, patients do not seek medical treatment until the disease is advanced, and remain infectious during this time. The difficulties of early case finding are an important limitation to the control of leprosy by chemotherapy. Other limitations include the considerable dropout rates from treatment with subsequent relapses and the development of dapsone resistance. Although ultimate control of leprosy by vaccination still seems far away, data from two recent African studies indicate that BCG vaccination confers some protection against leprosy in that part of the world [10, 27].

The clinical presentation and the histopathological and immunological features of the different types of leprosy according to the Ridley/Jopling classification are summarized in Tables 32.4 and 32.5. For details, the appropriate textbooks should be consulted. The diagnosis of leprosy can be made clinically in typical cases. Slit-skin smears identify the bacterial load of the individual patient, and, made serially, will monitor treatment by serial determination of the bacteriological and morphological index. In doubtful cases, histopathological examination of a skin biopsy is required for diagnosis. The histopathological examination also serves to correct classification of a patient in the disease spectrum of leprosy, which has an important bearing on the prognosis.

Treatment

This section intentionally deals exclusively with chemotherapy of leprosy. For other aspects of treatment, such as the various measures of orthopedic surgery, appropriate textbooks should be consulted.

Table 32.4. Clinical and histological features of the different types of leprosy. From [19]

Group	Clinical Features	Histologic Features
Tuberculoid (TT)	Single or a few anesthetic macules or plaques. Borders well defined. Peripheral nerve involvement common	Epithelioid-lymphocyte granulomas, with or without giant cells, in skin and nerves. No subepidermal clear zone. Bacilli rarely found in nerves
Borderline tuberculoid (BT)	Lesions similar to TT but more numerous. Borders of lesions less distinct. Satellite lesions sometimes present around larger lesions. Peripheral nerve involvement common	Granulomas similar to TT. Nerves are infiltrated. Bacilli frequently found in nerves
Borderline (BB)	More lesions than BT. Borders more vague. Satellite lesions often seen. Peripheral nerve involvement common	Epithelioid cells and histiocytic infitrations focalized by lymphocytes. Nerves show increased cellularity. Bacilli readily found in nerves
Borderline lepromatous (BL)	Lesions numerous and similar to BB. Some nerve damage	Histiocytic infiltrations show a tendency to evolve toward both epithelioid cells and foamy cells. Lymphocytes present. Nerves have less cellular infiltration. Bacilli plentiful in nerves
Lepromatous (LL)	Multiple, nonanesthetic, macular or papular, symmetrically distributed lesions. No neural lesions until late. Late complications of madarosis, leonine facies, testicular damage, etc.	Foamy histiocytes containing large numbers of bacilli. Bacilli in walls of blood vessels and arrector muscles. Few or no lymphocytes. Subepidermal clear zone. Numerous bacilli in nerves and perineurium without significant intraneural cellular infiltration
Indeterminate (I)	Vaguely defined hypopigmented or erythematous macule	Often indistinguishable from "chronic dermatitis!" Lymphocytes and histiocytes around skin appendages and nerves

Table 32.5. Immunological features of the different types of leprosy. From [3, 19]

Group	Lepromin reaction (Mitsuda)	Bacillary density in skin	Bacillary density in nose	Immuno-logical stability	Reversal reactions	Erythema nodosum leprosum
Tuberculoid (TT)	+ + +	− to ±	−	+ +	−	−
Borderline tuberculoid (BT)	+ + to −	− to +	−	±	+	−
Borderline (BB)	−	+ +	−	−	+ +	−
Borderline lepromatous (BL)	−	+ + +	±	±	+	±
Lepromatous (LL)	−	+ + + +	+ +	+[a]/+ +[b]	±[a]/−[b]	+ +
Indeterminate (I)	± to −	− to +	−	−	−	−

[a] Subpolar lepromatous.
[b] Polar lepromatous.

Efficacious chemotherapy of leprosy is being hampered, apart from other reasons, by the emergence of dapsone resistance, which is now widespread in many areas of the world and is more likely to occur in the case of low-dose treatment and irregular administration [31]. Secondary resistance of *Mycobacterium leprae* strains to other antileprosy drugs has also been reported [31].

Multidrug treatment of leprosy has therefore been recommended by a WHO study group [31] in order to prevent the continuing emergence and spread of resistant strains. The treatment schedules are given in Table 32.6. The significantly shortened duration of treatment may also reduce the dropout rates of existing programs. Whether the phenomenon of "persistence," the survival of viable, drug-sensitive *M. leprae* organisms despite prolonged regular treatment in adequate dosage, will be overcome by multidrug treatment is as yet not known. Similarly, it remains to be seen whether multidrug treatment will obviate the need for lifelong treatment in lepromatous leprosy. The recommended multidrug treatment is considerably more expensive than the previous dapsone monotherapy, which makes its implementation difficult in practice [26]. Studies to define the position of oral quinolones (pefloxacin, ofloxacin, ciprofloxacin) in the multidrug treatment of leprosy are under way.

Dapsone is well tolerated in the conventional dosage. Apart from mild hemolysis, more severe anemia may occur in the case of intercurrent infection and deficiency of glucose-6-phosphate dehydrogenase. Allergic rashes are uncommon, but cases of fatal toxic epidermal necrolysis have occurred. "Dapsone dermatitis," which may be associated with fever and jaundice, occurs 1-2 months after the onset of treatment.

Skin discoloration due to deposition of the drug, a red dye, in the subcutaneous fat is the major side effect of clofazimine. The discoloration is particularly noticeable in lighter-skinned people. Starting with a red discoloration, the skin later turns brownish to violet. Leprosy lesions stain darker than the surrounding normal

Table 32.6. Standard regimens for treatment of leprosy, as recommended by a WHO study group. From [31]

Multibacillary (BB, BL, and LL) leprosy		
Dapsone	100 mg daily, self-administered	for at least 2 years, or until the slit-skin smears are negative
Rifampicin	600 mg once monthly, supervised	
Clofazimine[a]	300 mg once monthly, supervised *and* 50 mg daily, self-administered	
Paucibacillary (TT, BT, and I) leprosy		
Dapsone	100 mg daily, self-administered	for 6 months
Rifampicin	600 mg once monthly, supervised	

The doses recommended apply to adults of approximately 60 kg body weight. According to weight, the doses are proportionately smaller in children.

BB, borderline; BL, borderline lepromatous; LL, lepromatous; TT, tuberculoid; BT, borderline tuberculoid; I, indeterminate.

[a] Where clofazimine is unacceptable, prothionamide or ethionamide may be given, as a single self-administered dose of 250-375 mg daily.

skin. The drug is also deposited in the intestines, and prolonged treatment at high dosage may cause an ileus-like syndrome [4].

Rifampicin, when given intermittently, can cause a "flulike-syndrome," which appears to be less common with monthly than with weekly doses [31]. In addition to considerable gastrointestinal intolerance, ethionamide and prothionamide may occasionally cause drug-induced hepatitis. Fatalities were observed after 3-4 months of combined treatment with dapsone, prothionamide, and rifampicin [15].

Antileprosy treatment should not be discontinued during reactions. In reversal (upgrading) reactions, nerve involvement with impairment or loss of function is common. Steroids (e.g., 30-40 mg prednisone) should be given at once if there is any nerve tenderness, since complete paralysis may develop over a few days. Early surgical decompression may restore function if paralysis has already supervened. Mild cases respond to nonsteroidal anti-inflammatory agents such as aspirin.

Aspirin or paracetamol also suffices for treatment of mild cases of erythema nodosum leprosum. Bed rest should always be advised. Severe cases require treatment with steroids and or clofazimine. The full anti-inflammatory effect of clofazimine (300 mg daily) is not achieved before 6 weeks. On the other hand, it is inadvisable to administer the drug for longer than 3 months at this increased dosage because of intestinal side effects. Thalidomide is also effective in the treatment of erythema nodosum leprosum. Thalidomide (100 mg 3-4 times daily, or one dose of 300-400 mg in the evening) should never be given to women of child-bearing age because of its teratogenic effects. Treatment with clofazimine or thalidomide helps to reduce the steroid dose.

Attention should be paid to any arising iridocyclitis, which is treated with a mydriatic and a topical steroid.

Buruli Ulcer

Buruli ulcers are known to occur in Central and West Africa, Southeast Asia, Papua New Guinea, Australia, and Central and South America [20, 22]. Infections typically occur in foci associated with swampy terrain and subtropical climate. A source of the infection in nature has not been discovered, and person-to-person spread is thought very unlikely. Infection is considered to result from inoculation of the pathogen into the skin.

The causative pathogen, *Mycobacterium ulcerans,* grows on ordinary mycobacterial culture media at 32 °C. Growth is slow and may take up to 8-14 weeks.

Clinically, the disease starts as a firm, painless, subcutaneous nodule that will ulcerate within weeks. A spreading necrosis of the deep dermis is the main histopathological feature. The ulcer may remain small and heal spontaneously or may spread rapidly to involve large parts of an extremity. The ulcer is characteristically painless, has widely undermined edges, and the surrounding skin is indurated by edema. If left untreated, the ulcer will eventually heal, often with extensive scarring resulting in gross contractures when adjacent to a joint.

The diagnosis is made by the finding of acid-fast bacilli in Ziehl-Neelsen-stained smears from the necrotic ulcer base or the edge of the ulcer.

Treatment

Treatment is largely surgical. In the early stage, cure can be achieved by excision of the subcutaneous nodule in total. When ulceration has supervened, the lesion must be excised widely. Skin grafting is usually required. As an adjunct to therapy, rifampicin, the most active drug in vitro, is given daily until healing has occurred [12]. Local application of heat may accelerate the healing process.

References

1. Aderele WI (1979) Pulmonary tuberculosis in childhood: an analysis of 263 cases seen at Ibadan, Nigeria. Trop Geogr Med 31: 41-51
2. Aluoch JA (1985) Practical therapeutics - chemotherapy for pulmonary tuberculosis. East Afr Med J 62: 612-620
3. Bloom BR, Godal T (1983) Selective primary health care: strategies for control of disease in the developing world. V. Leprosy. Rev Infect Dis 5: 765-780
4. Browne SG (1984) Leprosy. Acta Clinica. Ciba-Geigy, Basel
5. Daniel TM (1982) Selective primary health care: strategies for control of disease in the developing world. II. Tuberculosis. Rev Infect Dis 4: 1254-1265
6. Dutt AK, Stead WW (1985) Tuberculosis chemotherapy today. Ann Acad Med Singapore 14: 508-514
7. East African/British Medical Research Councils (1977) Tuberculosis in Tanzania: a follow-up of a national sampling survey of drug resistance and other factors. Tubercle 58: 55-78
8. East African/British Medical Research Councils (1978) Third study, first report: controlled clinical trial of four short-course regimens of chemotherapy for two durations in the treatment of pulmonary tuberculosis. Am Rev Respir Dis 118: 39-48
9. East African/British Medical Research Councils (1979) Tuberculosis in Kenya: follow-up of the second (1974) national sampling survey and a comparison with the follow-up data from the first (1964) national sampling survey. Tubercle 60: 125-149
10. Fine PEM, Maine N, Ponnighaus JM, Clarkson JA, Bliss L (1986) Protective efficacy of BCG against leprosy in northern Malawi. Lancet II: 499-502
11. Griffiths DL (1980) Tuberculosis of the spine: a review. Adv Tuberc Res 20: 92-110
12. Havel A, Pattyn SR (1975) Activity of rifampicin on *Mycobacterium ulcerans*. Ann Soc Belge Méd Trop 55: 105-108
13. Hong Kong Chest Service/British Medical Research Council (1981) Controlled trial of four thrice-weekly regimens and a daily regimen all given for 6 months for pulmonary tuberculosis. Lancet I: 171-174
14. Hopewell PC, Sanchez-Hermandez M, Baron RB, Ganter B (1984) Operational evaluation of treatment for tuberculosis. Results of a "standard" 12-month regimen in Peru. Am Rev Respir Dis 129: 439-443
15. Ji BH, Chen JK, Wang CM, Xia GA (1984) Hepatotoxicity of combined therapy with rifampicin and daily prothionamide for leprosy. Lepr Rev 55: 283-289
16. Joint IUAT/WHO study Group (1982) Tuberculosis control. Tech Rep Ser No 671. WHO, Geneva
17. Kilpatrick ME, Girgis NI, Yassin MW, Abu El Ella AA (1986) Tuberculous meningitis - clinical and laboratory review of 100 patients. J Hyg 96: 231-238
18. Leowski J (1982) Thioacetazone - a review. Indian J Chest Allied Sci 24: 184-189
19. Meyers WM (1984) Leprosy. In: Strickland GT (ed) Hunter's tropical medicine, 6th edition. Saunders, Philadelphia, pp 409-422
20. Meyers WM, Connor DH, McCullough B, Bourland J, Moris R, Proos L (1974) Distribution of *Mycobacterium ulcerans* infections in Zaire, including the report of new foci. Ann Soc Belge Méd Trop 54: 147-157

21. Nkinda SJ, Mulder DW, Styblo K (1984) Developments in the national tuberculosis control programme in Tanzania. Bull IUAT 59: 77-84
22. Oluwasanmi JO, Solankee TF, Olurin EO, Itayemi SO, Alabi GO, Lucas AO (1976) *Mycobacterium ulcerans* (Buruli) skin ulceration in Nigeria. Am J Trop Med Hyg 25: 122-128
23. Osuntokun BO, Adeuja AOG, Familusi JB (1971) Tuberculous meningitis in Nigerians - a study of 194 patients. Trop Geogr Med 23: 225-231
24. Pio A (1983) The future of the fight against tuberculosis in the world. Z Erkr Atmungsorgane 161: 96-99
25. Radhakrishna S (1983) National tuberculosis programme: relative merits of enhancing the operational efficiency of different components of the treatment programme. Indian J Tuberc 30: 3-8
26. Riley DN (1984) Treatment of leprosy in rural India as seen on a medical student elective. Lepr Rev 55: 397-402
27. Stanley SJ, Howland C, Stone MM, Sutherland J (1981) BCG vaccination of children against leprosy in Uganda. Final results. J Hyg 87: 233-248
28. Stott H (1978) The treatment of pulmonary tuberculosis in the developing countries. Trans R Soc Trop Med Hyg 72: 564-569
29. Styblo K (1983) The Kellersberger memorial lecture, 1982. Tuberculosis and its control: lessons to be learned from past experience, and implications for leprosy control programmes. Ethiop Med J 21: 101-122
30. Taqi AM, Abdurrahman MB, Ango SS (1982) Childhood tuberculosis as seen at the Ahmadu Bello University Teaching Hispital, Zaria. Niger J Pediatr 9: 99-103
31. WHO Study Group (1982) Chemotherapy of leprosy for control programmes. Tech Rep Ser No 675. WHO, Geneva

33 Actinomycosis

Actinomycosis is a chronic suppurative disease with a tendency toward abscess formation and formation of multiple draining sinuses. Histological examination of tissue or microscopic examination of pus will show the diagnostic sulfur granules, but many sections will have to be scrutinized before one or two granules are seen.

Actinomycosis is usually caused by the anaerobic bacterium *Actinomyces israelii;* other *Actinomyces* species and a related *Arachnia* species are occasionally responsible. Mixed infections with other oral saprophytes are common. In the early stage, gram-positive branching filaments may be seen, whilst the typical sulfur granules are usually found in advanced disease.

The organism belongs to the normal flora of the oral cavity, and cervicofacial lesions result when the mucous membranes are invaded, frequently following dental disease or dental manipulations. Cervicofacial actinomycosis is the commonest form of actinomycosis, while deep-seated actinomycosis, thoracic or abdominal, accounts for about 20% of cases.

Treatment

Benzylpenicillin is the drug of choice. Dosage and duration of treatment depend upon the severity and the extent of the disease. In severe cases, surgical measures such as resection, or incision and drainage may be required in addition to antibiotic treatment. A dose of 2–5 megaunits (MU) benzylpenicillin daily for 6 weeks may suffice in cervicofacial actinomycosis. Deep-seated actinomycosis is treated with parenteral benzylpenicillin at high dosage, 10–20 megaunits daily for 4–6 weeks. This is followed by 2–5 megaunits phenoxymethylpenicillin or oral tetracyclines daily for 6–12 months. Patients sensitive to penicillin can be treated with tetracyclines, erythromycin, or clindamycin. Metronidazole is not active against *Actinomyces israelii.*

34 Nocardiosis

Nocardiosis is an uncommon, acute or chronic, bacterial infection which occurs worldwide [1, 2]. Nocardiosis usually affects the lungs, with a tendency toward hematogenous dissemination into the soft tissues and the cerebral nervous system. Usually acquired by inhalation, the infection may remain transient and subclinical or produce bronchopulmonary lesions which are clinically similar to tuberculosis, mycotic infections, or carcinoma. Histologically, the lesions are suppurative with necrosis and abscess formation. Metastatic spread into the CNS with the formation of abscesses occurs in one third of cases. Opportunistic infection occurs in patients with lymphoreticular malignancy, patients on glucocorticosteroid therapy, patients with AIDS, and in chronic pulmonary diseases and tuberculosis [2].

Most cases are due to *Nocardia asteroides*, a soil-borne aerobic actinomycete, others are caused by *N. brasiliensis*, a cause of mycetoma in Mexico and South America, and *N. caviae*.

Whether isolation of *N. asteroides* from the sputum always signifies clinical infection, rather than respiratory tract colonization, is controversial, but repeated isolation in a consistent clinical setting should be considered diagnostic of nocardiosis.

The organism grows aerobically on blood agar. Growth is slow, and in mixed culture, for instance from sputum, overgrowth may occur. Microscopic examination of pus or sputum with Gram's stain and a modified Ziehl-Neelsen stain (1% sulfuric acid instead of acid alcohol) can be diagnostic when fine, gram-positive, beaded branching filaments are seen.

Treatment

Sulfonamides are considered the drugs of choice. Surgical procedures - drainage of abscesses or empyema, excision of sinus tracts - are also important. Sulfonamides like sulfadiazine are given for many months after apparent cure to prevent relapses. Metastatic lesions can occur during or after apparently efficacious treatment despite maintenance of sulfonamide plasma levels between 8 and 16 mg/dl. In case of sulfonamide allergy, minocycline or amikacin may be chosen.

References

1. Osoagbaka OU, Njoku-Obi ANU (1985) Nocardiosis in pulmonary diseases in parts of Nigeria. I. Preliminary observations on five cases. J Trop Med Hyg 88: 367-372
2. Randhawa HS, Mishra SK, Sandhu RS, Mazumdar PR, Jain RC, Rao PU, Parkash D, Viswanathan R, Subramaniam S, Sharma VN, Damodaran VN (1973) Prevalence of nocardiosis in bronchopulmonary diseases. Indian J Med Res 61: 689-699

35 Rare Bacterial Diseases: Bartonellosis, Erysipeloid, and Ratbite Fever

Bartonellosis, erysipeloid, and rat-bite fever are all rare bacterial diseases. Whereas bartonellosis is geographically restricted to the habitat of its sandfly vector (certain valleys in Peru, Ecuador, and Colombia), the other two infections occur worldwide. Erysipeloid is an occupational disease of people handling meat, poultry, or fish.

The antimicrobial treatment of the three infections and its duration are summarized in Table 35.1.

Table 35.1. Treatment of rate bacterial diseases

Disease	Pathogen(s)	Drug of choice (daily adult dose)	Alternative (daily adult dose)	Duration of treatment
Rat-bite fever	*Streptobacillus moniliformis* *Spirillum minor*	Procaine penicillin 0.6 MU twice	Tetracycline 2 g	At least 10-14 days
Erysipeloid	*Erysipelothrix rhusiopathiae*	Local disease: phenoxymethyl-penicillin 1.2 MU	Tetracycline 1-2 g	6-8 days
		Endocarditis: benzylpenicillin 12-20 MU	-	4-6 weeks
Bartonellosis	*Bartonella bacilliformis*	Oroya fever: chloramphenicol[a] 2-3 g	Benzylpenicillin, co-trimoxazole, ampicillin	At least 7 days

MU, megaunit.

[a] Chloramphenicol is preferred because of the high incidence of complicating *Salmonella* infections. Supportive treatment (transfusion!) is important.

36 Fungal Infections

Common fungal infections are very frequent in the tropics. The treatment of superficial infections (tinea, pityriasis, intertrigo, vaginitis, onychomycosis) is quite easy, but the high tendency toward recurrence may cause difficulties. Unusual fungi cause special forms of superficial (e.g., tinea imbricata), subcutaneous (mycetoma), or generalized (e.g., histoplasmosis, paracoccidioidomycosis) fungal infections in the tropics.

Diagnosis

Pathogenic fungi belong to very different biological groups. For practical purposes, they may be grouped as follows:

1. Dermatophytes (*Trichophyton, Microsporum, Epidermophyton,* etc.)
2. Facultative pathogenic yeasts (*Candida, Torulopsis,* etc.)
3. Facultative pathogenic molds (*Aspergillus, Mucor,* etc.)
4. Dimorphous fungi (*Histoplasma, Coccidioides,* etc.)

Filamentous bacteria *(Actinomyces, Nocardia)* should not be classified with the fungi. Despite morphological similarities, they are inhibited by antibacterial agents.

Common fungi are easily identified by microscopy and culture. Many species can be part of the normal body flora or may be present in the normal environment. The interpretation of positive isolates is often difficult. In superficial infections, the presence of fungi can be shown easily. A few infected hairs or scales scraped from the active edge of a lesion, placed in a drop of 10% potassium hydroxide solution on a slide covered with a coverslip and heated gently to clear the keratin should be examined microscopically (an oil emersion lens is not needed). Usually spores, hyphae, or yeast cells can be found in active infections. Culture and identification is important with yeasts and molds, since certain species, such as *Candida albicans* and *Aspergillus fumigatus,* are more important clinically than other species. Correlation with the clinical picture is necessary. An isolate in the absence of clinical manifestation does not justify antifungal therapy. The use of Wood's light may be helpful in the diagnosis of certain infections by dermatophytes like ringworm of the scalp.

The diagnostic procedures in fungal septicemia and deep tissue mycoses may be difficult. The culture of certain dimorphous fungi *(Histoplasma, Coccidioides)* may be hazardous; there is a considerable risk of laboratory infections. It is quite common to explain a nonresponse to blind antibiotic therapy by suspected fungal infection. This is relevant only in patients with bone marrow suppression.

Dermatophyte Infections

Many species of keratinolytic fungi (*Trichophyton, Epidermophyton,* etc.) may be pathogenic to man. They are usually transmitted by direct contact from infected patients, carriers, and animals. The culture and identification of the slow-growing fungi requires special mycological techniques; however, this is not essential when the clinical picture is unequivocal and fungal elements are seen in the direct preparation. The distinction between a candidal and a dermatophyte infection by culture is, however, important in inflammatory tinea. In intertrigo, mixed infections with *Candida* and bacteria are common. All skin areas may be infected by dermatophytes, but invasion of deep tissues or mucous membranes is practically unknown. Quite commonly, allergy to the infections can arise, with sterile blisters on noninfected areas (-id- reaction or "dermatophytid").

Treatment

The majority of dermatophyte infections will respond to local treatment. Oral treatment is only necessary in severe or nonresponsive cases. The great number of preparations for the topical therapy of dermatophyte infections are summarized in Table 36.1. In general, there is a choice between modern, expensive, broad-spectrum agents and older, cheap, less effective, and sometimes irritant preparations. Classical benzoic acid ointment is still useful for the treatment of sites where keratolytic activity is desired. Tolnaftate can be regarded as the drug of choice for local therapy in typical tinea. The imidazoles, such as miconazole and clotrimazole, have the additional advantage of activity against *Candida albicans.* The response of tinea to effective therapy can be seen within a few days. A longer course of local therapy is recommended. There is a considerable risk of relapse.

Oral treatment is necessary together with the local therapy in infections of the scalp or the nails or in intensive involvement. The classical drug of choice for the oral therapy of dermatophyte infections is griseofulvin in a single daily dose of 500–750 mg (10 mg/kg) with a fatty meal. The duration of therapy depends on the clinical manifestation – 3 weeks for skin involvement, 6 weeks for scalp involvement, and several months for onychomycoses.

A new alternative for oral treatment is ketoconazole. At a daily dose of 200 mg the drug is well tolerated, but hepatic injury may be possible. In long-term treatment, liver function tests must be controlled.

Tinea Capitis

Ringworm of the scalp is very common in children in the tropics. The classical form with scaly patches of the scalp is the most frequent. The drug of choice is griseofulvin. Various dose schedules have been advocated. A single massive dose of griseofulvin may cure up to 60% of cases of *Microsporum* ringworm.

The treatment should be continued for up to 12 days. Griseofulvin has only fungistatic activity, so that infected hair has to be extruded by growth. The effect of griseofulvin can be enhanced by shaving of the head. The treatment of superin-

Table 36.1. Topical antifungal agents. Modified from Speller [11]

Official name of preparation	Available formulations	Relative cost	Comments	Additional spectrum (besides dermatophytes)	
1. Benzoic acid compound ointment (Whitfield's ointment)	Straw-colored ointment	Cheap	Irritant on inflamed skin - dilute to half strength with emulsifying ointment for groins	Erythrasma	Not effective against *candida*
2. Undecenoic acid	Straw-colored ointment and white powder	Cheap	Mildly irritant. Numerous proprietary formulations	-	
3. Tolnaftate	White cream, colorless lotion	Fairly cheap	Lotion may sting on fissured or raw skin	-	
4. Pecilocin	White cream	Expensive	Apart from griseofulvin it is only anti-ringworm antibiotic. Non-irritant	-	
5. Clotrimazole	White cream	Expensive	Nonirritant. An imidazole compound	Erythrasma	Effective against *candida*
6. Miconazole	White cream, white powder	Expensive	Nonirritant. An imidazole compound	Erythrasma	
7. Bifonazole	White cream, gel, spray, powder	Expensive	Nonirritant. An imidazole compound	Erythrasma	
8. Econazole Oxyconazole Isoconazole	White cream, colorless solution		Probably as effective as other imidazoles but less well documented	Erythrasma	
9. Haloprogin	White cream		Nonirritant	-	
10. Ciclopiroxolamine	Cream, powder, solution	Expensive	Nonirritant, good penetration	-	
11. Naftifin	Cream, gel, solution	Expensive	Nonirritant	Erythrasma	
12. Magenta paint (Castellani's paint)	Magenta-coloured liquid	Cheap	Stains skin red, may sting. Start with half strength preparation (dilute with water)	Bacteria	Use in inflamed and infected tinea pedis
13. Aluminum chloride solution (20%-30%)	Colorless liquid	Cheap	May sting on raw surfaces and fissures. Effective drying preparation	Bacteria	
14. Potassium permanganate solution (1 in 8000)	Wine-red liquid	Cheap	Relatively non-irritating. Initial twice daily treatment may cause overdrying quickly, therefore reduce to daily treatment. Use as soaks	Bacteria	

fected tinea capitis (kerion) is difficult; it necessitates the addition of antibiotics (tetracyclines). Favus, as a malignant form of scalp ringworm (due to *Trichophyton schoenleinii*), necessitates greater duration of treatment with griseofulvin. Topical therapy is not necessary in scalp ringworm, but may be useful to reduce the infectivity.

Fungal Infections of the Nails

The most frequent causes of fungal infections of the nails are dermatophytes, although *Candida* species and molds can lead to similar infections. Mixed infections of *Candida* and a dermatophyte are not uncommon. Oral griseofulvin is effective in dermatophyte infections, but must be continued for a long time. A new alternative is oral ketoconazole (200 mg/day). Systemic treatment should be supplemented by topical antifungal agents (miconazole, ciclopiroxolamine, tolnaftate), mechanical filing, and occlusion. Infections of the toenails are difficult to treat. Sometimes avulsion of the nails may be necessary.

Tinea imbricata is a special variant of ringworm in Southeast Asia; the fungus grows in concentric rings, which may cover the whole body with gyrate patterns. Griseofulvin is regarded as the drug of choice, but all topical applications are also successful.

Tinea Versicolor

Tinea (pityriasis) versicolor is caused by a yeast-like organism, *Pityrosporon orbiculare* (formerly *Malassezia furfur*). The diagnosis is made from the clinical picture and woodlight; the pathogen can be seen in direct preparations from the skin. The occurrence of tinea versicolor is very dependent on host and environmental factors, such as heat, sweating, and poverty.

Many local agents have a certain activity. The drug of choice is a lotion containing 2.5% selenium sulfide for 3 days. Intermittent treatment with ketoconazole is a more modern and more expensive alternative. Treatment consisting of two 200-mg tablets given together followed by another tablet monthly for 6–8 months is recommended.

Candida Infections

The most important species is *Candida albicans*. Other species of *Candida* are less common and far less important, as are *Torulopsis* species. *Candida albicans* is a common component of the normal body flora (mouth, gut, vagina) and can be selected by antibiotics, oral contraceptives, pregnancy, diabetes, iron deficiency, and impaired host defenses. Severe nonresponsive *Candida* infection is a typical clinical symptom in AIDS. In many patients, the reason for the *Candida* infection is unknown. Clusters and partner infection may be possible.

Genital Candidiasis

Genital candidiasis occurs in females as vulvitis and vaginitis with redness, itching, white patches and a creamy discharge. Men may suffer from balanitis. Infection of sexual partners is common.

For local treatment, azoles, such as clotrimazole, miconazole, and also ciclopiroxolamine, and nystatin are effective. A cheap but messy alternative is the topical use of povidone-iodine.

With azoles, the duration of treatment should be 1–5 days. Sexual partners should also be treated, if possible. There is a high tendency toward relapse, but resistance almost never develops. Systemic therapy with ketoconazole is effective, but usually not necessary. Genital candidiasis is a common complication of antibiotic therapy.

Oral Candidiasis

Oral candidiasis (thrush) is common in newborns and in infants with impaired resistance. It is a common complication of antibiotic therapy in patients with lowered resistance and severe illnesses. The treatment of choice is oral nystatin given as a suspension instilled around the mouth every 3-6 h. The suspension should then be swallowed, since the esophagus is also often affected. Lozenges containing pimaricin, amphotericin, and clotrimazole are also available.

Oral candidiasis usually responds rapidly to oral nystatin. In patients with AIDS, response to treatment is usually poor, and long-term treatment with ketoconazole may be necessary.

Candida Esophagitis

Candida esophagitis is a relatively common and serious complication in patients with bone marrow suppression or AIDS. The clinical picture, with thrush and pain on swallowing, is typical. Exact diagnosis by esophagoscopy or barium swallow is difficult. Since *Candida* esophagitis may be the focus for disseminated infection, treatment is justified on suspicion. The treatment of choice is the oral suspension of nystatin or amphotericin B. Severely compromised hosts should be treated with systemic ketoconazole (400-600 mg/day).

Candida Enteritis

Candida enteritis is rare. Yeasts found in feces usually do not belong to the species *Candida albicans* and are usually of no clinical importance. *Candida albicans* may cause inflammatory changes in the gut of oncological patients. The intestinal tract may, however, serve as a reservoir for further infections with a recurrent genital candidiasis. In these cases, nystatin should be given orally in tablet form.

Candida Pneumonia

Candida pneumonia is rare and only affects patients with markedly lowered resistance. It usually results from hematogenous spread of *Candida* septicemia (intravenous catheters), but sometimes arises through aspiration. Yeasts found in expectorated sputum are usually oral contaminants and do not support the diagnosis of *Candida* infection usually of the lung. Yeasts are commonly found in tracheal secretions in patients who have been intubated or ventilated for long periods, usually in the absence of pneumonia. Established cases of *Candida* pneumonia should be treated systemically with amphotericin B + flucytosine; milder infections may be treated with i.v. miconazole or oral ketoconazole.

Candida Infections of the Urinary Tract

The presence of yeasts in normal urine is rare. *Candida albicans* is often present in small amounts in the urine of patients with *Candida* septicemia. The underlying conditions are multiple foci in the kidneys. The treatment of urinary tract infections by yeasts is difficult; ketoconazole gives only low urinary concentrations, but may treat renal microabscesses. Flucytosine is excreted in high concentrations, but encounters a marked problem of resistance. Intravenous amphotericin B is only justifiable in cases of *Candida* septicemia. Fluconazole (investigational) is promising for treatment of *Candida* urinary tract infections.

Candida Septicemia

Candida albicans sometimes enters the blood-stream in immunosuppressed patients. Intravenous lines may lead to secondary *Candida* infection. Generally, septicemia caused by *Candida* is rare. The most important symptom is fever; metastatic colonies are fairly commonly found in the retina, brain, and kidney.

The treatment of choice is a combination of amphotericin B + flucytosine. An infected foreign body should be removed whenever possible. For less severe cases, such as infection of a venous catheter, oral ketoconazole may be adequate.

Candida Infections of the Skin

Candida infections of the skin are common but not dangerous. Intertrigo, perianal eczema, diaper dermatitis, balanitis, chronic paronychia, and perlèche are typical clinical conditions caused by *Candida albicans*. Redness, itching, and peeling are suggestive clinical signs. Skin scrapings show a great number of budding yeast cells and pseudomycelia.

Treatment consists usually of local antifungal agents. Castellani's solution is effective and cheap, but has doubtful cosmetic effects. Polyenes (nystatin, amphotericin B) or imidazole derivatives (clotrimazole, bifonazole, miconazole) are used as topical preparations. Alternatives are ciclopiroxolamine, haloprogin, pimaricin, and povidone-iodine. Tolnaftate has no effect on yeast infections. Ketoconazole may be indicated as systemic treatment in severe infections or in underlying diseases. Griseofulvin is ineffective. The treatment of chronic paronychia is difficult, as topical antifungal agents do not reach pockets at the site of the nail, but the condition responds fairly well to treatment with ketoconazole.

Chronic Mucocutaneous Candidiasis

Chronic mucocutaneous candidiasis (candidiasis granulomatosa) is a quite rare condition in children which results from a primary immune defect. The treatment of long-lasting chronic granulomas is difficult, as topical therapy is often ineffective. Oral ketoconazole over a long period seems at present to be the treatment of choice [1].

Cryptococcal Infection

Cryptococcal infection is quite common in Africa. Infections by *Cryptococcus neoformans* are frequently associated with AIDS. Cryptococcal pneumonia and cryptococcal meningitis are most frequent, but involvement of many other organs is possible. The treatment is difficult. Only amphotericin B and flucytosine are active. Treatment consists usually of a combination of high-dose flucytosine with amphotericin B. Ketoconazole and miconazole are not effective in the treatment of cryptococcus infection. Fluconazole (investigational) is a new alternative.

Infections by Molds

Aspergillus Infections

Aspergillus fumigatus is the commonest cause. Other species of aspergillus are only occasionally involved *(A. nidulans, A. glaucus). A. fumigatus* is widespread in the inanimate environment (earth, wood, flower pots, dust). The spores are generally inhaled and expectorated with no ill effect. Isolated findings of aspergilli in sputum are not indicative of disease; the organisms must be present repeatedly or in large amounts and with associated clinical evidence.

Aspergillus fumigatus can cause several types of disease:

1. *Bronchopulmonary aspergillosis:* Allergic asthma may be due to inhalation of spores, but there is also an invasive disease in which the bronchi are attacked and the bronchial wall destroyed. Purely allergic forms should be treated with corticosteroids.
2. *Aspergilloma* is a noninvasive infection of preformed cavities. The chest X-Ray is typical and shows a fungus mass containing crescents of air. Hemoptysis often occurs. Aspergilloma does not respond to systemic or local chemotherapy and must be removed surgically.
3. *Invasive pulmonary aspergillosis* presents as pneumonia in patients with immunosuppresion (e.g., leukemia, AIDS); cavitation and hemoptysis sometimes occur.
4. *Aspergillus septicemia* is common in immunocompromised patients. The portal of entry may be lung, gut, or infected venous catheters, but is often unknown. Hematogenous spread leads to infected infarcts and metastatic abscesses, usually in the brain, skin, or kidneys, but sometimes in other organs. Blood culture is negative.
5. Eye infections, otitis externa, sinusitis, and colonization of chronic skin ulcers or burns are rare.

The treatment of *Aspergillus* infections is difficult; the relatively toxic combination of amphotericin B and flucytosine at full dosage as a long course is still recommended. Imidazole derivatives (miconazole i.v., ketoconazole orally) are only weakly active. The investigational azole derivative itraconazole has promising activity, also against aspergilli. Local treatment, such as inhalation of amphotericin B in bronchopulmonary aspergillosis, is seldom effective. The prognosis of invasive

Aspergillus infections in leukemia or related conditions is poor, even with the best treatment available.

Mycetoma

Choice of treatment depends upon whether one is dealing with a case of actinomycetoma or of eumycetoma. Actinomycetoma is caused by aerobic actinomycetes, whereas eumycetoma is caused by fungi. *Actinomadura madurae, Actinomadura pelletieri, Streptomyces somaliensis,* and *Nocardia brasiliensis* are the species most commonly responsible for actinomycetoma in the tropics. The most prevalent fungi are *Madurella mycetomatis, Madurella griseis, Leptosphaeria senegalensis, Pseudoallescheria boydii,* and *Aspergillus nidulans.*

Actinomycetoma. Medical treatment offers a chance of cure or improvement in the majority of patients. Combination of dapsone, 100 mg twice daily, or co-trimoxazole, two tablets (corresponding to 0.96 g) twice daily, with streptomycin, 1 g i.m. daily, for one month and on alternate days thereafter for about 3 to 4 months, is of proven value. When there is no clinical response to these regimens, combination treatment with rifampicin (450 mg daily) and co-trimoxazole or streptomycin, or with streptomycin and pyrimethamine-sulfadoxine, may be tried before resorting to surgery. In one study, duration of treatment was on average 9 months in responsive cases [3]. Clinical response is indicated by reduction of swelling and closure of sinus tracts.

Eumycetoma. Radical excision or amputation have so far been the only means of cure in advanced cases with bone involvement. Relapses will occur, unless all infected tissue is removed. Until recently, no effective chemotherapy was known. However, cure or improvement has been reported in five and four patients respectively who were treated with ketoconazole for eumycetoma due to *Madurella mycetomatis* [4]. The observation period after treatment was still short at the time of publication. Since the outcome appeared to vary with dosage, the authors recommend a daily dose of 400 mg ketoconazole [4]. Thus, initial medical management with ketoconazole, where available, may be tried before surgery is embarked upon (the lesions of eumycetoma are usually only slowly progressive).

Chromomycosis

Chromomycosis is due to several related species of fungi, the nomenclature of which is not yet settled. Irrespective of species, chromomycosis is diagnosed by the finding of brown, thick-walled spherical cells ("sclerotic cells") in tissue sections.

Surgical excision or cautery is suggested as treatment for early lesions of limited extent, since antifungal chemotherapy is not particularly effective. Oral flucytosine is usually chosen for treatment of advanced cases. Since relapse resistant to retreatment with flucytosine occurs, initial combined treatment with i.v. amphotericin B is suggested in advanced cases with extensive lesions. The use of amphotericin B is, however, controversial. Treatment with ketoconazole or thiabendazole are further, less toxic, alternatives. Local heat therapy may also be effective.

Rhinosporidiosis

There is no effective antifungal chemotherapy for rhinosporidiosis, and treatment is surgical by excision or electrocoagulation.

Sporotrichosis

Cutaneous sporotrichosis is treated orally with saturated (1 g/ml) potassium iodide solution. One usually starts at 1 ml thrice daily and increases by increments of three to five drops to a dose of 3–4 ml thrice daily. If symptoms of iodism occur (salivation, lacrimation, nausea), treatment should be discontinued for several days and reinstituted thereafter at a somewhat lower dosage. Treatment is continued until 4–6 weeks after cure has been achieved, usually for 3 months altogether. Local heat treatment is apparently also successful. Ketoconazole, 400 mg daily, may also be considered in patients with allergy to iodide.

Potassium iodide is ineffective in extracutaneous or disseminated sporotrichosis. Intravenous amphotericin B is then the drug of choice.

Subcutaneous Phycomycosis

Subcutaneous phycomycosis, also known as subcutaneous zygomycosis, is an entomophtoromycosis due to *Basidiobolus haptosporus*. This tropical fungal infection is usually treated with a saturated solution of potassium iodide (see sporotrichosis above) given orally in doses up to 3 (-4) ml thrice daily. Patients who fail to respond may be treated with ketoconazole, amphotericin B, or a combination of co-trimoxazole, two 0.48-g tablets twice daily, and potassium iodide [8].

Rhinoentomophthoromycosis

Treatment of entomophthoromycosis (due to *Conidiobolus*) is the same as that for subcutaneous phycomycosis. Monotherapy with co-trimoxazole at high dosage, two 0.48-g tablets 4 times daily (this is a daily dose of 3.48 g!), has also been reported to be efficacious in a small series of patients [6].

Histoplasmosis

Histoplasmosis due to Histoplasma capsulatum. The USA National Institutes of Health Mycosis Study Group came to the following conclusion after a prospective, randomized clinical trial of treatment of blastomycosis and histoplasmosis with ketoconazole: "The success rate for all patients with histoplasmosis [chronic pulmonary, localized, or disseminated histoplasmosis] treated for 6 months or more was 85%. ... ketoconazole is effective for immunocompetent patients with non-life-threatening, nonmeningeal forms of blastomycosis and histoplasmosis" [5]. The recommended dose is 400 mg ketoconazole daily taken as a single oral dose for a minimum of 6 months, depending on the clinical course.

Amphotericin B remains the drug of choice for histoplasmosis occurring in immunocompromised patients, in acutely life-threatening disease, and in cases with meningeal involvement. When amphotericin B is given, the initial 1-mg test dose is usually increased by 5-10 mg each day up to 40-50 mg daily (equal to 0.6 mg/kg/day) for a total dose of 2-2.5(-3) g. The higher total dose of 3 g may be needed in disseminated disease. Infants and children tolerate up to 1 mg/kg/day.

Histoplasmosis due to Histoplasma duboisii (African Histoplasmosis). Amphotericin B is still the drug of choice. Cure is usually achieved with total doses of 1-2 g, although higher amounts may be needed in selected patients. Treatment with ketoconazole daily, 400 mg in adults, and 200 mg in a 6-year-old girl who had not responded to a course of amphotericin B, has also been successful [2]. Surgical excision of localized, cutaneous lesions has been curative, but relapses may occur.

Blastomycosis

Amphotericin B is still considered the drug of choice for all forms of blastomycosis. This may, however, change with wider recognition of the results of a prospective, randomized clinical trial of treatment of blastomycosis with ketoconazole. The authors, the Mycosis Study Group of the American National Institute of Allergy and Infectious Diseases, state that "our findings indicate that ketoconazole provides an effective alternative to amphotericin B in patients with mild to moderately severe disease" [5]. Thirty-three patients were treated with 400 mg ketoconazole daily, 32 patients received 800 mg daily. Both groups were treated for 6 months or more. The success rates were 79% and 100% respectively. Despite the better results achieved with the high dose of 800 mg, initial treatment with 400 mg ketoconazole daily is suggested because of the high frequency of side effects in patients on 800 mg ketoconazole daily. If the disease progresses in patients on 400 mg ketoconazole daily, the dose should be increased to 800 mg daily (this is the upper limit) when tolerated by the patient. Otherwise, amphotericin B has to be used. In the rare case with involvement of the central nervous system, amphotericin B should be used from the outset.

Amphotericin B is given up to a total dose of 1.5-2.5 g. Treatment on alternate days, usually with 50 mg, may be considered in patients who are not severely ill, and is often more convenient than daily treatment with 0.5-0.6 mg/kg/day.

Paracoccidioidomycosis

Ketoconazole is now considered the drug of choice for treatment of paracoccidioidomycosis (South American blastomycosis). Treatment with a dose of 200-600 mg daily for at least 6 months is suggested. Besides being highly effective, this regimen also apparently reduces posttreatment relapse rates and mortality [9, 10].

Severe cases and patients not responding to other treatment should be treated with amphotericin B, which is given up to a total dose of 1-3 g. Thereafter, maintenance therapy, usually with sulfonamides or co-trimoxazole, must continue for years in order to prevent relapse.

Coccidioidomycosis

When treatment is indicated, as in progressive pulmonary disease, disseminated disease, or in the case of high complement fixation titers suggesting impending or occult dissemination, amphotericin B is the drug of choice. The drug is usually given up to a total dose of 1-2.5 g according to the clinical course. In coccidioidal meningitis, amphotericin B has to be given both intrathecally and intravenously. Prolonged intrathecal treatment is commonly required.

Ketoconazole has shown promising results in early studies, but high relapse rates somewhat subdue the initial optimism.

Systemic Treatment of Fungal Infections

The treatment of fungal infections is quite difficult. There are a remarkable number of topical agents (see Table 36.1). In general, the modern antifungal agents are more expensive, but cover a broad spectrum of different fungi. The older agents are cheap, but are generally active only against dermatophytes. A limited number of agents are available for systemic therapy of fungal infections. Amphotericin B, flucytosine, and griseofulvin are included in the WHO list. Amphotericin B has a broad spectrum of activity. It is active against yeasts, molds, and the agents of systemic fungal infections. It has very intensive side effects; the nephrotoxicity puts the patient at risk of acute renal failure. Treatment has to be started at a low daily dosage and increased by increments up to a maximal dose of 1 mg/kg/day. A daily dose of 50 mg and a total dose of 2-4 g should not be exceeded. Febrile reactions by the administration of amphotericin B are common. In general, the use of this agent is justified only in hospital by specialists.

Flucytosine has good activity against *Candida,* cryptococci, aspergilli, and the pathogens of chromomycosis. There is synergy with amphotericin B against *Candida,* cryptococci, and aspergilli. The main problem is high primary or secondary resistance. Usually flucytosine should be given in a high daily dose of 100-150 mg/kg (=6-10 g/day) together with amphotericin B. It can be given orally or by intravenous infusion. Flucytosine is fairly well tolerated, but bone marrow depression, rise in liver enzymes, and other complications are possible.

Griseofulvin is a classical selective agent only for dermatophytes. It is quite ef-

Table 36.2. Systemic antifungal agents

Substance	Application	Daily dosage	Main activity	Side effects, comments
Amphotericin B	i.v. as slow infusion	10-50 mg (0.1-1 mg/kg) in increasing doses	*Candida* spp., *Cryptococcus neoformans, Aspergillus* spp., *Histoplasma* spp., *Blastomyces dermatitidis, Coccidioides immitis, Paracoccidioides brasiliensis, Mucor, Sporotrichum schenckii*	Marked nephrotoxicity, fever, thrombophlebitis. Use only under strict clinical control by experts
Flucytosine	p.o., i.v. as 1% infusion	100-150 mg/kg in four doses	*Candida* spp., *Cryptococcus neoformans, Aspergillus* spp., agents of chromomycosis	Neutropenia, rise in liver enzymes. Primary resistance is frequent and secondary resistance may develop rapidly
Miconazole	i.v.	0.6 g as infusion	*Candida* spp., *Cryptococcus neoformans, Histoplasma* spp., *Blastomyces dermatitidis, Paracoccidioides brasiliensis, Coccidioides immitis*	Shock, allergy, hyperlipemia, mainly due to solvent
Ketoconazole	p.o.	200-600 mg	*Candida* spp., *Histoplasma* spp., *Paracoccocidioides immitis, Blastomyces dermatitidis, Malassezia furfur,* dermatophytes	Liver damage, gynecomastia, decrease in serum testosterone and ACTH-stimulated cortisol response. Poor absorption
Fluconazole (investigational)	p.o., i.v.	50-400 mg	*Candida* spp., *Cryptococcus neoformans*	Good tolerance, only limited experience
Itraconazole (investigational)	p.o.	100-200 mg	Similar to ketoconazole, but also active against *Aspergillus* spp.	Liver damage. Only limited experience
Griseofulvin	p.o.	0.5 g prolonged treatment	Dermatophytes only	Liver damage, photosensitivity, disorders of central nervous system

fective in tinea, but poor absorption and the lack of more than fungistatic activity are disadvantages. The oral imidazole derivative, ketoconazole, has good activity against *Candida* and in systemic mycoses. It is generally quite well tolerated, but may lead to severe hepatic reactions in some patients. Ketoconazole cannot be given intravenously, but there exists a poorly tolerated i.v. preparation of the related agent miconazole.

Itraconazole is a still investigational oral antifungal agent with a somewhat broader spectrum of activity. Fluconazole (investigational) can be given orally and i.v. It is only active against *Candida* and *Cryptococcus* spp.. In contrary to the other azole derivatives, it is not metabolized in the liver. It is excreted in the urine in active form.

The topical and systemic agents available for treatment of common fungal infections are summarized in Tables 36.1 and 36.2.

References

1. Drouhet E, Dupont B (1983) Laboratory and clinical assessment of ketoconazole in deep-seated mycoses. Am J Med 74 (1 B): 30-47
2. Mabey DCW, Ajdukiewicz AB, Hay RJ (1983) Treatment of African histoplasmosis with ketoconazole. Trans R Soc Trop Med Hyg 77: 219-220
3. Mahgoub ES (1976) Medical management of mycetoma. Bull WHO 54: 303-310
4. Mahgoub ES, Gumaa SA (1984) Ketoconazole in the treatment of eumycetoma due to *Madurella mycetomii.* Trans R Soc Trop Med Hyg 78: 376-379
5. National Institute of Allergy and Infectious Diseases Mycoses Study Group (1985) Treatment of blastomycosis and histoplasmosis with ketoconazole. Results of a prospective randomized clinical trial. Ann Intern Med 103: 861-872
6. Okafor BC, Gugnani HC (1983) Nasal entomophthoromycosis in Nigerian Igbos. Trop Geogr Med 35: 53-57
7. Pettit JHS (1986) Treatment of superficial fungal infections. Trop Doct 16: 105-112
8. Piens MA, Garin C, Richard-Lenoble C, Garin JP (1985) Premier cas de phycomycose souscoutanée observée au Gabon et traitée avec succès par kétoconazole. Bull Soc Pathol Exot Filiales 78: 170-178
9. Restrepo A, Gómez I, Cano LE, Aranyo MD, Gutierrez F, Sanin A, Robledo MA (1983) Treatment of paracoccidioidomycosis with ketoconazole: a three-year experience. Am J Med 74 (1 B): 48-52
10. Restrepo A, Gómez I, Cano LE, Aranyo MD, Gutierrez F, Sanin A, Robledo MA (1983) Posttherapy status of paracoccidioidomycosis treated with ketoconazole. Am J Med 74 (1 B): 53-57
11. Speller DCE (ed) (1980) Antifungal chemotherapy. Wiley, Chichester

37 Treatment of AIDS

Safe and efficient treatment of the underlying infection with the HIV virus (HTLV III-LAS virus) is not yet possible. Any chemotherapeutic agents used have to suppress the virus proliferation and probably must be given for very long periods - maybe for life. The investigational drug azidothymidine (AZT) apparently has activity in HIV infection; however, it is not free from severe side effects (anemia, neutropenia, central nervous system toxicity). It is an open question whether a safe, simple, cheap and curative therapy of HIV infection can ever be developed.

AIDS is the final stage of HIV infection. The patients display a marked decrease of defense mechanisms (T-cell deficiency, macrophage depletion, neutropenia). A remarkable number of quite different secondary infections may arise, the treatment of which may be difficult.

The common opportunistic infections in AIDS are classified according to their response to antimicrobial therapy (Table 37.1). The possible methods of treatment are summarized in Table 37.2. Rapidly developing areactive tuberculosis is a main manifestation of AIDS in the tropics. Severe lung infiltration or abscesses can develop within a few days. Tuberculosis in AIDS responds well to the usual combination therapy with isoniazid, rifampicin, and streptomycin.

A typical manifestation of AIDS in African patients is diarrhea leading to a wasting syndrome (slim disease). The underlying infection of this condition may be *Cryptosporidium* infection; other intestinal pathogens may also be present. In the majority of patients with wasting and diarrhea no pathogen is found (infection by HIV virus, cytomegalovirus infection?).

Table 37.1. Secondary infections in AIDS

Treatable	Difficult to treat	Practically nontreatable
Pneumocystis carinii	*Cryptococcus neoformans*	*Cryptosporidium*
Toxoplasma gondii	*Aspergillus* spp.	*Mycobacterium avium intracellulare*
Candida spp.	Epstein-Barr virus	Cytomegalovirus
Mycobacterium tuberculosis	Molluscum contagiosum	
Salmonella spp.	*Histoplasma capsulatum*	
Entamoeba histolytica	*Coccidioides immitis*	
Giardia lamblia		
Strongyloides stercoralis		
Herpes virus		
Zoster/varicella virus		

Table 37.2. Treatment of opportunistic infections in AIDS

Pathogen	Treatment of choice	Nonresponse or intolerance	Prevention of recurrence
Pneumocystis carinii (pneumonia)	Co-trimoxazole, 3-4 times the usual dose (=5.76-7.68 g), p.o. or as infusion for a minimum of 14 days	Pentamidine 4 mg/kg/day i.v.	If necessary 1. Two tablets Fansidar/week *or* 2. 1.92 g co-trimoxazole/day
Toxoplasma gondii (encephalitis)	Pyrimethamine 50-100 mg/day + 500 mg sulfamethoxydiazine for 14 days-3 weeks	Clindamycin in high dose	If necessary 1. Two tablets Fansidar/week *or* 2. Pyrimethamine 50 mg/day
Cryptosporidium (enteritis)	Trial with spiramycin 4 g/day p.o. (unsafe)	Symptomatic treatment	Unknown
Candida albicans (thrush, esophagitis)	Ketoconazole 400-600 mg/day	Miconazole gel Nystatin p.o.	1. Ketoconazole 200 mg/day *or* 2. Nystatin p.o.
Aspergillus spp. (pneumonia)	Amphotericin B 0.5 mg/kg/day + flucytosine 150 mg/kg/day	Itraconazole (investigational)	Unknown
Cryptococcus neoformans (meningitis)	Amphotericin B 0.5 mg/kg/day + flucytosine 150 mg/kg/day	Fluconazole (investigational)	Fluconazole (investigational)
Tuberculosis	Combination of isoniazid, streptomycin, rifampicin	Other antituberculous agents	Apparently prolonged therapy not necessary
Mycobacterium avium intracellulare	Ansamycin 150-300 mg/day p.o. + clofazimine 3 × 100 mg/day p.o.	Unknown	Unknown
Salmonella spp. (septicemia)	Quinolones (ofloxacin, ciprofloxacin)	Cefotaxime, ceftriaxone	Co-trimoxazole
Herpes I + II (herpes encephalitis)	Acyclovir 10 mg/kg/day Acyclovir 20-30 mg/kg/day	None Vidarabin 15 mg/kg/day	Sometimes long-term acyclovir therapy necessary
Zoster	Acyclovir 10 mg/kg/day Duration of treatment: until disappearance of vesicles	Unknown	Usually unnecessary, but sometimes necessary
Cytomegalovirus infections	Ganciclovir, (DHPG), (investigational)	Hyperimmune gammaglobulin	If necessary, DHPG, e.g., 5 times/week (?)

The treatment of diarrhea in AIDS is difficult. Co-trimoxazole in normal doses or quinolones may be tried. *Entamoeba histolytica* or *Strongyloides stercoralis* infections should be treated, if present. *Cryptosporidium* infection is practically nontreatable.

Pneumocystis pneumonia is a main severe manifestation in AIDS patients in the USA and Europe. It must be suspected if pneumonia does not respond to the usual antibiotics. The treatment of choice is co-trimoxazole in very high dosage (3-4 times the normal daily dose). Delay of treatment deteriorates the result. Many patients will develop allergy to sulfonamides. An alternative is pentamidine; however, it has considerable side effects. In sulfonamide intolerance, treatment with trimethoprim alone (300 mg/day), possibly together with dapsone, is a further alternative. For prophylaxis of relapses, Fansidar (pyrimethamine/sulfadoxine) or low-dose co-trimoxazole is advocated.

Brain abscesses are the main manifestation of CNS toxoplasmosis. Presumptive diagnosis can be made only by computed axial tomography. Treatment decisions must be made on the basis of clinical evidence, using a combination of pyrimethamine and sulfonamides. Patients who do not tolerate sulfonamides may be given clindamycin in high dosage in combination with pyrimethamine. Brain toxoplasmosis has a high tendency to relapse. Treatment with two tablets of Fansidar a week is necessary to prevent recurrences.

Nearly all patients with AIDS suffer from severe oral *Candida* infection. The response to oral nystatin is poor. The treatment of choice is ketoconazole orally (400-600 mg/day). Fluconazole is a new alternative.

Cryptococcal meningitis and other cryptococcal infections are typical manifestations of AIDS in African patients. Therapy is difficult. The standard treatment with amphotericin B + flucytosine is often ineffective. Fluconazole is a valuable agent.

Histoplasmosis and coccidioidomycosis may be activated in AIDS. Infections with *Aspergillus* and *Mucor* are also possible. The treatment of severe mycotic infections is very difficult. The fungal infection usually is the final complication.

Common bacterial infections (staphylococci, streptococci) are quite rare in AIDS. Salmonella infections quite often lead to septicemia. It is common for a patient with slight diarrhea and high fever to have blood cultures positive for salmonellae. Quinolones (ofloxacin, ciprofloxacin) are the treatment of choice. Co-trimoxazole and chloramphenicol have only poor efficacy.

Besides tuberculosis, infections with atypical mycobacteria are quite common. Infections by *Mycobacterium avium intracellulare* are practically nontreatable. Sometimes they respond only to the investigational agent ansamycin and the antileprosy compound clofazimine.

Often chronic latent virus infections are activated by AIDS. Severe herpes infection (herpes labialis, herpes analis) can be treated with acyclovir. Usually the infection responds well to treatment, but relapses are common. Zoster and chickenpox also respond well to acyclovir. The treatment of severe cytomegalovirus (CMV) infection is very difficult; in CMV retinitis the investigational compound ganciclovir, (DHPG), can be tried. In tropical areas, modified courses of malaria, leishmaniosis, and *Strongyloides* infection may lead to additional secondary problems in AIDS patients.

The high frequency of secondary tumors (Kaposi's sarcoma, lymphoma) complicates the clinical picture. The best available therapy of full-blown AIDS is the effective treatment of a series of different complications, but in the end the patients die from nontreatable infection, encephalopathy, or tumor.

Subject Index

Abortion, septic 66, 226
Abscess, brain 101
–, lung 134
–, peritonsillar 110
–, pyogenic liver 165
–, renal 236
–, retropharyngeal 110
Abscess-forming pneumonia 129
Absorption of antibiotics 39
Accident, traffic 201
Actinobacillus actinomycetemcomitans 71
Actinomadura madurae 313
Actinomyces 306
- *israelii* 302
Actinomycetoma 313
Actinomycosis 302
Acyclovir 321
Acylaminopenicillins 21
Administration of antibiotics 39 ff.
Adsorbents 154
Aeromonas hydrophila 61, 145, 160, 193
AIDS 249, 309, 319–321
Alcohol, intolerance 45
Allergic side effects 47
Allergy to penicillin 47, 78
Ameboma 161
Amikacin 15
Aminoglycosides 15
Amoxicillin 11, 24
Amphotericin B 310, 315, 316
Ampicillin 11
Anaphylactic shock 47
Angina, Ludwig's 110
Animal bites 201
Anthrax 276
–, cutaneous 276
Antibiotics, absorption of 39
–, administration of 39 ff.
–, choice of 26 ff.
–, combinations of 49 ff.
–, costs of 51 ff.
–, CSF penetration of 87
–, dosage of 41 ff.
–, pregnancy 228
Appendicitis 161
ARI (acute respiratory infection) 105
Arthritis 178, 179
–, gonococcal 179
–, septic 178 ff.
Aspergilloma 312
Aspergillus 306
Aspergillus fumigatus 306, 312
Aspiration pneumonia 129
Atropine 154
Augmentin 25
Azidamphenicol 252
Azidocillin 10
Azidothymidine (AZT) 319
Azlocillin 21

Bacampicillin 11
Bacillus anthracis 276
- *cereus* 148
Bacteremia, infusion 68
Bacterial endocarditis 71
- infections 187
- keratitis 256
- meningitis 83 ff.
- resistance 34
Bacteriuria 231
Bacteroides fragilis 132, 163
- *melaninogenicus* 129
Bartholinitis 222
Bartonella bacilliformis 305
Bartonellosis 305
Basidiobolus haptosporus 314
Benzathine penicillin 11
Benzoic acid 307
Benzylpenicillin 10
Bifonazole 307
Biliary tract infections 166
Biological side effects 48
Bites 201, 204
–, animal 201
Blastomycosis 315
Blepharitis, ulcerating 251
Blood cultures 61
Bordetella pertussis 135
Borrelia recurrentis 274
- *vincenti* 193
Brain abscess 101
Bronchiectasis 134
Bronchiolitis 112
Bronchitis 106, 111, 112

-, acute 111
-, chronic 111
Bronchopneumonia 106
Brucella 71, 177
Brucellosis 271
Burn wounds 192
Burns 201
Buruli ulcer 299

Campylobacter fetus 65
- *jejuni* 145, 160
Cancrum oris 248
Candida 306 ff.
- *albicans* 306
- enteritis 310
- esophagitis 310
- infections 309 ff.
- pneumonia 310
- septicemia 310
- stomatitis 247
-, urinary tract infection 311
Candidiasis 223, 310
-, oral 310
Capreomycin 286
Cardiobacterium hominis 71
Carfecillin 239
Carindacillin 239
Cataract extraction 258
Cefaclor 21
Cefadroxil 21
Cefalexin 21
Cefamandole 21
Cefazedone 21
Cefazolin 21
Cefmenoxime 21
Cefmetazole 21
Cefoperazone 21
Cefotaxime 21
Cefotetan 21
Cefotiam 21
Cefoxitin 21
Cefradin 21
Ceftazidime 21
Ceftizoxime 21
Ceftriaxone 21
Cefuroxime 21
Cellulitis 199
-, lid 251
-, orbital 252
Cephaloridine 21
Cephalosporins 21
Cephalothin 21
Cervical lymphadenitis 249
Chancroid 210
Charcoal 154
Chemoprophylaxis 198
-, tuberculosis 294
Chemotherapy, topical 255
Chlamydia trachomatis 210, 219
Chlamydial ophthalmia 254
Chloramphenicol 13, 260
Chlorhexidine 197
Cholangitis 67, 166
Cholecystitis 67
Cholelithiasis 166
Chorioamnionitis 227
Chromobacterium violaceum 61, 193
Chromomycosis 313
Ciclopiroxolamine 308
Cilastatin 24
Cinoxacin 19
Ciprofloxacin 19
Clavulanic acid 24
Clindamycin 23
Clioquinol 154
Clofazimine 298, 321
Clonorchis sinensis 166, 261
Clostridium difficile 23, 48, 148, 160
- *perfringens* 148, 206
- *tetani* 205
Clotrimazole 307
Cloxacillin 12, 78
Coccidioides 306
Coccidioidomycosis 316
Combinations of antibiotics 49
Common bacterial infections 187
Common cold 108
Community-acquired pneumonia 119
Conidiobolus 314
Conjunctivitis 252, 254
Contraindications 46
Corneal ulcers 256
Corynebacterium diphtheriae 136
- *pyogenes* 193
Costs of antibiotics 51 ff.
Co-trimoxazole 14, 320
Coxiella burnetii 71
Croup 110
Cryptococcal infection 312
- meningitis 95, 321
Cryptosporidium 145, 320
CSF, penetration of antibiotics 87
Cutaneous anthrax 276
- diphtheria 193
Cycloserine 286
Cystitis 236, 237
-, acute, single-dose therapy 237

Dacryocystitis 251
Dapsone 298
Dehydration 149
Demeclocycline 12
Dermatophyte infections 306, 307
Diarrheal diseases 142

Dibekacin 15
Dicloxacillin 12
Diloxanide furoate 154
Diphenoxylate 154
Diphtheria 136 ff.
-, cutaneous 193
-, nasal 108
Donovanosis 210
Dosage, Antibiotics 41 ff.
-, neonates 43
-, pediatric 43
Doxycycline 12
Dracontiasis 200
Dracunculus medinensis 180, 205
Drugs, list of essential, WHO 5
Dysentery 148, 155

Econazole 308
Eczema 200
Empyema 131
-, subdural 102
Encephalitis, herpes 320
Endocarditis, bacterial 71
-, enterococcal 71
-, infective 71 ff.
-, prevention 79
-, staphylococcal 77
-, streptococcal 75
Endophthalmitis 257
Enoxacin 19
Entamoeba histolytica 145
ENT infections 243
Enteric fever 259
Enteritis, *Candida* 310
- necroticans 160
Enterococcal endocarditis 71
Enterocolitis, pseudomembranous 48
Enterotoxigenic *Escherichia coli* 144
Entomophthoromycosis 314
Eumycetoma 313
Epidermophyton 306
Epididymitis 240
Erysipelas 199
Erysipeloid 305
Erysipelothrix rhusiopathiae 305
Erythrasma 308
Erythromycin 17
Escherichia coli, enterotoxigenic 144
Esophagitis, *Candida* 310
Essential drugs, WHO list
Ethambutol 286
Ethionamide 286, 298
Exsiccosis 149
Eye drops 253
- infection 251 ff.
- ointment 252

Favus 308
Fever, enteric 259
-, Herxheimer's 219, 220
-, Oroya 305
-, paratyphoid 259
-, Q- 265
-, ratbite 305
-, relapsing 274
-, rheumatic 109
-, spotted 265
-, typhoid 259 ff.
Flucloxacillin 12
Fluconazole 95, 311, 317
Flucytosine 316
Fracture, open 201
Fungal infections 33, 306 ff.
- -, nails 308
- keratitis 257
- meningitis 95
Fusobacterium fusiforme 193

Ganciclovir 321
Gardnerella vaginalis 223
Gas gangrene 206
Gastrointestinal tract infections 142 ff.
Genital herpes simplex 210
- ulcers 212
Genitourinary infections 231 ff.
Gentamicin 16, 75
Giardia lamblia 145
Gingivitis, necrotizing ulcerative 247
Gonococcal arthritis 179
- ophthalmia 253
- - neonatorum 218
- vulvovaginitis 219
Gonorrhea 210
-, oropharyngeal 217
Granuloma inguinale (donovanosis) 210
Griseofulvin 307, 316
Gynecological infections 222 ff.
Gyrase inhibitors 19

Haloprogin 308
Hemophilus aphrophilus 71
- *ducreyi* 213
- *influenzae* 83, 93, 106, 175
- - meningitis 93
Herpes encephalitis 320
- simplex, genital 210
Herpetic keratitis 256
Herxheimer's fever 219, 220
Histoplasma 306, 315
- *capsulatum* 315
- *duboisii* 315
Histoplasmosis 315
HIV infection 319
- virus 319

Hordeolum 251
Hospital-aquired pneumonia 130
Hydroxyquinolines 154
Hypersensitivity 47
Hypopyon 256

Imipenem 24
Impetigo 199
Inclusion conjunctivitis 254
Infection, bacterial 187
-, biliary tact 166
-, *Candida* 309
-, cryptococcal 312
-, dermatophyte 207
-, ENT 243 ff.
-, eye 251 ff.
-, fungal 23, 306 ff.
-, gastrointestinal tract 142 ff.
-, genitourinary 231 ff.
-, gynecological 222 ff.
-, HIV 319
-, joint 178 ff.
-, *Listeria* 280
- by molds 312
-, mycobacterial 285 ff.
-, nail 308
-, nosocomial 36
-, obstetric 222 ff.
-, odontogenic 248
-, respiratory tract 105 ff.
-, *Salmonella* 259 ff.
-, skin 187 ff.
-, urinary tract 231, 311
-, - -, *Candida* 311
-, viral 247
-, wound 187 ff.
Infective endocarditis 71 ff.
Influenza 131
Infusion bacteremia 68
Injection, intramuscular 40
Interstitial pneumonia 130
Intramuscular injection 40
Isoconazole 308
Isoniazid 286
Itraconazole 317

Jaw, osteomyelitis 248
Joint infection 178 ff.
Josamycin 16

Kanamycin 15, 286
Kaolin 154
Kaposi's sarcoma 321
Keratitis 256 ff.
-, bacterial 256
-, fungal 257
-, herpetic 256
-, measles 257
-, phlyctenular 257
Ketoconazole 307, 313, 317
Klebsiella pneumoniae 129

Legionella pneumophila 117, 130
Legionnaire's disease 130
Leprosy 200, 295
Leptospirosis 269
Leukemia 68
Lid, cellulitis 251
- infection 251
Lincomycin 23
Listeria infections 280
- meningitis 95
- *monocytogenes* 83, 280
Liver abscess, pyogenic 165
Loperamide 154
Ludwig's angina 110
Lung abscess 134
Lyell's syndrome 15
Lymphadenitis 199
-, cervical 249
Lymphangitis 66, 199
Lymphogranuloma venereum 210

Macrolide antibiotics 16
Madurella mycetomatis 313
Mafenide 197
Malassezia furfur 309
Mastitis 227
Mastoiditis 245
Measles keratitis 257
- pneumonia 127
Melioidosis 282
Meningococcal meningitis 91, 96
Meningitis 83 ff.
-, bacterial 83 ff.
-, cryptococcal 95, 321
-, fungal 95
-, *Hemophilus influenzae* 93
-, *Listeria* 95
-, meningococcal 91, 96
-, pneumococcal 92
-, *Salmonella* 94
-, tuberculous 292
Methacycline 12
Metronidazole 17, 126, 223
Mezlocillin 21
Miconazole 307, 317
Microsporum 306
- ringworm 307
Minocycline 12
Molds 306
-, infections 312
Mucor 306
Mycetoma 313

Mycobacterial infections 285 ff.
Myobacterium avium-intracellulare 321
- *balnei* 193
- *leprae* 298
- *tuberculosis* 285
- ulcerans 193, 299
Mycoplasma pneumoniae 119

Nafcillin 12
Naftifin 308
Nail infection 308
Nalidixic acid 19, 155
Nasal diphtheria 108
Neisseria gonorrhoeae 35, 210
Neisseria meningitidis 83
Neomycin 15
Neonatal septicemia 68
Neonates, dosage 43
Nephrotoxicity 16
Netilmicin 15
Neurosurgery 202
NGU (nongonococcal urethritis) 219
Nitrofurans 18
Nitrofurantoin 18, 236
Nitroimidazole 17
Nocardia 303, 306
- *asteroides* 303
- *brasiliensis* 313
Nocardiosis 303
Noma 193, 248
Norfloxacin 19
Nosocomial infections 36
Nystatin 309

Obstetric infections 222 ff.
Odontogenic infection 248
Ofloxacin 19
Oleandomycin 16
Ophthalmia neonatorum 253 ff.
- -, gonococcal 218, 253
Opium 154
Oral candidiasis 310
Orbital cellulitis 252
Ornidazole 17
Oropharyngeal gonorrhea 217
Oroya fever 305
ORS (oral rehydration solution) 151
Osteomyelitis 170, 175, 176, 248
-, chronic 176
-, *Hemophilus* 175
-, jaw 248
-, *Salmonella* 175
Otitis 243-245
-, external 243
-, -, malignant 243
- media 243-245
- -, acute 244
- -, tuberculous 245
Ototoxicity 16
Oxacillin 12
Oxolinic acid 19
Oxyconazole 308
Oxytetracycline 12

Pancytopenia 13
Para-aminosalicylic acid 286
Paracoccidioidomycosis 316
Paratyphoid fever 259
Paromomycin 15
Pasteurella multocida 65, 201
Pecilocin 308
Pectine 154
Pediatric dosage 43
Pefloxacin 19
Pelvic inflammatory disease (PID) 217, 222, 225
Pelvioperitonitis 225
Penicillin 10, 11, 21, 78
-, Acylamino- 21
-, allergy 47, 78
-, Benzathine 11
-, Benzyl- 10
-, Phenoxymethyl- 10
-, Procain 11
-, Repository 11
Penicillinase-producing *Neisseria gonorrhoeae* *PPNG* 35, 210, 216
Pentamidine 320
Pericarditis 81 ff.
Periodontal disease 248
Peritonitis 163
Pertussis 135
Pharyngitis 108
Phenoxymethylpenicillin 10
Phlyctenular keratitis 257
Phycomyosis 314
PID (pelvic inflammatory disease) 217, 222, 225
Pinta 284
Pipemidic acid 19, 239
Piperacillin 21
Piromidic acid 19
Pityrosporon orbiculare 309
Pivampicillin 11
Plague 278
Plesiomonas shigelloides 145
Pneumococcal meningitis 92
Pneumocystis pneumonia 321
Pneumonia 106, 112 ff., 127, 130, 131, 310
-, abscess-forming 129
-, aspiration 129
-, *Candida* 310
-, community-acquired 119

–, hospital-acquired 130
–, interstitial 130
–, *Klebsiella* 129
–, *Legionella* 130
–, measles 127
–, postoperative 131
–, staphyloccocal 127
Povidone-iodine 197, 309
PPNG (penicillinase-producing *Neisseria gonorrhoeae*) 35, 36, 213, 216
Prednisone 293
Pregnancy, antibiotics 228
Prevention, endocarditis 79
Prices/costs of antibiotics 51ff.
Procaine penicillin 11
Prophylaxis, wound infections 198ff.
Propicillin 10
Prostatitis 240
Prothionamide 286, 298
Pseudomembranous enterocolitis 48
Pseudomonas pseudomallei 61, 129, 282
Puerperal sepsis 66, 225, 226
Pyelonephritis 231
Pylephlebitis 166
Pyoderma 187, 199
Pyogenic liver abscess 165
Pyomyositis 183
Pyrazinamide 286
Pyrexia during labor 227
Pyrimethamine 15, 321

Q-fever 265
Quinolones 19

Ratbite fever 305
Rehydration 151
- solution, oral (ORS) 151
Reiter's syndrome 180
Relapsing fever 274
Renal abscess 236
Repository penicillins 11
Resistance 34, 159
–, bacterial 34
–, multiple (streptococcus pneumoniae) 35
–, staphylococcus aureus 78
Respiratory tract infections 105ff.
Rheumatic fever 109
Rhinitis 108
Rhinoentomophthoromycosis 314
Rhinosporidiosis 314
Rickettsial diseases 265
Rickettsia conorii 265
- *prowazekii* 265
Rifampicin 19, 126, 286, 298
Rifampin 19
Rosoxacin 216
Rotavirus 144

Salmonella carriers 261
- infections 259ff.
- meningitis 94
- osteomyelitis 175
- spp. 144
- *typhimurium* 262
Salmonellosis 262
Salpingitis 225
Scabies 189, 200
Schistosoma haematobium 261
Scrub typhus 265
Sepsis, puerperal 66, 225, 226
Septic abortion 66, 226
- arthritis 178ff.
Septicemia 55ff., 67, 68
–, *Candida* 310
–, neonatal 68
–, postoperative 67
Sexually transmitted disease (STD) 210ff.
Shigella spp. 144
Shock, anaphylactic 47
Sickle-cell disease 100, 175
Side effects 45ff.
- –, allergic 47
- –, biological 48
- –, toxic 45
Silver sulfadiazine 197
Single-dose treatment, acute cystitis 237
–, chancroid 214
–, gonorrhea 214, 216
–, meningococcal meningitis 96
–, shigellosis 159
–, typhus 267
Sinusitis 245
Sisomicin 15
Skin infections 187ff.
Spectinomycin 216
Spiramycin 16
Spirillum minor 305
Splenectomy 100
Sporotrichosis 314
Spotted fever 265
Staphyloccocal endocarditis 77
- pneumonia 127
Staphylococcus aureus, cloxacillin-resistant 78
- *epidermidis* 78, 193
- *saprophyticus* 235
STD (sexually transmitted disease) 210ff.
Stomatitis 247
–, *Candida* 247
Streptobacillus moniliformis 305
Streptococcal endocarditis 75
Streptococcus milleri 101
- *pneumoniae* (multiple resistance) 35, 83, 106
Streptomycin 15, 75, 286
Subdural empyema 102

Sulfadiazine 15
–, silver 197
Sulfadimidine 15
Sulfadoxine 15
Sulfafurazole 15
Sulfaguanidine 15, 154
Sulfaguanole 15
Sulfalene 15
Sulfamethoxazole 15
Sulfamoxole 15
Sulfmethoxydiazine 15
Sulfonamides 15
Syndrome, Lyell's 15
–, Reiter's 180
Syphilis 210

Teichoplanin 23
Tetanus 205, 206
Tetracycline 12
Tetroxoprim 14
Thalidomide 299
Thiacetazone 286
Thiamphenicol 13
Thrush 247, 310
Typhoid fever 259
Tifomycine 96
Tinea 306–309
– imbricata 306
– versicolor 309
Tinidazole 17
Tobramycin 15
Tolnaftate 307
Tonsillitis 108
Topical administration 40
– treatment 196, 255
Torulopsis 306, 309
Toxic side effects 45
Toxoplasma gondii 101
Toxoplasmosis 321
Trachoma 254, 255
Treatment, duration 41
–, single-dose 213
–, –, typhus 267
–, topical 196, 255
Treponematoses, nonvenereal 284
Trichomoniasis 213, 223
Trichophyton 306
Trimethoprim 14
Tropical ulcer 193, 199
Tuberculosis 285, 291, 294
–, chemoprophylaxis 294
–, chemotherapy, short-course 291
Tuberculous meningitis 292
– otitis media 245
Tungiasis 200
Typhoid fever 259 ff.
Typhus, epidemic 265
–, scrub 265
–, single-dose treatment 267

Ulcer, Buruli 299
–, corneal 256
–, genital 212
–, tropical 193, 199
Ulcerating blepharitis 251
Undecenoic acid 308
Urethral discharge 211
Urethritis 219, 240
–, nongonococcal (NGU) 219
Urinary tract infections (UTI) 231, 311
– – –, *Candida* 311

Vaginal discharge 223
– disease 223
Vaginitis 213, 223
Vancomycin 23
Vibrio cholerae 144, 159, 160
– – *non-01* 160
– *parahaemolyticus* 144, 160
– *vulnificus* 61, 193
Viomycin 286
Viral infections 247
Vulvitis 223
Vulvovaginitis in children 222
–, gonococcal 219

Weil's disease 269
WHO, list of essential drugs 5
Whooping cough 135
Wound infections 187 ff., 198
– –, prophylaxis 198 ff.
Wounds, burn 192

Yaws 284
Yeasts 306
Yersinia enterocolitica 145
Yersinia pestis 278

Zoster 320

GPSR Compliance

The European Union's (EU) General Product Safety Regulation (GPSR) is a set of rules that requires consumer products to be safe and our obligations to ensure this.

If you have any concerns about our products, you can contact us on ProductSafety@springernature.com

In case Publisher is established outside the EU, the EU authorized representative is:

Springer Nature Customer Service Center GmbH
Europaplatz 3
69115 Heidelberg, Germany

Batch number: 10704763

Printed by Printforce, the Netherlands